FUNDAMENTAL IDENTITIES [Section 3.1]

PYTHAGOREAN

$$\cos^2 t + \sin^2 t = 1$$

$$1 + \tan^2 t = \sec^2 t$$

$$\cot^2 t + 1 = \csc^2 t$$

RECIPROCAL

$$\sec t = \frac{1}{\cos t}$$

$$\csc t = \frac{1}{\sin t}$$

$$\cot t = \frac{1}{\tan t}$$

EVEN-ODD

$$\sin(-t) = -\sin t$$

$$\cos(-t) = \cos t$$

$$\tan(-t) = -\tan t$$

QUOTIENT

$$\tan t = \frac{\sin t}{\cos t}$$

$$\cot t = \frac{\cos t}{\sin t}$$

Cofunction Formulas [Section 3.1]

$$\sin\left(\frac{\pi}{2} - t\right) = \cos t \qquad \cos\left(\frac{\pi}{2} - t\right) = \sin t$$

$$\tan\left(\frac{\pi}{2} - t\right) = \cot t$$

Addition Formulas [Section 4.2]

$$\sin(u + v) = \sin u \cos v + \cos u \sin v$$

$$\cos(u + v) = \cos u \cos v - \sin u \sin v$$

$$\tan(u + v) = \frac{\tan u + \tan v}{1 - \tan u \tan v}$$

Subtraction Formulas [Section 4.2]

$$\sin(u - v) = \sin u \cos v - \cos u \sin v$$

$$\cos(u - v) = \cos u \cos v + \sin u \sin v$$

$$\tan(u - v) = \frac{\tan u - \tan v}{1 + \tan u \tan v}$$

Double Angle Form

$$\sin 2t = 2 \sin t \cos$$

$$\cos 2t = \cos^2 t - \sin^2 t = 1 - 2 \sin^2 t$$
$$= 2 \cos^2 t - 1$$

$$\tan 2t = \frac{2 \tan t}{1 - \tan^2 t}$$

Half Angle Formulas [Section 4.3]

$$\sin \frac{t}{2} = \pm\sqrt{\frac{1 - \cos t}{2}} \qquad \cos \frac{t}{2} = \pm\sqrt{\frac{1 + \cos t}{2}}$$

$$\tan \frac{t}{2} = \frac{1 - \cos t}{\sin t} = \frac{\sin t}{1 + \cos t}$$

$$\sin^2 t = \frac{1 - \cos 2t}{2}$$

$$\cos^2 t = \frac{1 + \cos 2t}{2}$$

$$\tan^2 t = \frac{1 - \cos 2t}{1 + \cos 2t}$$

Product Formulas [Section 4.4]

$$\sin u \sin v = \tfrac{1}{2}[\cos(u - v) - \cos(u + v)]$$

$$\cos u \cos v = \tfrac{1}{2}[\cos(u - v) + \cos(u + v)]$$

$$\sin u \cos v = \tfrac{1}{2}[\sin(u + v) + \sin(u - v)]$$

$$\cos u \sin v = \tfrac{1}{2}[\sin(u + v) - \sin(u - v)]$$

Sum Formulas [Section 4.4]

$$\sin u + \sin v = 2 \sin \frac{u + v}{2} \cos \frac{u - v}{2}$$

$$\sin u - \sin v = 2 \cos \frac{u + v}{2} \sin \frac{u - v}{2}$$

$$\cos u + \cos v = 2 \cos \frac{u + v}{2} \cos \frac{u - v}{2}$$

$$\cos u - \cos v = -2 \sin \frac{u + v}{2} \sin \frac{u - v}{2}$$

Trigonometry

Trigonometry

SECOND EDITION

Dennis G. Zill
Loyola Marymount University

Jacqueline M. Dewar
Loyola Marymount University

Portions of this work also appear in
Algebra and Trigonometry.

McGRAW-HILL PUBLISHING COMPANY

New York   St. Louis   San Francisco   Auckland   Bogotá
Caracas   Hamburg   Lisbon   London   Madrid   Mexico
Milan   Montreal   New Delhi   Oklahoma City   Paris
San Juan   São Paulo   Singapore   Sydney   Tokyo   Toronto

Trigonometry

234567890 DOH DOH 943210

ISBN 0-07-557099-8

This book was set in Times Roman by York Graphic Services, Inc.
The editors were Robert A. Weinstein and John M. Morriss;
the designer was Geri Davis;
the production supervisor was Michael Weinstein.
The cover was designed by Hardy House.
R. R. Donnelley & Sons Company was printer and binder.

Library of Congress Cataloging-in-Publication Data

Zill, Dennis G., (date).
　Trigonometry / Dennis G. Zill, Jacqueline M. Dewar—2nd ed.
　　p. cm.
　Rev. ed. of: College algebra and trigonometry, 1988. Now published
　　in two separate volumes.
　Reprint. Originally published: New York: Random House, 1989.
　Includes index.
　ISBN 0-07-557099-8
　　1. Trigonometry. I. Dewar, Jacqueline M. II. Zill, Dennis G., (date).
　　College algebra and trigonometry. III. Title.
　QA531.Z55　1990b
　516.24—dc20　　　　　　　　　　　　　　　　　　　89-7984

Preface

PHILOSOPHY

This text reflects our philosophy that a mathematics text at the beginning college level should be readable, straightforward, and loaded with motivation. But ultimately students can learn mathematics only by doing mathematics. Therefore, we have emphasized problem-solving as a means of understanding. The examples are designed to motivate, instruct, and guide students. The exercises then give students the opportunity to test their comprehension, challenge their understanding, and apply their knowledge to real-world situations.

AUDIENCE AND FLEXIBILITY

We intend this text to provide a treatment of trigonometry, vectors, logarithms, analytic geometry, polar coordinates, and parametric equations that is accessible to a college student with two years of high-school mathematics. We have provided more than sufficient material here for a standard one-semester or one-quarter course. This wealth of topics allows the instructor to choose those best suited to the objectives of the course and the backgrounds and abilities of the students. Thus we believe that the text provides the prerequisites for a course in calculus.

FEATURES
Pedagogy

- *Examples*   It has been our experience that examples and exercises are the primary learning sources in a mathematics text. We have found that students rely on examples, not theorems and proofs. Therefore we have included numerous examples to illustrate both the theoretical concepts and the computational techniques covered in the text.

- *Exercises*   We feel that students can learn only by doing. Therefore, in order to promote active participation in problem-solving, the exercises are extensive and varied. The exercise sets include an abundance of drill problems, true-false questions, fill-in-the-blank questions, applications, challenging problems, graphing problems, and problems that require interpretation of graphs. This variety gives students the opportunity to solidify their understanding of basic concepts, see practical uses for abstract mathematical ideas, and test their ingenuity. For this edition each exercise set has been reorganized and expanded; the text now contains over 2700 exercises. Appendix B is a special section of challenging problems in trigonometry and involves concepts from several different parts of the text. Topics covered include distance and area measurement, space science, geometry, inequalities, and even tool-making. Answers to all odd-numbered exercises are given at the back of the book.

- *Applications*   We have added many new applications culled from journals, newspapers, and scientific texts. These "real-life" problems show students the power and usefulness of the mathematics they learn in this course. The applications in this edition span a wide variety of disciplines, including astronomy, biology, business and economics, chemistry, ecology, engineering, geology, medicine, meteorology, navigation, optics, and physics.

- *Motivation*   While a number of proofs are included, we have typically motivated concepts in an intuitive or geometric manner. In addition, wherever possible we have used figures to illustrate an idea or aid in a solution.

- *Emphasis on Functions*   Since functions are an essential concept in this course and in mathematics as a whole, we have increased the emphasis on functions and function notation in this edition. (See *Changes in Topic Coverage in the Second Edition*.)

- *Emphasis on Graphing*   There is also a greater emphasis on graphing equations and functions. We have stressed symmetry, use of shifted graphs, reflections, intercepts, and interpretation of graphs throughout the text.

- *Graphics and Photographs*   As an aid for comprehension and solving problems, we have included over 800 carefully labeled figures in this edition. In addition, photographs are located throughout the text to enhance the text's appeal and to increase student interest in the applications.

For Each Chapter

- *Chapter Opening Material* Each chapter opens with a table of its contents, a motivational discussion of the material, and a brief historical account of a mathematician who had a great influence on the development of the mathematics in that chapter.

- *Key Concepts* Each chapter concludes with a checklist of important concepts introduced in the chapter that students can use to review the material.

- *Review Exercises* In addition, at the end of each chapter there is an extensive set of review exercises so that a student can test his or her understanding.

Special Help for Students

- *Notes of Caution* Common mistakes and misinterpretations are pointed out to students in **Notes of Caution.** These notes range from brief reminders of simple errors to longer discussions of plotting enough points for a graph, using a calculator correctly, and eliminating extraneous solutions. These numerous reminders will alert students to common errors and clarify hidden assumptions.

- *Use of a Calculator* Our experience is that most, if not all, students own a calculator and can perform simple arithmetic operations using it. They may be less familiar, however, with other keys and capabilities of the calculator, especially for exponential, logarithmic, and trigonometric functions. Therefore, where appropriate, we include descriptions and examples on **Use of a Calculator.**

COVERAGE OF TOPICS

- *Prerequisite Topics* Concepts that are essential for the understanding of trigonometry, such as the Cartesian coordinate system, functions, and graphing, are presented in Chapter 1.

- *Trigonometry* In this revision the discussion of trigonometry has been expanded and reorganized into three chapters.

- *Triangle Trigonometry* The trigonometric functions are introduced in Chapter 2 using right triangles. This approach builds on the students' intuition and knowledge of geometry. Applications of right triangle trigonometry, including the law of sines and the law of cosines, are discussed in this chapter.

- *Analytic Trigonometry* In Chapter 3 the unit circle definition of the trigonometric functions of real numbers is presented. Throughout this chapter the emphasis is on the circular functions and their graphs and applications.

- *Trigonometric Identities and Equations* The original section of special trigonometric formulas has been expanded to two sections: one on addition and subtraction formulas (Section 4.2) and one on multiple angle formulas (Section 4.3). A new section (Section 4.4) on the product and sum formulas has been added to this chapter.

- *Vectors and Complex Numbers* Chapter 5 presents applications of trigonometry involving vectors and complex numbers. The discussion of the dot product has been expanded to a new section (Section 5.2).

- *Calculators vs. Logarithms* The topic of logarithms (Chapter 6) as a computational tool has been deleted from the text in favor of increased emphasis on the use of the calculator. Since calculators are used so extensively and are now relatively inexpensive, we feel that it is not unreasonable to expect a college student to possess a calculator. However, for those wishing to acquaint the student with logarithmic tables, we have included a discussion of interpolation and the use of such tables in Appendix C.

- *Analytic Geometry* Topics in analytic geometry are given a thorough treatment. In Chapter 7 we discuss straight lines, conic sections, and translation and rotation of axes.

- *Polar Coordinates and Parametric Equations* Chapter 8 contains two new sections: on special polar graphs and parametric equations and on curves.

- *Linear and Angular Velocity* These applied concepts, which are briefly introduced in Section 2.1 on angle measurement, are given fuller treatment in Appendix A.

ACCURACY

From our many years of teaching, we realize how frustrating it is to students and instructors when they encounter errors in the text or in the answers to exercises. To ensure accuracy, not only have we checked all answers ourselves, but the answers have also been checked independently by both Barry A. Cipra and Warren S. Wright. If, however, you do find any errors we would greatly appreciate it if you would advise our publisher so that these glitches may be eliminated from subsequent printings.

SUPPLEMENTS

This text is accompanied by a well-rounded supplements package, including a Student's Solutions Manual, Instructor's Resource Manual, Answer Manual, and Computerized and Print Test Bank.

STUDENT'S SOLUTIONS MANUAL The Student's Solutions Manual was prepared by Warren S. Wright and contains detailed solutions to odd-numbered exercises.

INSTRUCTOR'S RESOURCE MANUAL The Instructor's Resource Manual includes Sample Tests, prepared by Linda Hawley, and numerous transparency masters for classroom use.

ANSWER MANUAL The Answer Manual was prepared by Barry Cipra and includes answers to the even-numbered exercises.

TEST BANK The questions found in the sample tests, and additional test questions, are also available in test bank form which corresponds to a computerized version for IBM-PC and compatibles.

ACKNOWLEDGMENTS

We would like to take this opportunity to express our appreciation to Barry A. Cipra for supplying many of the applied problems that appear in the exercise sets, to Mary Margaret Grady for retyping the entire manuscript, and to Warren S. Wright for giving us the use of his material from the first edition. It was also our good fortune to have the following individuals read all or part of this second edition in manuscript form. Their criticisms and many fine suggestions are gratefully acknowledged:

Wayne Andrepoint	*University of Southwestern Louisiana*
Nancy Angle	*Colorado School of Mines*
James E. Arnold	*University of Wisconsin—Milwaukee*
Judith Baxter	*University of Illinois—Chicago Circle*
Margaret Blumberg	*Southwestern Louisiana University*
Robert A. Chaffer	*Central Michigan University*
Daniel Drucker	*Wayne State University*
Chris Ennis	*Carleton College*
E. John Hornsby	*University of New Orleans*
Don Johnson	*New Mexico State University*
Jimmie Lawson	*Louisiana State University*
Gerald Ludden	*Michigan State University*
Stanley M. Lukawecki	*Clemson University*
Richard Marshall	*Eastern Michigan University*
Glenn Mattingly	*Sam Houston State University*
Michael Mays	*West Virginia University*
Phillip R. Montgomery	*University of Kansas*
Bruce Reed	*Virginia Polytechnic Institute and State University*

Jean Rubin	*Purdue University*
Helen Salzberg	*Rhode Island College*
George L. Szoke	*University of Akron*
Darrell Turnbridge	*Kent State University*

We would also like to thank the following who responded to a detailed market survey sent out by our publisher:

Carol Achs	*Mesa Community College*
Joseph Altinger	*Youngstown State University*
Phillip Barker	*University of Missouri—Kansas City*
Wayne Britt	*Louisiana State University*
Kwang Chul Ha	*Illinois State University*
Duane Deal	*Ball State University*
Richard Friedlander	*University of Missouri—St. Louis*
August Garver	*University of Missouri—Rolla*
Irving Katz	*George Washington University*
Janice Kilpatrick	*University of Toledo*
Barbara Meininger	*University of Oregon*
Eldon Miller	*University of Mississippi*
Judith Rollstin	*University of New Mexico*
Monty J. Strauss	*Texas Tech University*
Faye Thames	*Lamar University*
Waldemar Weber	*Bowling Green State University*

Lastly, we are indebted to the staff at Random House and McGraw-Hill, especially Alexa Barnes, developmental editor; Margaret Pinette, project manager; John Martindale, senior editor; Robert Weinstein, sponsoring editor; and John Morriss, editing manager, for their constant flow of ideas, encouragement, support, and for an occasional but necessary prodding.

Series

The Zill and Dewar Precalculus Series

College Algebra

A text designed for a one-term course covering topics such as equations and inequalities, algebraic functions, exponential and logarithmic functions, matrices, analytic geometry, and probability.

Trigonometry

This text includes coverage of triangle and analytic trigonometry, exponential and logarithmic functions, vectors, analytic geometry, and polar coordinates.

Algebra and Trigonometry

This title combines the content of the two texts described above. It includes sufficient material for a standard one-semester or two-quarter or even a slower-paced full-year course.

Contents

3 Analytic Trigonometry   129

4 Trigonometric Identities and Equations   165

5 Vectors and Complex Numbers   207

6 Exponential and Logarithmic Functions   243

7 Topics in Analytic Geometry   283

8 Polar Coordinates and Parametric Equations   343

Appendixes

CONTENTS

Trigonometry

1

Prerequisites for Trigonometry

Gottfried Wilhelm Leibniz

If the question "What is the most important mathematical concept?" were posed to a group of mathematicians, teachers, and scientists, certainly the term *function* would appear near or even at the top of the list of their responses. In Chapters 3 and 4, we will focus primarily on the definition and the graphical interpretation of a function.

The word "function" was probably introduced by the German mathematician and "co-inventor" of calculus, Gottfried Wilhelm Leibniz, in the late seventeenth century and stems from the Latin word "functo," meaning to act or perform. In the seventeenth and eighteenth centuries, mathematicians had only the most intuitive notion of a function. To many of them, a functional relationship between two variables was given by some smooth curve or by an equation involving the two variables. Although formulas and equations play an important role in the study of functions, we will see in Section 3.4 that the "modern" interpretation of a function (dating from the middle of the nineteenth century) is that of a special type of correspondence between the elements of two sets.

1.1

The Real Number Line

SET THEORY

Set theory enables us to describe in a very precise way collections of numbers that share a common property. This can be very useful in stating the solutions to certain types of problems. Therefore, we begin with a brief review of basic concepts from set theory. A **set** is a collection of distinct objects. An object in a set is called an **element** of the set.

A set can be specified in two ways: by listing the elements in the set or by stating a property that determines the elements in the set. In each case, braces { } are used. For example, the set consisting of the numbers 5, 10, and 15 can be denoted by either

$$\{5, \ 10, \ 15\} \quad \text{or} \quad \{x \mid x = 5n, \ n = 1, \ 2, \ 3\}.$$

The latter is read "the set of all numbers x such that $x = 5n$, where $n = 1, 2, 3$."

If every element of a set B is also an element of a set A, we say that B is a **subset** of A. Two sets A and B are said to be **equal,** written $A = B$, if A and B contain precisely the same elements.

NUMBERS: INTEGER, RATIONAL, IRRATIONAL, AND REAL

Recall that the set of **natural numbers,** or **positive integers,** is

$$N = \{1, \ 2, \ 3, \ 4, \ . \ . \ .\}.$$

This set N is a subset of the set of **integers:**

$$Z = \{. \ . \ . \ -3, \ -2, \ -1, \ 0, \ 1, \ 2, \ 3, \ . \ . \ .\}.$$

The set Z includes both the positive and the negative integers and the number zero, which is neither positive nor negative. In turn, the set of integers is a subset of the set of **rational numbers:**

$$Q = \left\{ \frac{p}{q} \ \middle| \ p \text{ and } q \text{ are integers, } q \neq 0 \right\}.$$

The set Q consists of all quotients of two integers, provided that the denominator is nonzero; for example,

$$\frac{-1}{2}, \quad \frac{17}{5}, \quad \frac{10}{-2} = -5, \quad \frac{6}{1} = 6, \quad \frac{0}{8} = 0.$$

Note of Caution: The quotient a/b is undefined if $b = 0$. For example, 8/0 and 0/0 are undefined.

The set of rational numbers is not sufficient to solve certain elementary algebraic and geometric problems. For example, there is no rational number p/q for which

$$\left(\frac{p}{q}\right)^2 = 2.$$

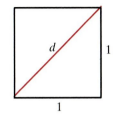

FIGURE 1

(See Problem 109.) Thus we cannot use rational numbers to describe the length of a diagonal of a unit square (see Figure 1). By the Pythagorean theorem, we know that the length of the diagonal d must satisfy

$$d^2 = (1)^2 + (1)^2 = 2.$$

We write $d = \sqrt{2}$ and call d "the square root of 2." As we have just indicated, $\sqrt{2}$ is not a rational number. It belongs instead to the set of **irrational numbers,** that is, the set of numbers that cannot be expressed as a quotient of two integers. Other examples of irrational numbers are π, $-\sqrt{3}$, and $\sqrt{5}/4$.

The set of numbers that are either rational or irrational is called the set of **real numbers** and is denoted by R.

DECIMALS

Every real number can be written in **decimal form.** For example,

$$\tfrac{1}{4} = 0.25,$$
$$\tfrac{131}{99} = 1.323232\ldots,$$
$$\pi = 3.14159265\ldots,$$

and
$$\sqrt{2} = 1.41421356\ldots.$$

Numbers such as 0.25 are said to be **terminating decimals,** whereas numbers such as

repeats

$$1.32\ 32\ 32$$

are called **repeating decimals.** A repeating decimal, such as $1.323232\ldots$, is often written as $1.\overline{32}$, where the bar indicates the number or numbers that repeat. It can be shown that every rational number has either a repeating or a terminating decimal representation. Conversely, every repeating or terminating decimal is a rational number. It is also a basic fact that every decimal number is a real number. It follows, then, that the set of irrational numbers consists of all decimals that neither terminate nor repeat. Thus π and $\sqrt{2}$ have nonrepeating and nonterminating decimal representations. The notation $\pi \approx$ 3.14 is used to indicate that π is *approximately equal* to 3.14.

THE REAL NUMBER LINE

Real numbers can be represented geometrically as points on a **number line,** as follows. Given any horizontal line, we choose a point O on the line to represent the number 0. This point is called the **origin.** The side to the right of the origin is designated as the **positive direction** and is indicated by an arrowhead. If we now select a line segment of unit length, we can locate the integers by laying off successive unit lengths on either side of O, as shown in Figure 2. Points corresponding to rational numbers, such as $\frac{1}{2}$ and $\frac{-8}{3}$, can be determined by subdividing the equal line segments. Points corresponding to irrational numbers, such as π, can be located as accurately as desired if we consider the successive rational approximations 3, 3.1, 3.14, 3.141, and so on.

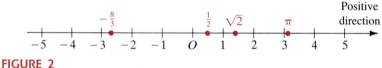

FIGURE 2

This association results in a one-to-one correspondence between the set of real numbers and the set of points on a straight line, called the **real number line, number line,** or **coordinate line.** For any given point P on the number line, the number p, which corresponds to this point, is called the **coordinate** of P. Numbers that correspond to points to the right of O in Figure 2 are called **positive,** whereas those that correspond to points to the left are **negative.**

In general, we will not distinguish between a point on the number line and its coordinate. Thus, for example, we will sometimes refer to the point on the real number line with coordinate 5 as "the point 5."

LESS THAN AND GREATER THAN

Two real numbers a and b, $a \neq b$, can be compared by the order relation **less than,** designated by the symbol $<$. We say that

a is less than b if b − a is positive.

If a is less than b, we write $a < b$.

Equivalently, we can say that b is **greater than** a and write $b > a$. For example, $-7 < 5$, since $5 - (-7) = 12$ is positive. Alternatively, we can write $5 > -7$.

The real number line is useful in demonstrating the order relation less than. Geometrically, $a < b$ means that the point corresponding to a on the number line in Figure 3 lies to the left of the point corresponding to b.

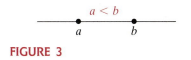

FIGURE 3

EXAMPLE 1

Using the order relation greater than, compare the real numbers π and $\frac{22}{7}$.

Solution From $\pi = 3.1415 \ldots$ and $\frac{22}{7} = 3.1428 \ldots$, we find that

$$\tfrac{22}{7} - \pi = (3.1428 \ldots) - (3.1415 \ldots) = 0.001 \ldots$$

Since this difference is positive, we conclude that $\frac{22}{7} > \pi$.

Two additional order relations are of importance:

1. a **is less than or equal to** b, given by

$$a \leq b \text{ if and only if either } a < b \text{ or } a = b;$$

2. a **is greater than or equal to** b, given by

$$a \geq b \text{ if and only if either } a > b \text{ or } a = b.$$

For example, since $2 = \sqrt{4}$, we can write $2 \geq \sqrt{4}$ or $2 \leq \sqrt{4}$. We can also write $4 \leq 9$, since $4 < 9$.

INEQUALITIES

The relations $a < b$, $a > b$, $a \leq b$, and $a \geq b$ are called **inequalities,** and the symbols $<$, $>$, $\leq$, and $\geq$ are **inequality signs.** If $b > 0$, then b is **positive,** whereas if $b < 0$, then b is **negative.** If $b \geq 0$, then b is not negative and we say that b is **nonnegative.**

The **simultaneous inequality**

$$a < x < b$$

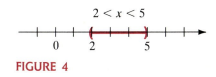

FIGURE 4

means that both $a < x$ *and* $x < b$. For example, the set of real numbers that satisfy $2 < x < 5$ is graphed in Figure 4. To **graph** a set of real numbers, we darken the points on the real number line that correspond to the given set.

Using basic inequalities and simultaneous inequalities, we can describe certain sets of real numbers called **intervals.** Corresponding to these intervals is a special interval notation and terminology shown in the following table.

INTERVAL	INTERVAL NOTATION	NAME	GRAPH
$\{x \mid a < x < b\}$	(a, b)	open interval	
$\{x \mid a \le x \le b\}$	$[a, b]$	closed interval	
$\{x \mid a < x \le b\}$	$(a, b]$	half-open interval	
$\{x \mid a \le x < b\}$	$[a, b)$	half-open interval	
$\{x \mid a < x\}$	(a, ∞)	open half-line	
$\{x \mid x < b\}$	$(-\infty, b)$	open half-line	
$\{x \mid x \le b\}$	$(-\infty, b]$	closed half-line	
$\{x \mid a \le x\}$	$[a, \infty)$	closed half-line	

In the table above, the symbols ∞, read "infinity," and $-\infty$, read "minus infinity," represent certain kinds of intervals called half-lines. These symbols *do not* represent real numbers. For example, the notation $(7, \infty)$ simply means all real numbers greater than 7. Note that a square bracket indicates that the respective endpoint is included in the interval, whereas a parenthesis indicates that the respective endpoint is not included in the interval.

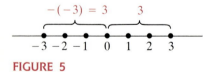

FIGURE 5

ABSOLUTE VALUE

We can also use the real number line to picture distance. As shown in Figure 5, the distance from the point 3 to the origin is 3 units, and the distance from the point -3 to the origin is 3, or $-(-3)$, units. In general, the distance from any number to the origin is the "unsigned value" of that number.

More precisely, as shown in Figure 6, for any positive real number x, the distance from the point x to the origin is x, but for any negative number y, the distance from the point y to the origin is $-y$. Of course, for $x = 0$, the distance to the origin is 0. The concept of the distance from a point on the number line to the origin is described by the **absolute value.**

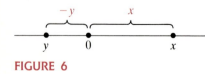

FIGURE 6

DEFINITION 1

For any real number a, the **absolute value** of a, denoted by $|a|$, is

$$|a| = \begin{cases} a, & \text{if } a \geq 0. \\ -a, & \text{if } a < 0. \end{cases}$$

Since 3 and $\sqrt{2}$ are positive numbers,

$$|3| = 3 \quad \text{and} \quad |\sqrt{2}| = \sqrt{2}.$$

But since -3 and $-\sqrt{2}$ are negative numbers,

$$|-3| = -(-3) = 3 \quad \text{and} \quad |-\sqrt{2}| = -(-\sqrt{2}) = \sqrt{2}.$$

EXAMPLE 2

(a) $|2 - 2| = |0| = 0$
(b) $|2 - 5| = |-3| = -(-3) = 3$
(c) $|2| - |-5| = 2 - [-(-5)] = 2 - 5 = -3$

EXAMPLE 3

Find $|\sqrt{2} - 3|$.

Solution To find $|\sqrt{2} - 3|$, we must first determine whether $\sqrt{2} - 3$ is positive or negative. Since $\sqrt{2} \approx 1.4$, we see that $\sqrt{2} - 3$ is a negative number. Thus,

$$|\sqrt{2} - 3| = -(\sqrt{2} - 3) = -\sqrt{2} + 3 = 3 - \sqrt{2}.$$

Note of Caution: It is a common mistake to think that $-y$ represents a negative number because the symbol y is preceded by a minus sign. We emphasize that if y represents a negative number, then $-y$ is a positive number. Hence, if y is *negative,* then $|y| = -y$.

EXAMPLE 4

Find $|x - 6|$ if (a) $x > 6$, (b) $x = 6$, and (c) $x < 6$.

Solution

(a) If $x > 6$, then $x - 6$ is positive. From the definition of absolute value, we conclude that $|x - 6| = x - 6$.

(b) If $x = 6$, then $x - 6 = 0$; hence, $|x - 6| = |0| = 0$.

(c) If $x < 6$, then $x - 6$ is negative and we have that $|x - 6| = -(x - 6) = 6 - x$.

For any real number x and its negative, $-x$, the distance to the origin is the same. That is, $|x| = |-x|$. This is one of several special properties of the absolute value, which we now list.

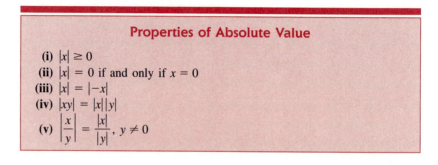

Properties of Absolute Value

(i) $|x| \geq 0$

(ii) $|x| = 0$ if and only if $x = 0$

(iii) $|x| = |-x|$

(iv) $|xy| = |x||y|$

(v) $\left|\dfrac{x}{y}\right| = \dfrac{|x|}{|y|}$, $y \neq 0$

Restating these properties in words is one way of increasing your understanding. For example, property (i) states that the absolute value of a quantity is always nonnegative. Property (iv) says that the absolute value of a product equals the product of the absolute values of the two factors.

Another important property of the absolute value is the **Triangle Inequality.**

Triangle Inequality

$$|a + b| \leq |a| + |b|.$$

DISTANCE BETWEEN POINTS

The concept of absolute value describes the distance from a point to the origin. It is also useful in finding the distance between two points on the number line. Since we want to describe distance as a positive quantity, we subtract one coordinate from the other and take the absolute value of the difference.

DEFINITION 2

If a and b are two points on the real number line, the **distance** from a to b is

$$d(a, b) = |b - a|.$$

See Figure 7.

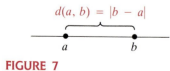

$d(a, b) = |b - a|$

$a \qquad b$

FIGURE 7

■■■ EXAMPLE 5

(a) The distance from -5 to 2 is

$$d(-5, 2) = |2 - (-5)| = 7.$$

(b) The distance from 3 to $\sqrt{2}$ is

$$d(3, \sqrt{2}) = |\sqrt{2} - 3| = 3 - \sqrt{2}.$$ ■■■

We see that the distance from a to b is the same as the distance from b to a, since by property (iii),

$$d(a, b) = |b - a| = |-(b - a)| = |a - b| = d(b, a).$$

Thus,

$$d(a, b) = d(b, a).$$

We have already found it convenient to use letters, such as x or y, to represent numbers. Such a symbol is called a **variable.** Unless otherwise specified, throughout this text variables will represent real numbers. The **domain** of the variable in an expression such as $\sqrt{x}$ consists of all real numbers x for which the expression represents a real number. Thus, for $\sqrt{x}$, the domain is the set of all nonnegative real numbers $\{x \,|\, x \geq 0\}$. Since in the expression $3/(x + 1)$ we must avoid dividing by zero, the domain is the set of all real numbers except $x = -1$, that is, $\{x \,|\, x \neq -1\}$.

When two expressions are set equal, an **equation** is obtained. For example,

$$\sqrt{x - 1} = 2, \quad x^2 - 1 = (x + 1)(x - 1), \quad \text{or} \quad |x + 1| = 5$$

are equations in the variable x.

A **solution,** or **root,** of an equation is any number that when substituted into the equation makes it a true statement. A number is said to **satisfy an equation** if it is a solution of the equation. To **solve an equation** means to find its solution.

■■■ EXAMPLE 6

The number 2 is a solution of $3x - 2 = x + 2$ since

$$3(2) - 2 = 2 + 2.$$

As we will see later, there are no other values of x that satisfy this equation. ■■■

An equation is called an **identity** if it is satisfied by all numbers in the domain of the variable. If there is at least one number in the domain of the variable for which the equation is *not* satisfied, then it is said to be a **conditional equation.**

▮ EXAMPLE 7

(a) The equation

$$\frac{x^2 - 1}{x - 1} = x + 1$$

is satisfied by the set of all real numbers except $x = 1$. Since 1 is not in the domain of the variable, the equation is an identity.

(b) The number 3 is in the domain of the variable in the equation

$$4x - 1 = 2,$$

but it does not satisfy this equation since $4(3) - 1 \neq 2$. Thus, $4x - 1 = 2$ is a conditional equation. ▮

We say that two equations are **equivalent** if they have the same solutions. For example,

$$2x - 1 = 0, \qquad 2x = 1, \quad \text{and} \quad x = \tfrac{1}{2}$$

are equivalent equations. Generally, we solve an equation by finding an equivalent equation with solutions that are easily determined.

We assume that you are familiar with the basic techniques for solving linear and quadratic equations in one variable. For reference, we state the **quadratic formula** here.

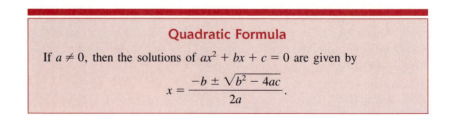

Quadratic Formula

If $a \neq 0$, then the solutions of $ax^2 + bx + c = 0$ are given by

$$x = \frac{-b \pm \sqrt{b^2 - 4ac}}{2a}.$$

As illustrated in the following example, **factoring** is another important technique for solving equations.

▮ EXAMPLE 8

Solve $4x^3 + 8x^2 - 4x = 0$.

Solution Factoring, we write the equation as

$$4x(x^2 + 2x - 1) = 0$$

or

$$x(x^2 + 2x - 1) = 0.$$

It follows that $x = 0$ or $x^2 + 2x - 1 = 0$.

Solving the second equation by the quadratic formula, we have

$$x = \frac{-2 \pm \sqrt{4 - 4(1)(-1)}}{2(1)}$$

$$= \frac{-2 \pm \sqrt{4 + 4}}{2}$$

$$= \frac{-2 \pm 2\sqrt{2}}{2}$$

$$= -1 \pm \sqrt{2}.$$

Therefore, the solutions are 0, $-1 - \sqrt{2}$, and $-1 + \sqrt{2}$.

When $b^2 - 4ac \geq 0$, the roots of a quadratic equation $ax^2 + bx + c = 0$ are real numbers. When $b^2 - 4ac < 0$, the quadratic equation has no real roots. This last case will be considered in greater detail in Section 5.3.

EXERCISE 1.1

For Problems 1–14, answer true or false.

1. $\frac{1}{3}$ is an element of Z. ___
2. $-\frac{1}{2}$ is an element of Q. ___
3. $\sqrt{3}$ is an element of R. ___
4. $\sqrt{2}$ is a rational number. ___
5. $0.1333\ldots$ is an irrational number. ___
6. 1.5 is a rational number. ___
7. $0.121212\ldots$ is a rational number. ___
8. $\frac{8}{0}$ is an element of Q. ___
9. -4 is an element of Z, but -4 is not an element of N. ___
10. π is an element of R, but π is not an element of Q. ___
11. Every irrational number is a real number. ___
12. Every integer is a rational number. ___
13. Every decimal number is a real number. ___
14. Every decimal can be written as a quotient of two integers. ___
15. Construct a number line and locate on it the points

$$0, \ -\tfrac{1}{2}, \ 1, \ -1, \ 2, \ -2, \ \tfrac{4}{3}, \ 2.5.$$

16. Construct a number line and locate on it the points

$$0, \ 1, \ -1, \ \sqrt{2}, \ -3, \ -\sqrt{2} + 1.$$

In Problems 17–24, write the given statement as an inequality.

17. x is positive
18. y is negative
19. $x + y$ is nonnegative
20. a is less than -3
21. b is greater than or equal to 100
22. $c - 1$ is less than or equal to 5
23. $|t - 1|$ is less than 50
24. $|s + 4|$ is greater than or equal to 7

In Problems 25–30, compare the given pair of numbers using the order relation "less than."

25. $15, \ -3$
26. $-9, \ 0$
27. $\dfrac{4}{3}, \ 1.33$
28. $\dfrac{-7}{15}, \ \dfrac{-5}{11}$
29. $\pi, \ 3.14$
30. $1.732, \ \sqrt{3}$

In Problems 31–36, compare the given pair of numbers using the order relation "greater than or equal to."

31. $-2, -7$

32. $\dfrac{-1}{7}, -0.143$

33. $2.5, \dfrac{5}{2}$

34. $0.333, \dfrac{1}{3}$

35. $\dfrac{423}{157}, 2.6$

36. $\sqrt{2}, 1.414$

In Problems 37–44, write the inequality using interval notation and graph the interval.

37. $x < 0$

38. $0 < x < 5$

39. $x \geq 5$

40. $-1 \leq x$

41. $8 < x \leq 10$

42. $-5 < x \leq -3$

43. $-2 \leq x < 1$

44. $x > -4$

In Problems 45–66, find the absolute value.

45. $|7|$

46. $|-7|$

47. $|22|$

48. $\left|\dfrac{22}{7}\right|$

49. $\left|\dfrac{-22}{7}\right|$

50. $|\sqrt{5}|$

51. $|-\sqrt{5}|$

52. $|0.13|$

53. $|\pi - 4|$

54. $|2 - 6|$

55. $|6 - 2|$

56. $||2| - |6||$

57. $|-6| - |-2|$

58. $|\sqrt{5} - 3|$

59. $|3 - \sqrt{5}|$

60. $|8 - \sqrt{7}|$

61. $|\sqrt{7} - 8|$

62. $|-(\sqrt{7} - 8)|$

63. $|\sqrt{5} - 2.3|$

64. $\left|\dfrac{\pi}{2} - 1.57\right|$

65. $|6.28 - 2\pi|$

66. $|\sqrt{17} - 4.123|$

In Problems 67–78, write the expression without using absolute value symbols.

67. $|h|$, if h is negative

68. $|-h|$, if h is negative

69. $|x - 2|$, if $x < 2$

70. $|x - 2|$, if $x = 2$

71. $|x - 2|$, if $x > 2$

72. $|5 - x|$, if $x < 5$

73. $|5 - x|$, if $x = 5$

74. $|5 - x|$, if $x > 5$

75. $|x - y| - |y - x|$

76. $\dfrac{|x - y|}{|y - x|}$, $x \neq y$

77. $\dfrac{|t|}{|t|}$, $t \neq 0$

78. $\dfrac{|z|}{-|z|}$, $z \neq 0$

79. For what values of x is it true that $x \leq |x|$?

80. For what values of x is it true that $x = |x|$?

81. Use Definition 1 to prove that $|xy| = |x||y|$ for any real numbers x and y.

82. Use Definition 1 to prove that

$$\frac{|x|}{|y|} = \left|\frac{x}{y}\right|$$

for any real number x and any nonzero real number y.

In Problems 83–90, find the distance between the given points.

83. $7, 3$

84. $2, 5$

85. $0.6, 0.8$

86. $-100, 255$

87. $-5, -8$

88. $6, -4.5$

89. $\dfrac{3}{2}, -\dfrac{3}{2}$

90. $-\dfrac{1}{4}, \dfrac{7}{4}$

91. Find c so that $3(y - c) = 3y + 7$ is an identity.

92. Find a so that $(x - 1)(x + a) = x^2 - 2x - a$ is an identity.

In Problems 93–106, solve the equation.

93. $3s^2 - 13s + 4 = 0$

94. $4x^2 + 8x + 4 = 0$

95. $s^2 - 4s - 4 = 0$

96. $2.4 + 1.0y + 0.1y^2 = 0$

97. $8t^2 + 10t + 5 = 0$

98. $r^2 + 2r = 35$

99. $24t^3 - 3t = 0$

100. $9u^2 + 25 = 30u$

101. $4p^2 = 60$

102. $5(c + 1)^2 = 25$

103. $x^3 - 9x = 0$

104. $16p^4 - p^2 = 0$

105. $4q^5 - 25q^3 = 0$

106. $x^4 - 18x^2 + 32 = 0$

107. The Rhind papyrus (circa 1700 B.C.) indicates that the Egyptians used $(16/9)^2$ as the value of π.
   (a) Is this approximation larger or smaller than π?
   (b) Show that the error in using this approximation is less than 1% of π.

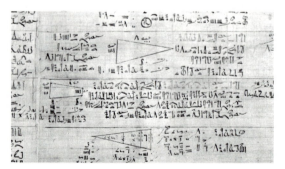

108. Using the fact that the circumference of a circle is π times the diameter, determine what value of π is implied from the following biblical quotation. "Also he made the molten sea of ten cubits from brim to brim, round in compass, and the height thereof was five cubits; and a line of thirty cubits did compass it round about." (This is taken from II Chronicles 4:2 and I Kings 7:23, which date from the tenth century B.C.)

109. Show that $\sqrt{2}$ cannot be written as a quotient of integers. [*Hint:* Assume that there is a fraction p/q, reduced to lowest terms, such that $(p/q)^2 = 2$. This simplifies to $p^2 = 2q^2$, which implies that p^2, and hence p, is an even integer, say $p = 2r$. Make this substitution and consider $(2r/q)^2 = 2$. You should arrive at a contradiction to the fact that p/q was reduced to lowest terms.]

110. A company that owned one manufacturing plant next to a river bought two additional manufacturing plants, one x miles upstream and the other y miles downstream. Now the company wants to build a processing plant located so that the total shipping distance from the processing plant to the three manufacturing plants is a minimum. Use the Triangle Inequality to show that the processing plant should be built at the same location as the original manufacturing plant. [*Hint:* Think of the plants as being located at 0, x, and $-y$ on a number line. (See Figure 8.) Using absolute values, find an expression for the total shipping distance if the processing plant is located at point d.]

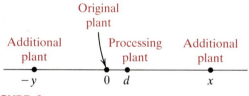

FIGURE 8

111. Some secret codes work by shifting letters of the alphabet. Figure 9 shows a shift of 2. Each letter in a message can be shifted by a different amount. Such a coding scheme can be represented by the digits in a decimal number. For example, the decimal number .12121212 . . . codes the message STUDY MATH into TVVFZ OBVI. If using 9/37 produces the coded message RCWJEJQVDU PLXIV, what was the original message?

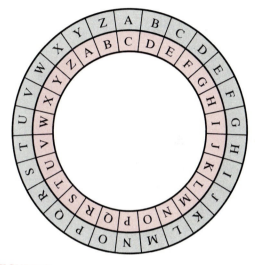

FIGURE 9

112. Is the product of two irrational numbers necessarily irrational? Explain.

113. Is the quotient of two irrational numbers necessarily irrational? Explain.

114. Determine under what conditions equality holds in the Triangle Inequality. That is, when is it true that $|a + b| = |a| + |b|$?

115. Use the Triangle Inequality to show $|a - b| \leq |a| + |b|$.

116. Use the Triangle Inequality to show $|a - b| \geq |a| - |b|$. [*Hint:* $a = (a - b) + b$.]

The Cartesian Coordinate System, Relations, and Graphs

1.2

Earlier we saw that each real number can be associated with exactly one point on the number line. We now examine a correspondence between points in the plane and ordered pairs of real numbers.

CARTESIAN COORDINATE SYSTEM

A **Cartesian,** or **rectangular, coordinate system*** is formed in a plane by two perpendicular number lines that intersect at the point corresponding to the number 0 on each line. This point of intersection is called the **origin** and is denoted by O. The horizontal and vertical number lines are often called the **x-axis** and the **y-axis,** respectively. The axes divide the plane into four regions, called **quadrants,** which are numbered as shown in Figure 10(a). As we can see in Figure 10(b), the scales on the x- and y-axes need not be the same. A plane containing a rectangular coordinate system is called a **Cartesian plane,** a **coordinate plane,** or an **xy-plane.**

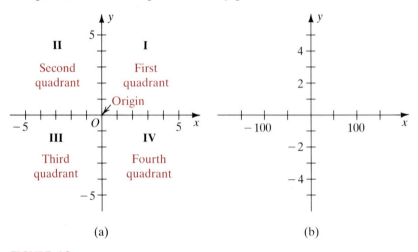

(a)   (b)

FIGURE 10

ABSCISSA AND ORDINATE

FIGURE 11

Let P be a point in a Cartesian plane. We associate an ordered pair of real numbers with P by drawing a vertical line from P to the x-axis and a horizontal line from P to the y-axis. If the vertical line intersects the x-axis at a and the horizontal line intersects the y-axis at b, we associate the ordered pair (a, b) with the point P (see Figure 11). Conversely, to each ordered pair (a, b) of real numbers, there corresponds a point P in the plane. This point lies at the intersection of the vertical line passing through a on the x-axis and the horizontal line passing through b on the y-axis. Hereafter we will refer to an ordered pair as a **point** and denote it by $P(a, b)$ or, simply, (a, b).† We call

---

*This system was named after the French philosopher, soldier, and mathematician René Descartes (1596–1650).
†This is the same notation used to denote an open interval. It should be clear from the context of the discussion whether we are considering a point (a, b) or an open interval (a, b).

a the **x-coordinate,** or **abscissa,** of *P* and *b* the **y-coordinate,** or **ordinate,** of *P* (see Figure 11).

The algebraic signs of the *x*-coordinate and the *y*-coordinate of any point (x, y) in each of the four quadrants are indicated in Figure 12. Points on either of the axes are not considered to be in any quadrant. When we locate a point in a Cartesian plane corresponding to an ordered pair of numbers and represent it using a dot, we say that we **plot** the point.

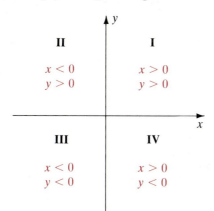

FIGURE 12

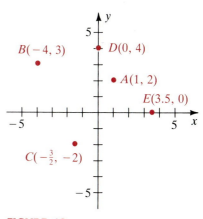

FIGURE 13

EXAMPLE 1

Plot the points $A(1, 2)$, $B(-4, 3)$, $C(-\frac{3}{2}, -2)$, $D(0, 4)$, and $E(3.5, 0)$. Specify in which quadrant each point lies.

Solution   The five points are plotted in the Cartesian plane in Figure 13. Point *A* is in quadrant I, *B* in quadrant II, and *C* is in quadrant III. Points *D* and *E*, which lie on the *y*- and *x*-axes, respectively, are not in any quadrant.

RELATIONS AND GRAPHS

In general, any set of ordered pairs of real numbers is called a **relation,** and the corresponding set of points in the plane is called the **graph of the relation.**

EXAMPLE 2

Graph the relation $S = \{(-1, 2), (0, 4), (3, -\frac{5}{2}), (-3, -1)\}$.

Solution   In Figure 14 we have plotted the four points that correspond to the ordered pairs in the relation *S*.

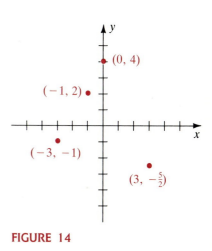

FIGURE 14

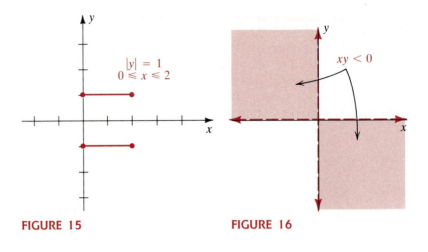

FIGURE 15   **FIGURE 16**

■ EXAMPLE 3

Graph the relation $T = \{(x, y) \mid 0 \le x \le 2, |y| = 1\}$.

Solution   First, recall that $|y| = 1$ implies that $y = 1$ or $y = -1$. Thus, to graph the relation T, we plot the points whose x-coordinates are numbers in the interval $[0, 2]$ and whose y-coordinates are either 1 or -1. See Figure 15.   ■

■ EXAMPLE 4

Graph the set of points (x, y) in the plane that satisfy the condition $xy < 0$.

Solution   A product of two numbers is negative when one of the numbers is positive and the other is negative. Thus, $xy < 0$ when $x > 0$ and $y < 0$ or when $x < 0$ and $y > 0$. We see from Figure 12 that $xy < 0$ for all points (x, y) in quadrants II and IV. Thus we can represent the set of points (x, y) for which $xy < 0$ by the shaded regions in Figure 16. The coordinate axes are shown as dashed lines to indicate that the points on these lines are not included in the solution.   ■

THE GRAPH OF AN EQUATION

In the preceding two examples we examined graphs of relations defined by inequalities. We now consider relations defined by an equation relating two variables x and y. We call the set of points in the plane corresponding to the ordered pairs (x, y) in the relation the **graph of the equation.**

■■ **EXAMPLE 5**

Graph the equation $y = x^2$.

Solution   Since, in general, the graph of an equation consists of infinitely many points, we plot a few of them and join these with a smooth curve, as shown in Figure 17.

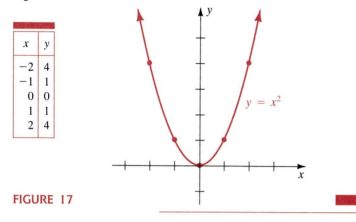

x	y
-2	4
-1	1
0	0
1	1
2	4

$y = x^2$

FIGURE 17

Note of Caution: Several problems can arise with the technique of point plotting. You must plot enough points to be able to discern the shape of the graph. But what is "enough points"? Six points? Twenty points? See Figure 18. Of course, the answer depends on both the equation that you are graphing and your experience. As you gain experience in mathematics you will find that there are often ways to graph an equation while keeping the plotting of points to a minimum. Also, not every graph is a "smooth curve." See Figure 19. Because of the sharp peak at the origin, the curve joining the points in Figure 19(b) is not considered "smooth." Thus when plotting points, proceed with caution.

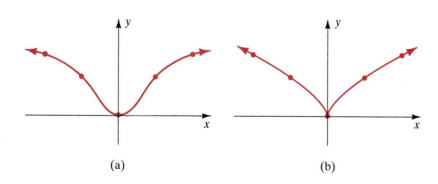

(a)   (b)

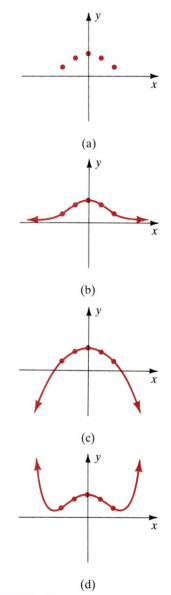

(a)

(b)

(c)

(d)

FIGURE 18

Five points joined by different smooth curves.

FIGURE 19

Five points joined by a curve: (a) smooth; (b) not smooth.

■ EXAMPLE 6

Graph the equation $y = \sqrt{x}$.

Solution   In order to obtain real values for y, we observe that x cannot be negative. We plot the points corresponding to the ordered pairs listed in the accompanying table, and, as in Example 5, join them with a smooth curve (see Figure 20).

x	y
0	0
1	1
2	$\sqrt{2} \approx 1.41$
3	$\sqrt{3} \approx 1.73$
4	2

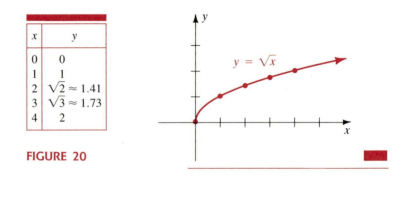

FIGURE 20

■ EXAMPLE 7

Graph the equation $x = 1$.

Solution   The x-coordinate of each point that satisfies this equation must equal 1. Since y does not appear explicitly in the equation, it is understood that the y-coordinate of a point satisfying the equation can be any real number. As we see in Figure 21, the graph of $x = 1$ is a vertical line one unit to the right of the y-axis.

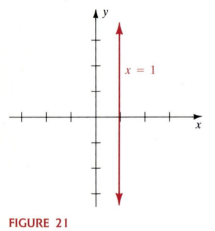

FIGURE 21

INTERCEPTS

Locating the points at which the graph of an equation crosses the coordinate axes can be helpful in sketching its graph. The **x-intercepts** of a graph of an equation are the x-coordinates of the points at which the graph crosses the x-axis. Since every point on the x-axis has y-coordinate 0, the x-intercepts (if there are any) can be determined from the given equation by setting $y = 0$ and solving for x. In turn, the **y-intercepts** of the graph of an equation are the y-coordinates of the points at which the graph crosses the y-axis. These values can be found by setting $x = 0$ in the equation and solving for y. See Figure 22.

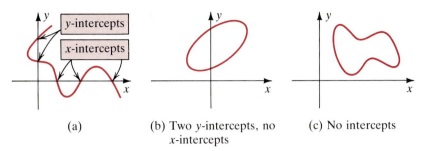

(a)

(b) Two *y*-intercepts, no
x-intercepts

(c) No intercepts

FIGURE 22

SYMMETRY

A graph can also possess **symmetry.** Figure 17 shows that the graph of $y = x^2$ is symmetric with respect to the *y*-axis since the portion of the graph that lies in the second quadrant is the mirror image of that portion of the graph in the first quadrant. As Figure 23 illustrates, a graph is **symmetric with respect to the y-axis** if whenever (x, y) is a point on the graph, $(-x, y)$ is also a point on the graph. A graph is **symmetric with respect to the x-axis** if whenever (x, y) is a point on the graph, $(x, -y)$ is also a point on the graph. Finally, a graph is **symmetric with respect to the origin** if whenever (x, y) is a point on the graph, $(-x, -y)$ is also a point on the graph. When graphing an equation, we can determine whether its graph possesses any of these symmetries *before* plotting points.

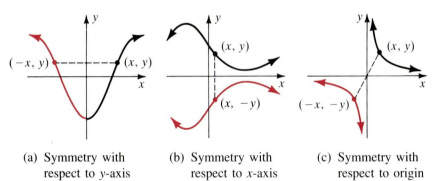

(a) Symmetry with
respect to *y*-axis

(b) Symmetry with
respect to *x*-axis

(c) Symmetry with
respect to origin

FIGURE 23

Tests for Symmetry

The graph of an equation is symmetric with respect to:

(i) the **y-axis** if replacing *x* by $-x$ results in an equivalent equation;
(ii) the **x-axis** if replacing *y* by $-y$ results in an equivalent equation;
(iii) the **origin** if replacing *x* by $-x$ and *y* by $-y$ results in an equivalent equation.

The advantage of using symmetry in graphing should be apparent: If, say, the graph of an equation is symmetric with respect to the y-axis, then we need only plot points for $x \geq 0$ since points on the graph for $x < 0$ are obtained by taking the mirror images, through the y-axis, of the points in the first and fourth quadrants.

■■■ EXAMPLE 8

By replacing x by $-x$ in $y = x^2$ and using $(-x)^2 = x^2$, we see that

$$y = (-x)^2 \quad \text{is equivalent to} \quad y = x^2.$$

This proves what is apparent in Figure 17: The graph of $y = x^2$ is symmetric with respect to the y-axis.   ■■

The next example illustrates how to use these new tools as aids in graphing.

■■■ EXAMPLE 9

Graph the equation $x + y^2 = 10$.

Solution

Intercepts: Setting $y = 0$ in the equation immediately gives $x = 10$. Thus the x-intercept is 10. When $x = 0$, we get $y^2 = 10$, which implies that $y = -\sqrt{10}$ or $y = \sqrt{10}$. The y-intercepts are then $-\sqrt{10}$ and $\sqrt{10}$.

Symmetry: The graph is symmetric with respect to the x-axis since, replacing y by $-y$, we find that

$$x + (-y)^2 = 10 \quad \text{is equivalent to} \quad x + y^2 = 10.$$

Plotting points: The entries in the accompanying table were obtained by assigning values to y. We note that because of the symmetry we need only consider $y \geq 0$. The color points in Figure 24(a) in the third and fourth quadrants were obtained by taking the mirror images of the points that were plotted in the first and second quadrants.

Using all the information above, we obtain the graph in Figure 24(b).

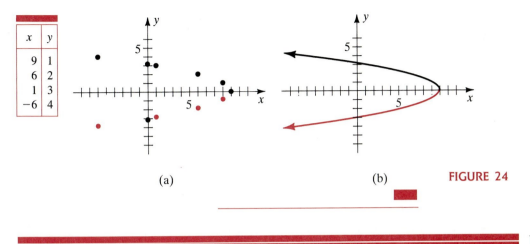

x	y
9	1
6	2
1	3
−6	4

(a)

(b)

FIGURE 24

EXERCISE 1.2

In Problems 1–4, plot the given points.

1. (2, 3), (4, 5), (0, 2), (−1, −3)
2. (1, 4), (−3, 0), (−4, 2), (−1, −1)
3. (−½, −2), (0, 0), (−1, ⁴⁄₃), (3, 3)
4. (0, 0.8), (−2, 0), (1.2, −1.2), (−2, 2)

In Problems 5–16, determine the quadrant in which the given point lies if (a, b) is in quadrant I.

5. (−a, b) **6.** (a, −b)
7. (−a, −b) **8.** (b, a)
9. (−b, a) **10.** (−b, −a)
11. (a, a) **12.** (b, −b)
13. (−a, −a) **14.** (−a, a)
15. (b, −a) **16.** (−b, b)
17. Plot the points given in Problems 5–16 if (a, b) is the point shown in Figure 25.

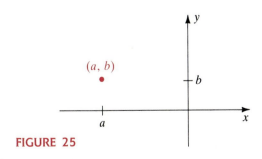

(a, b)

b

a

x

FIGURE 25

18. Give the coordinates of the points shown in Figure 26.

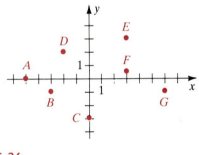

E

D

A 1

F

B

C

G

x

FIGURE 26

In Problems 19–22, graph the given relation.

19. $\{(x, y) \mid xy = 0\}$ **20.** $\{(x, y) \mid xy \le 0\}$
21. $\{(x, y) \mid xy > 0\}$ **22.** $\{(x, y) \mid -1 \le x \le 1, y = 2\}$

In Problems 23–30, test for symmetry with respect to the x- and y-axes and the origin. Do not graph.

23. $y = x^5 - x^3 + 2x$ **24.** $y = 7x^3 + 8x - 2$
25. $y = (x^3 - x)^2$ **26.** $y = x\sqrt{x^2 + 1}$
27. $x^{2/3} + y^{2/3} = 9$ **28.** $x^2y + 3x^2 = y$
29. $xy = 4$ **30.** $x^2y = 1$

In Problems 31–68, graph the given equation. Find intercepts. Use symmetry when possible.

31. $x = -3$

32. $y = 4$

33. $y = x$

34. $y = -x$

35. $x - y = 1$

36. $2x + 3y = 6$

37. $y = -x^2$

38. $y^2 = x$

39. $y = x^2 - 3$

40. $y = x^2 + 1$

41. $y = (x + 1)^2$

42. $y = (x - 1)^2$

43. $y^2 - x = 9$

44. $y^2 = 16(x + 4)$

45. $y = x^3$

46. $y = -x^3$

47. $y^3 = x$

48. $y = x^4$

49. $y^2 = x^2$

50. $y^2 = x^3$

51. $y = |x|$

52. $|y| = x$

53. $|x + y| = 0$

54. $|x - y| = 0$

55. $|x - y| = 2$

56. $x = |y| - 2$

57. $y^2 = 9$

58. $x^2 - 16 = 0$

59. $(x - 1)(y + 2) = 0$

60. $(2x + 9)(5 - y) = 0$

61. $x^2 = x$

62. $y^2 - 3y - 10 = 0$

63. $x^2 + y^2 = 0$

64. $x^2 + y^2 = 4$

65. $x^2 + 4y^2 = 16$

66. $\dfrac{x^2}{25} + \dfrac{y^2}{9} = 1$

67. $x^2 - y^2 = 1$

68. $y^2 - x^2 = 9$

In Problems 69–72, use symmetry to complete the graph.

69. The graph is symmetric with respect to the y-axis.

70. The graph is symmetric with respect to the x-axis.

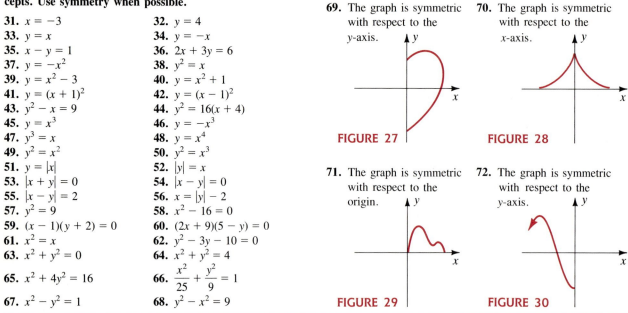

FIGURE 27   FIGURE 28

71. The graph is symmetric with respect to the origin.

72. The graph is symmetric with respect to the y-axis.

FIGURE 29   FIGURE 30

1.3

The Distance Formula and the Circle

In this section we develop a formula for finding the distance between two points in the plane whose coordinates are known. We then apply this formula to the problem of finding the equation of a circle.

THE DISTANCE FORMULA

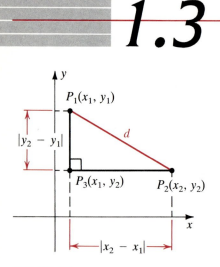

FIGURE 31

Suppose that $P_1(x_1, y_1)$ and $P_2(x_2, y_2)$ are two distinct points not on a vertical or a horizontal line. Then, as shown in Figure 31, P_1, P_2, and $P_3(x_1, y_2)$ are vertices of a right triangle. The length of the side P_3P_2 is $|x_2 - x_1|$, and the length of the side P_1P_3 is $|y_2 - y_1|$. If we denote the length of P_1P_2 by d, we have

$$d^2 = |x_2 - x_1|^2 + |y_2 - y_1|^2 \tag{1}$$

from the Pythagorean theorem. Since the square of any real number is equal to the square of its absolute value, we can drop the absolute value signs in (1). It then follows from (1) that

$$d = \sqrt{(x_2 - x_1)^2 + (y_2 - y_1)^2}.$$

Although we derived this equation for two points not on a vertical or a horizontal line, it holds in these cases as well.

Distance Formula

The distance $d(P_1, P_2)$ between any two points $P_1(x_1, y_1)$ and $P_2(x_2, y_2)$ is given by

$$d(P_1, P_2) = \sqrt{(x_2 - x_1)^2 + (y_2 - y_1)^2}. \tag{2}$$

Since $(x_2 - x_1)^2 = (x_1 - x_2)^2$, it makes no difference which point is used first in the distance formula; that is,

$$d(P_1, P_2) = d(P_2, P_1).$$

■ EXAMPLE 1

Find the distance between $A(8, -5)$ and $B(3, 7)$.

Solution

$$\begin{aligned}
d(A, B) &= \sqrt{(3 - 8)^2 + (7 - (-5))^2} \\
&= \sqrt{(-5)^2 + (12)^2} \\
&= \sqrt{25 + 144} \\
&= \sqrt{169} = 13
\end{aligned}$$

This distance is illustrated in Figure 32.   ■

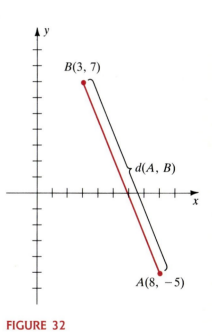

FIGURE 32

■ EXAMPLE 2

Determine whether the points $P_1(7, 1)$, $P_2(-4, -1)$, and $P_3(4, 5)$ are the vertices of a right triangle.

Solution From plane geometry we know that a triangle is a right triangle if and only if the sum of the squares of the lengths of two of its sides is equal to the square of the length of the remaining side. Now, from the distance formula, we find

$$d(P_1, P_2) = \sqrt{(-4 - 7)^2 + (-1 - 1)^2}$$

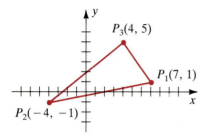

FIGURE 33

$$= \sqrt{121 + 4} = \sqrt{125},$$
$$d(P_2, P_3) = \sqrt{(4 - (-4))^2 + (5 - (-1))^2}$$
$$= \sqrt{64 + 36}$$
$$= \sqrt{100} = 10,$$

and
$$d(P_3, P_1) = \sqrt{(7 - 4)^2 + (1 - 5)^2}$$
$$= \sqrt{9 + 16}$$
$$= \sqrt{25} = 5.$$

Since
$$[d(P_3, P_1)]^2 + [d(P_2, P_3)]^2 = 25 + 100$$
$$= 125$$
$$= [d(P_1, P_2)]^2,$$

it follows that P_1, P_2, and P_3 are the vertices of a right triangle with the right angle at P_3 (see Figure 33).

CIRCLES

The distance formula can be used to find an equation of the set of all points equidistant from a given point.

DEFINITION 3

> A **circle** is the set of all points P in the plane that are a given fixed distance r, called the **radius,** from a given fixed point C, called the **center.**

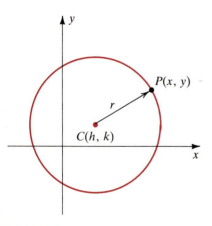

FIGURE 34

In Figure 34 we have sketched a circle of radius r centered at the point $C(h, k)$. From Definition 3 we know that a point $P(x, y)$ is on this circle if and only if

$$d(P, C) = r, \quad \text{or} \quad \sqrt{(x - h)^2 + (y - k)^2} = r.$$

Since $(x - h)^2 + (y - k)^2$ is always nonnegative, we obtain an equivalent equation when both sides are squared.

Equation of a Circle

A circle of radius r with center $C(h, k)$ has the equation

$$(x - h)^2 + (y - k)^2 = r^2. \tag{3}$$

Equation (3) is called the **standard form** of the equation of a circle. When $h = 0$ and $k = 0$, we see from equation (3) that the equation of a circle of radius r with center at the origin is given by

$$x^2 + y^2 = r^2.$$

See Figure 35.

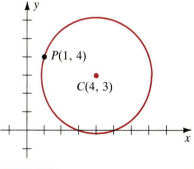

■ EXAMPLE 3

Find the center and the radius of the circle whose equation is $(x - 3)^2 + (y + 2)^2 = 49$.

Solution If we write this equation in the standard form (3),

$$(x - 3)^2 + (y - (-2))^2 = 7^2,$$

we see that $h = 3$, $k = -2$, and $r = 7$. Thus the circle is centered at $(3, -2)$ and has radius 7.

FIGURE 35

■ EXAMPLE 4

Find an equation of the circle centered at $C(-5, 4)$ with radius $\sqrt{2}$.

Solution Using the standard form (3) with $h = -5$, $k = 4$, and $r = \sqrt{2}$, we obtain

$$(x - (-5))^2 + (y - 4)^2 = (\sqrt{2})^2,$$

or

$$(x + 5)^2 + (y - 4)^2 = 2.$$

■ EXAMPLE 5

Find an equation of the circle centered at $C(4, 3)$ passing through $P(1, 4)$.

Solution Identifying $h = 4$ and $k = 3$, we have from (3)

$$(x - 4)^2 + (y - 3)^2 = r^2. \tag{4}$$

Since $P(1, 4)$ lies on the circle (see Figure 36), its coordinates must satisfy (4). Thus,

$$(1 - 4)^2 + (4 - 3)^2 = r^2, \quad \text{or} \quad 10 = r^2,$$

and so the equation in standard form is

$$(x - 4)^2 + (y - 3)^2 = 10.$$

FIGURE 36

In the next example we use the technique of *completing the square* to find the center and the radius of a circle. Recall that adding $(B/2)^2$ to the expression $x^2 + Bx$ yields $x^2 + Bx + (B/2)^2$, which is the perfect square $(x + B/2)^2$.

■ EXAMPLE 6

Find the center and the radius of the circle with equation

$$x^2 + y^2 + 10x - 2y + 17 = 0.$$

Solution We can find the center and the radius of the circle by rewriting the given equation in the standard form $(x - h)^2 + (y - k)^2 = r^2$. By rearranging terms, we have

$$(x^2 + 10x \quad) + (y^2 - 2y \quad) = -17.$$

Now we complete the square of each quantity in parentheses by adding $(10/2)^2$ in the first and $(-2/2)^2$ in the second. Note that we must be careful to add these numbers to both sides of the equation.

$$[x^2 + 10x + (\tfrac{10}{2})^2] + [y^2 - 2y + (\tfrac{-2}{2})^2] = -17 + (\tfrac{10}{2})^2 + (\tfrac{-2}{2})^2$$

$$(x^2 + 10x + 25) + (y^2 - 2y + 1) = 9$$

$$(x + 5)^2 + (y - 1)^2 = 3^2$$

From this equation it follows that the circle is centered at $(-5, 1)$ and has radius 3 (see Figure 37). ■

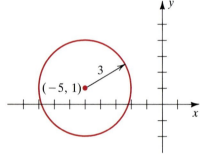

FIGURE 37

We note that not every equation of the form

$$Ax^2 + Ay^2 + Bx + Cy + D = 0$$

is necessarily a circle (see Problems 27 and 28).

EXERCISE 1.3

In Problems 1–6, find the distance between the points.

1. $A(1, 2)$, $B(-3, 4)$ **2.** $A(-1, 3)$, $B(5, 0)$

3. $A(2, 4)$, $B(-4, -4)$ **4.** $A(-12, -3)$, $B(-5, -7)$

5. $A(-\tfrac{3}{2}, 1)$, $B(\tfrac{5}{2}, -2)$ **6.** $A(-\tfrac{5}{3}, 4)$, $B(-\tfrac{2}{3}, -1)$

In Problems 7–10, determine whether the points A, B, and C are vertices of a right triangle.

7. $A(8, 1)$, $B(-3, -1)$, $C(10, 5)$

8. $A(-2, -1)$, $B(8, 2)$, $C(1, -11)$

9. $A(2, 8)$, $B(0, -3)$, $C(6, 5)$

10. $A(4, 0)$, $B(1, 1)$, $C(2, 3)$

11. Determine whether the points $A(0, 0)$, $B(3, 4)$, and $C(7, 7)$ are vertices of an isosceles triangle.

12. Find all points on the y-axis that are 5 units from the point $(4, 4)$.

13. Consider the line segment joining $A(-1, 2)$ and $B(3, 4)$.
   (a) Find an equation that expresses the fact that a point $P(x, y)$ is equidistant from A and from B.
   (b) Describe geometrically the set of points described by the equation in part (a).

14. Use the distance formula to determine whether the points $A(-1, -5)$, $B(2, 4)$, and $C(4, 10)$ lie on a straight line.

15. Kansas City and Chicago are not directly connected by an interstate highway, but each city is connected to St. Louis and Des Moines (see Figure 38). Des Moines is approximately 40 mi east and 180 mi north of Kansas City, St. Louis is approximately 230 mi east and 40 mi south of Kansas City, and Chicago is approximately 360 mi east and 200 mi north of Kansas City. Assume that this part of the midwest is a flat plane and that the connecting highways are straight lines. Which route from Kansas City to Chicago, through St. Louis or through Des Moines, is shorter?

FIGURE 38

16. The coordinates of the **midpoint** of the line segment joining the points $P_1(x_1, y_1)$ and $P_2(x_2, y_2)$ are

$$\left(\frac{x_1 + x_2}{2}, \frac{y_1 + y_2}{2}\right).$$

Find the distance from the midpoint of the line segment joining $A(-1, 3)$ and $B(3, 5)$ to the midpoint of the line segment joining $C(4, 6)$ and $D(-2, -10)$.

In Problems 17–26, find the center and the radius of the given circle.

17. $(x - 1)^2 + (y - 3)^2 = 49$
18. $(x + 3)^2 + (y - 5)^2 = 25$
19. $(x - \frac{1}{2})^2 + (y - \frac{3}{2})^2 = 5$
20. $(x + 5)^2 + (y + 8)^2 = \frac{1}{4}$
21. $x^2 + y^2 + 8y = 0$
22. $x^2 + y^2 + 2x - 4y - 4 = 0$
23. $x^2 + y^2 - 18x - 6y - 10 = 0$
24. $x^2 + y^2 - 16y + 3x + 63 = 0$
25. $8x^2 + 8y^2 + 16x + 64y - 40 = 0$
26. $5x^2 + 5y^2 + 25x + 100y + 50 = 0$

In Problems 27 and 28, show that the given equation does not represent a circle.

27. $x^2 + y^2 + 2y + 9 = 0$
28. $2x^2 + 2y^2 - 2x + 6y + 7 = 0$

In Problems 29–38, find an equation of the circle that satisfies the given conditions.

29. Center $(0, 0)$, radius 1   **30.** Center $(1, -3)$, radius 5
31. Center $(0, 2)$, radius $\sqrt{2}$ **32.** Center $(-9, -4)$, radius $\frac{3}{2}$
33. Endpoints of a diameter at $(-1, 4)$ and $(3, 8)$ [*Hint:* See Problem 16.]
34. Endpoints of a diameter at $(4, 2)$ and $(-3, 5)$
35. Center $(0, 0)$, passing through $(-1, -2)$
36. Center $(4, -5)$, passing through $(7, -3)$
37. Center $(5, 6)$, tangent to the x-axis
38. Center $(-4, 3)$, tangent to the y-axis

In Problems 39–44, graph the given relation.

39. $x^2 + y^2 \geq 9$ **40.** $(x - 1)^2 + (y + 5)^2 \leq 25$
41. $1 < x^2 + y^2 < 4$ **42.** $x^2 + y^2 > 2y$
43. $(x - 2)^2 + (y - 6)^2 = 0$ **44.** $x^2 = -y^2$

Functions and Function Notation   *1.4*

DEFINITION OF A FUNCTION

Using the objects and persons in the world around us, it is easy to make up a *rule of correspondence* that associates, or pairs, the members of one set with the members of another set. For example, to each name in the Los Angeles telephone directory there is a number, to each baby there corresponds a mother, to each car registered in the state of California there is a license plate number, to each book there corresponds an author, to each pitcher for the Dodgers there corresponds a won/lost record, and so on. In mathematics we are interested in a very special type of correspondence called a **function.**

DEFINITION 4

> A **function** from a set X to a set Y is a rule of correspondence that assigns to each element x in X one and only one element y in Y. The set X is called the **domain** of the function.

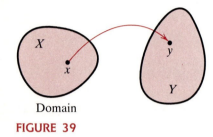

Domain

FIGURE 39

We often use a diagram, such as Figure 39, to illustrate the pairing of all the elements in the set X with some, or all, of the elements in the set Y. Figure 39 also suggests the fundamental characteristic of a function: To each element x in the set X, there corresponds a *unique* element y in the set Y.

For the remainder of this chapter we will assume that both X and Y are subsets of the set R of real numbers.

EXAMPLE 1

x	y
1 → 5
2 → 7
3 → 9
4 → 11

(a)

x	y
1 → 4
2 → 5
3 → 6
↘ 7

(b)

(a) Table (a) shown at left here defines a rule of correspondence between the numbers in the set $\{1, 2, 3, 4\}$ and the numbers in the set $\{5, 7, 9, 11, 13\}$. This correspondence is a function since one and only one number y is associated with a number x. The set $X = \{1, 2, 3, 4\}$ is the domain of the function. Note that we paired all the elements of X with only some of the elements in the set $Y = \{5, 7, 9, 11, 13\}$.

(b) Table (b) defines a rule of correspondence between the set $\{1, 2, 3\}$ and the set $\{4, 5, 6, 7\}$. This correspondence is not a function since there are two numbers—namely, 6 and 7—in the set $\{4, 5, 6, 7\}$ associated with the number 3 in the set $\{1, 2, 3\}$.

ALTERNATIVE DEFINITION

Since a rule of correspondence will generate pairs of elements, we can define a function in an alternative manner:

> A **function** is a set of ordered pairs (x, y) such that no two distinct ordered pairs in the set have the same first element.

▩ EXAMPLE 2

The set of ordered pairs $\{(1, 3), (3, 5), (6, 7), (8, 7)\}$ is equivalent to the correspondence shown in the accompanying table. Since to each value of x there corresponds one and only one value of y, the set of ordered pairs describes a function. Note that it is perfectly acceptable to correspond the same value of y to more than one value of x. ▩

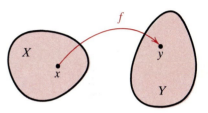

x		y
1	→	3
3	→	5
6	→	7
8	↗	

A function is usually denoted by a letter such as f or g. We can then represent a function f from a set X to a set Y by the notation $f: X \rightarrow Y$ or by a diagram (see Figure 40). The number y in the set Y shown in Figure 40 that is associated with x by the function f is written

$$y = f(x),$$

which is read "y equals f of x." The number $f(x)$ is also said to be the **value** of the function f at x or the **image** of x under f.

A function f is often defined by an explicit formula, such as $f(x) = x^2$. If the domain of f is the set R of real numbers, then $f: R \rightarrow R$, since the square of a real number is a real number. To find values of f, we substitute real numbers for x in the formula $f(x) = x^2$. For example, the value of the function corresponding to $x = 3$ is

$$f(3) = (3)^2 = 9,$$

and the value of the function at $x = -4$ is

$$f(-4) = (-4)^2 = 16.$$

In terms of interval notation, the domain of $f(x) = x^2$ is written as $(-\infty, \infty)$.

Strictly speaking, the function f is the rule given by $y = f(x)$, whereas $f(x)$ is simply the number associated with x. However, we will frequently ignore this distinction and refer to "the function $f(x)$."

A function can also be compared to a computing machine. A number x is the *input* to the "machine," and the corresponding **functional value** $f(x)$ is the *output* obtained after the "machine" has acted on x, as illustrated in Figure 41.

FIGURE 40

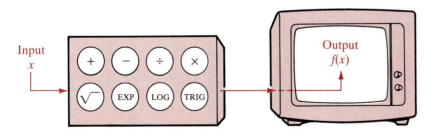

FIGURE 41

EXAMPLE 3

Find the values of $f(x) = \sqrt{x + 4}$ corresponding to $x = 0, 5, 8,$ and 12.

Solution This function "machine" takes a value of x, such as $x = 0$, adds the number 4 to it, and then extracts the square root of this sum. This process is illustrated in Figure 42. Thus

$$f(0) = \sqrt{0 + 4} = \sqrt{4} = 2,$$
$$f(5) = \sqrt{5 + 4} = \sqrt{9} = 3,$$
$$f(8) = \sqrt{8 + 4} = \sqrt{12} = \sqrt{4 \cdot 3} = \sqrt{4}\sqrt{3} = 2\sqrt{3}, \quad \text{and}$$
$$f(12) = \sqrt{12 + 4} = \sqrt{16} = 4.$$

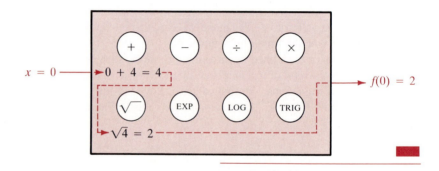

FIGURE 42

DEPENDENT AND INDEPENDENT VARIABLES

Since the value of the variable y in $y = f(x)$ always depends on the choice of x, we say that y is the **dependent variable.** By contrast the choice of x is independent of y, hence x is called the **independent variable.**

EXAMPLE 4

If $f(x) = x^2 - x + 1$, find $f(-1)$, $f(x + h)$, and $f(x^2 + 1)$.

Solution We replace the independent variable x by -1, $x + h$, and $x^2 + 1$ in turn. To emphasize this replacement, we write the original function as

$$f(\quad) = (\quad)^2 - (\quad) + 1.$$

Thus, for the given inputs, we have

$$f(-1) = (-1)^2 - (-1) + 1$$
$$= 1 + 1 + 1$$
$$= 3;$$

$$f(x + h) = (x + h)^2 - (x + h) + 1$$
$$= x^2 + 2xh + h^2 - x - h + 1;$$

and
$$f(x^2 + 1) = (x^2 + 1)^2 - (x^2 + 1) + 1$$
$$= x^4 + 2x^2 + 1 - x^2 - 1 + 1$$
$$= x^4 + x^2 + 1.$$

RANGE

In our analogy with the machine, the domain of a function is the set of all real inputs that result in real outputs. The set of outputs is called the *range* of the function. Formally, we define the **range** of a function f with domain X to be the set $\{f(x) \mid x \text{ in } X\}$. For example, the range of the function $f(x) = x^2$ is the set of nonnegative real numbers.

When a function is defined by a formula,

the domain is understood to be the set of real numbers for which the formula makes sense in the real number system.

This set is sometimes referred to as the *implicit,* or *natural,* domain of the function. For example, the domain of $f(x) = \sqrt{x}$ is the set of all nonnegative real numbers.

EXAMPLE 5

Find the domain of the function $f(x) = \sqrt{4 - x}$.

Solution Since the radicand $4 - x$ must be nonnegative, the domain is determined by the inequality $4 - x \geq 0$; that is, the domain is the set $\{x \mid x \leq 4\}$. Using interval notation we write the domain as $(-\infty, 4]$.

EXAMPLE 6

Find the domain of the function

$$f(x) = \frac{x}{x^2 - 4}.$$

Solution A quotient of two real numbers is a real number unless the denominator is zero. Therefore, the domain of the function f consists of all real numbers x *except* those satisfying

$$x^2 - 4 = 0.$$

The solutions of this equation are $x = 2$ and $x = -2$. Thus the domain of the function is the set $\{x \mid x \neq \pm 2\}$.   ▰▰

▰▰ EXAMPLE 7

The domain of the function

$$g(x) = \frac{x}{x^2 + 4}$$

is the set R of real numbers, since there is no real number x for which $x^2 + 4 = 0$.   ▰▰

▰▰ EXAMPLE 8

Find the domain and the range of $g(x) = 5 + \sqrt{x - 3}$.

Solution The domain, determined by the requirement $x - 3 \geq 0$, is $[3, \infty)$. Since $\sqrt{x - 3} \geq 0$ for $x \geq 3$, we have $5 + \sqrt{x - 3} \geq 5$ for these same values of x. Thus the range of g is $[5, \infty)$.   ▰▰

APPLICATIONS

Many formulas from geometry and science define functions. For example, the area A of a square is a function of the length of one side: If x denotes the length of a side of a square, then $A = x^2$. The area A and the circumference C of a circle are functions of its radius r: $A = \pi r^2$ and $C = 2\pi r$. The distance s that a body falls under the influence of gravity is a function of time t: $s = 16t^2$.

In applications, it is often necessary to set up a functional relationship between two variables by interpreting written data. Consider the following example.

▰▰ EXAMPLE 9

Water is being pumped into a conical tank whose height is 12 ft and whose radius is 4 ft. Express the volume of the water at any time as a function of its depth.

Solution The conical tank is illustrated in Figure 43(a), and a cross-sectional view of the tank is shown in Figure 43(b). We have introduced the variables r and h to denote the radius and the depth of the water, respectively. Now we see that the volume of the water is the volume of a right circular

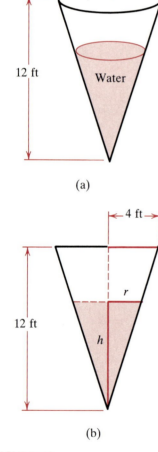

(a)

(b)

FIGURE 43

(a) Conical tank; (b) cross-sectional view.

cone. From geometry we know that the volume of such a cone is given by

$$V = \frac{\pi}{3} r^2 h. \qquad (5)$$

Since the right triangles shown in Figure 43(b) are similar, the lengths of their sides are proportional: $r/h = 4/12$. Substituting $r = \frac{1}{3}h$ into (5) then gives the volume of the water at any time as a function of its depth h:

$$V = \frac{\pi}{3}\left(\frac{1}{3}h\right)^2 h, \quad \text{or} \quad V = \frac{\pi}{27}h^3. \qquad ■$$

■ **EXAMPLE 10**

Express the area A of a circle as a function of its circumference C.

Solution The area A and the circumference C of a circle are functions of the radius r of the circle:

$$A = \pi r^2 \quad \text{and} \quad C = 2\pi r.$$

Now from the second of these equations we get $r = C/2\pi$. Substituting this expression into the first equation gives A as a function of C:

$$A = \pi(C/2\pi)^2, \quad \text{or} \quad A = \frac{1}{4\pi}C^2. \qquad ■$$

EXERCISE 1.4

In Problems 1–6, determine whether the correspondence given by the set of ordered pairs (x, y) is a function.

1. $\{(1, 2), (2, -3), (3, 4), (-4, -1), (1, 5)\}$
2. $\{(-1, 5), (7, 2), (3, -4)\}$
3. $\{(0, 0), (1, 1), (2, 2)\}$
4. $\{(4, 2), (-4, 3), (8, 6), (5, 4)\}$
5. $\{(0, 1), (1, 1), (2, 1)\}$
6. $\{(3, 2), (-6, 2), (-3, 9), (-6, 9)\}$
7. If $f(x) = x^2 - 1$, find $f(0)$, $f(1)$, $f(\sqrt{2})$, $f(-2)$, $f(x + h)$, and $\dfrac{f(2 + h) - f(2)}{h}$.
8. If $f(x) = x^2 + x$, find $f(0)$, $f(1)$, $f(\sqrt{2})$, $f(-2)$, $f(x + h)$, and $\dfrac{f(3 + h) - f(3)}{h}$.

9. If $f(x) = \sqrt{x + 1}$, find $f(0)$, $f(3)$, $f(-1)$, and $f(5)$.
10. If $f(x) = \sqrt{2x + 4}$, find $f(0)$, $f(4)$, $f(\frac{1}{2})$, and $f(-\frac{1}{2})$.
11. If $f(x) = 3x/(x^2 + 1)$, find $f(0)$, $f(1)$, $f(\sqrt{2})$, and $f(-1)$.
12. If $f(x) = x^2/(x^3 - 1)$, find $f(0)$, $f(-1)$, $f(\sqrt{2})$, and $f(\frac{1}{2})$.
13. If $f(x) = 3x^3 - x$, find $f(a)$, $f(a + 1)$, $f(a^2)$, $f(1/a)$, and $f(-a)$.
14. If $f(x) = 2x^4 - x^2$, find $f(b)$, $f(b + 1)$, $f(b^2)$, $f(1/b)$, and $f(-b)$.

In Problems 15–24, find the domain of the function.

15. $f(x) = \sqrt{2x + 3}$
16. $f(x) = \sqrt{15 - 5x}$
17. $f(x) = \dfrac{x^2}{x^4 - 16}$
18. $f(x) = \dfrac{3}{x^2 + 6x + 5}$

19. $f(x) = \dfrac{x}{x^2 + 25}$

20. $f(x) = \dfrac{1}{\sqrt{x - 2}}$

21. $f(x) = \dfrac{\sqrt{x + 1}}{x^2}$

22. $f(x) = \dfrac{5}{x\sqrt{x + 4}}$

23. $f(x) = \dfrac{\sqrt{x + 1}}{\sqrt{2 - x}}$

24. $f(x) = \sqrt{x} + \sqrt{x + 6}$

In Problems 25–32, find the domain and the range of the function.

25. $f(x) = 3x - 15$

26. $f(x) = x^2 + 4$

27. $f(x) = x^3$

28. $f(x) = \sqrt{x - 5}$

29. $f(x) = -1 + \sqrt{2x - 6}$

30. $f(x) = \sqrt{3x + 2}$

31. $f(x) = \sqrt{16 - x}$

32. $f(x) = |x| - 10$

33. For what values of x is $f(x) = \sqrt{x - 4}$ equal to 4?

34. For what values of x is $f(x) = \sqrt{x^2 - 1}$ equal to 0?

35. For what values of x is $f(x) = 4/(x + 2)$ equal to -8? to 4?

36. Determine whether the numbers 5 and -5 are in the range of the function $g(x) = x(x - 4)$.

37. Express the perimeter P of a square as a function of its area A.

38. Express the area A of a circle as a function of its diameter d.

39. Express the area A of an equilateral triangle as a function of the length s of one side.

40. Express the area A of an equilateral triangle as a function of the height h of the triangle.

41. Express the volume V of a cube as a function of the area A of its base.

42. Express the surface area S of a solid right circular cylinder of volume 1 m^3 as a function of its radius r.

43. An open box is made from a rectangular piece of cardboard by cutting a square of length x from each corner and then bending up the sides. If the cardboard measures 2 ft by 3 ft (Figure 44), express the volume V of the box as a function of x.

44. A rain gutter with a rectangular cross section is made from a 1-ft by 20-ft piece of metal by bending up equal amounts x from the 1-ft side (see Figure 45). Express the volume V of the gutter as a function of x.

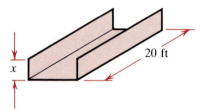

FIGURE 45

45. An open rectangular box is to be constructed with a square base of length x and a volume of 16,000 cm^3. Express the surface area S of the box as a function of x.

46. A closed container is made in the shape of a right circular cylinder of radius r. The container is to have a volume of 4π m^3. If the cost per square meter of the material for the lateral side is twice the cost of that used for the top and the cost per square meter of the material for the bottom is four times the cost of that used for the top, express the total cost C of construction of the container as a function of r.

47. A rectangular piece of pasture land is to be fenced and divided into two equal portions by an additional fence parallel to two sides. The piece of land contains 3000 m^2. Express the amount of fencing F in terms of the length x shown in Figure 46.

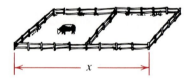

FIGURE 46

48. A person 6 ft tall walks toward a 20-ft street lamp, as shown in Figure 47. Express the length L of his shadow as a function of his distance x from the street lamp.

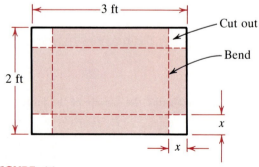

FIGURE 44

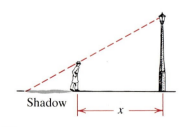

FIGURE 47

49. The window shown in Figure 48 consists of a rectangle surmounted by a semicircle. Express the area A of the window as a function of the indicated width x if it is known that the perimeter of the window is 20 m.

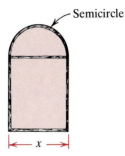

Semicircle

FIGURE 48

50. A wire of length L is cut x units from one end. One piece of the wire is bent into the shape of a circle and the other is bent into the shape of a square. Express the sum S of the areas as a function of x.

51. Two streets intersect as shown in Figure 49. At 12 P.M. car A crosses the intersection heading south at a constant speed of 50 mph. At the same time, car B is 4 mi east of the intersection and is traveling west at a constant speed of 30 mph. Let $t = 0$ represent the time 12 P.M. Express the distance d between the two cars as a function of time $t > 0$.

52. The ends of a 12-ft-long water trough form isosceles triangles. The equal sides of the triangles are 4 ft long, and the remaining side is x ft long. (See Figure 50.) Express the volume V of the trough as a function of x.

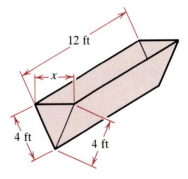

12 ft

x

4 ft   4 ft

FIGURE 50

53. A jogging track is to be constructed with two straight segments and two semicircular segments, as shown in Figure 51. The radius of each semicircular segment is r. The length of the track is to be 1 km. Express the area A enclosed by the track as a function of r.

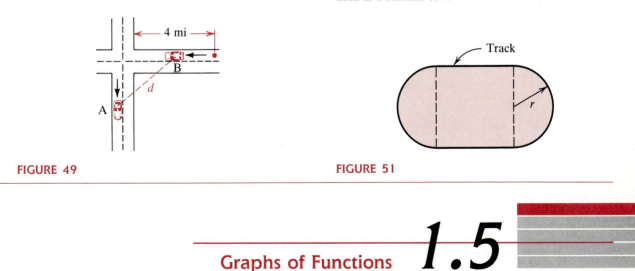

Track

r

4 mi

B

d

A

FIGURE 49   **FIGURE 51**

Graphs of Functions *1.5*

A function is often used to describe problems or phenomena in fields such as science, engineering, and business. In order to interpret and utilize data obtained from such a function, we find that it is useful to display the data in the form of a graph (see Figure 52).

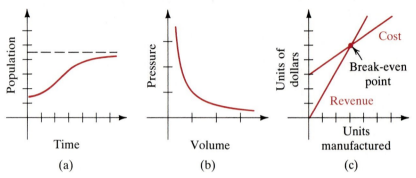

FIGURE 52   (a)   (b)   (c)

In an *xy*-plane the **graph of a function** $y = f(x)$ is defined to be the graph of the relation

$$\{(x, y)\,|\,y = f(x),\ x \text{ in the domain of } f\}.$$

In other words, the graph of a function f is the set of points (x, y) in the plane whose coordinates satisfy $y = f(x)$.

EXAMPLE 1

The graph of the function f defined by the accompanying table consists of the four points shown in Figure 53.

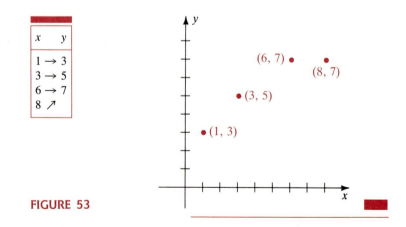

x	y
1 →	3
3 →	5
6 →	7
8 ↗	

FIGURE 53

INTERCEPTS

To graph a function defined by an equation $y = f(x)$, it is usually a good idea to determine first whether the graph of f has any intercepts. Recall that the

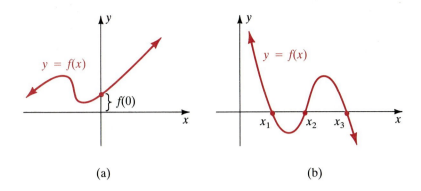

(a)   (b)

FIGURE 54

y-axis is the line $x = 0$. Thus, if 0 is in the domain of f, the **y-intercept** of its graph is the number $f(0)$. (See Figure 54(a)). Similarly, the *x*-axis is the line $y = 0$. Thus, to find the **x-intercepts** of the graph of $y = f(x)$, we must solve the equation $f(x) = 0$. The numbers satisfying this equation are also called the *zeros* of f. The *real* zeros of f are the *x*-intercepts of its graph. In Figure 54(b), we see that $f(x_1) = 0$, $f(x_2) = 0$, and $f(x_3) = 0$. Therefore, x_1, x_2, and x_3 are the *x*-intercepts of the graph of the function f.

PLOTTING POINTS

To obtain other points on the graph of a function $y = f(x)$, we can choose numbers $x_1, x_2, \ldots , x_n$ in its domain, compute $f(x_1), f(x_2), \ldots , f(x_n)$, and then plot the corresponding points $(x_1, f(x_1))$, $(x_2, f(x_2))$, $\ldots$, $(x_n, f(x_n))$. Keep in mind that a functional value $f(x)$ is a directed distance from the *x*-axis (see Figure 55).

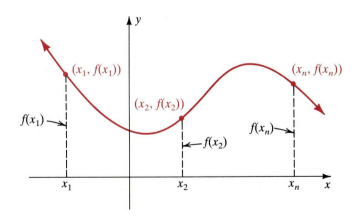

FIGURE 55

▦ EXAMPLE 2

Graph the function $f(x) = x^2 - 2x - 3$.

Solution   The y-intercept is $f(0) = -3$. To find the x-intercepts, we solve

$$x^2 - 2x - 3 = 0,$$

or $$(x - 3)(x + 1) = 0.$$

Thus, since $f(x) = 0$ when $x = -1$ or $x = 3$, the x-intercepts are -1 and 3. By plotting the points from the accompanying table, we obtain the graph shown in Figure 56.

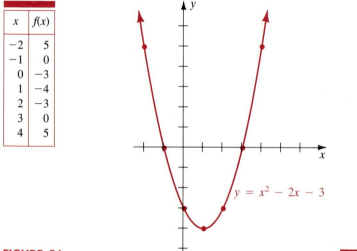

x	$f(x)$
-2	5
-1	0
0	-3
1	-4
2	-3
3	0
4	5

$$y = x^2 - 2x - 3$$

FIGURE 56

If the graph of a function $y = f(x)$ is drawn accurately, it is usually possible to *see* the domain and the range of f (Figure 57). The domain of f is some interval, or other set of real numbers, on the x-axis; and the range of f is some interval, or other set of real numbers, on the y-axis. In Example 2, the domain

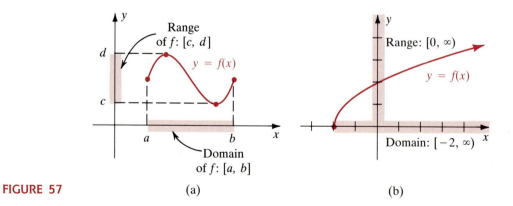

Range of f: $[c, d]$

$y = f(x)$

Domain of f: $[a, b]$

Range: $[0, \infty)$

$y = f(x)$

Domain: $[-2, \infty)$

FIGURE 57   (a)   (b)

of the given function is the set R of real numbers, that is, the interval $(-\infty, \infty)$. The range of the function appears from the graph in Figure 56 to be $[-4, \infty)$.

THE VERTICAL LINE TEST

From the definition of a function we know that for each x in the domain of f, there corresponds a *unique* value $f(x)$ in the range. This means that any vertical line intersecting the graph of f can do so in at most one point. Conversely, if each vertical line intersects the graph of a relation in at most one point, then the relation is a function. This last statement is called the **vertical line test** for a function.

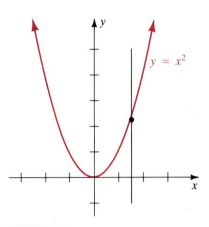

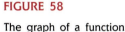

FIGURE 58

The graph of a function

■ EXAMPLE 3

(a) From Figure 58 we see that any vertical line intersects the graph of the relation defined by $y = x^2$ in at most one point. Thus, by the vertical line test, the relation defines a function $y = f(x) = x^2$.

(b) As Figure 59 shows, a vertical line can intersect the graph of the relation defined by $x^2 + y^2 = 4$ in more than one point. Thus the relation does not define a function $y = f(x)$. ■

LINEAR FUNCTIONS

One of the simplest yet most important types of functions is the linear function. Any function of the form

$$f(x) = ax + b, \quad a \neq 0, \tag{6}$$

where a and b are constants, is called a **linear function.** Since $f(x)$ is a real number for any choice of x, we conclude that the domain of (6) is the set of real numbers. As its name suggests, the graph of a linear function is a *straight line*. Lines will be considered in greater detail in Section 7.1.

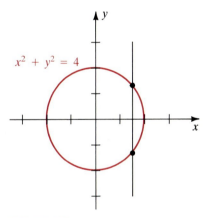

FIGURE 59

Not the graph of a function

■ EXAMPLE 4

Graph the linear function $f(x) = \frac{1}{2}x - \frac{3}{2}$.

Solution The y-intercept of the graph is $b = -\frac{3}{2}$, that is, $f(0) = -\frac{3}{2}$. To graph a straight line, we need only two points. Although we could substitute any value of x into $f(x)$ to obtain another point, we will determine the x-intercept of the graph. Setting $f(x) = 0$, we have

$$\tfrac{1}{2}x - \tfrac{3}{2} = 0,$$

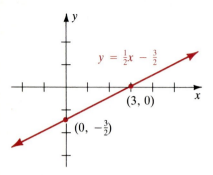

FIGURE 60

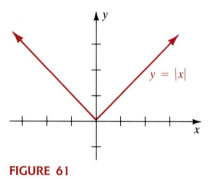

FIGURE 61

which gives $x = 3$. Thus the x-intercept is 3. The graph of the line in Figure 60 is drawn through the points $(0, -\frac{3}{2})$ and $(3, 0)$. ▨

▨ **EXAMPLE 5**

Graph the absolute value function $f(x) = |x|$.

Solution   Using Definition 1, we have

$$f(x) = |x| = \begin{cases} x, & x \geq 0, \\ -x, & x < 0. \end{cases}$$

Hence the graph of f consists of two pieces. We draw, in turn, the line $y = x$ for $x \geq 0$ and the line $y = -x$ for $x < 0$. The graph is shown in Figure 61. ▨

ONE-TO-ONE FUNCTIONS

For each element x in the domain of a function, there is only one element y in its range. But an element y in the range of a function f may correspond to *several* elements in its domain. For example, for the function $f(x) = x^2$, the number 9 in its range corresponds to the numbers -3 and 3 in its domain, that is, $f(-3) = f(3) = 9$. In contrast, for the function $g(x) = 2x - 1$, the number 9 in its range corresponds to the *single* number 5 in its domain: $g(5) = 2(5) - 1 = 9$. Functions, such as g, for which each element in its range corresponds with precisely one element in the domain is called a **one-to-one function.**

DEFINITION 5

> A function f is said to be **one-to-one** if and only if each element in the range of f is associated with exactly one element in the domain.

To show that a function f is *not* one-to-one, we need only find two elements x_1 and x_2 in its domain such that $f(x_1) = f(x_2)$.

▨ **EXAMPLE 6**

The absolute value function of Example 5 is not one-to-one since $f(-2) = f(2) = 2$. ▨

We can determine whether a function f is one-to-one by looking at its graph.

THE HORIZONTAL LINE TEST

Let $y = f(x)$ be a one-to-one function and consider its graph

$$\{(x, y)|y = f(x), \ x \text{ in the domain } X \text{ of } f\}.$$

Since each y-value corresponds to at most one x-value, any horizontal line intersects the graph of $y = f(x)$ in at most one point. Conversely, if each horizontal line intersects the graph of a function in at most one point, then the function is one-to-one. The **horizontal line test** is illustrated in Figure 62.

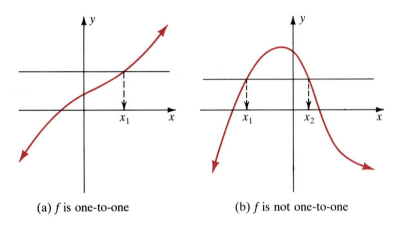

(a) f is one-to-one   (b) f is not one-to-one

FIGURE 62

▪ EXAMPLE 7

Determine whether the function $f(x) = x^2 - 2x$ is one-to-one.

Solution From Figure 63 we see that a horizontal line intersects the graph of the function f at more than one point. It follows from the horizontal line test that f is not one-to-one.

Inspection of the graph of the linear function $f(x) = \frac{1}{2}x - \frac{3}{2}$ in Example 4 reveals that the function is one-to-one.

SYMMETRY

In Section 1.2 we discussed symmetry of a graph with respect to the y-axis, the x-axis, and the origin. The graph of a function can be symmetric with respect to the y-axis or the origin, but the graph of a nonzero function *cannot* be symmetric with respect to the x-axis. (Why not?) If the graph of a function f is symmetric with respect to the y-axis, we say that f is an **even function.** A function whose graph is symmetric with respect to the origin is said to be an

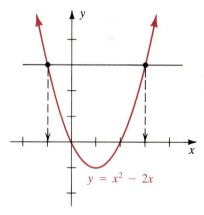

$y = x^2 - 2x$

FIGURE 63

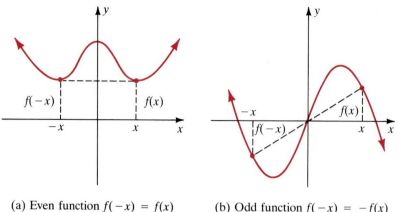

(a) Even function $f(-x) = f(x)$ (b) Odd function $f(-x) = -f(x)$

odd function. For functions, the following two tests for symmetry are equivalent to tests (i) and (iii), respectively, on page 19 (see Figure 64).

Tests for Symmetry

The graph of a function f with domain X is symmetric with respect to:

(i) the **y-axis** if $f(-x) = f(x)$ for every x in X, and
(ii) the **origin** if $f(-x) = -f(x)$ for every x in X.

Inspection of Figure 56 in Example 2 shows that f is neither even nor odd. We also note that $f(-x)$ does not equal either $f(x)$ or $-f(x)$. On the other hand, in Figure 61 in Example 5, we see that the absolute value function is an even function since its graph is symmetric with respect to the y-axis. In this case, $f(-x) = |-x| = |x| = f(x)$.

■ EXAMPLE 8

Graph the function $f(x) = x^3$.

Solution   Since $f(0) = 0$ and since $f(x) = x^3 = 0$ implies that $x = 0$, we see that the y- and x-intercepts are the same, namely, 0. This means that the graph of f passes through the origin. Also,

$$f(-x) = (-x)^3$$
$$= -x^3$$
$$= -f(x)$$

shows that the graph of f is symmetric with respect to the origin. Therefore, f

is an odd function. Using these results about the intercepts and symmetry and plotting the points from the accompanying table gives us the graph shown in Figure 65.

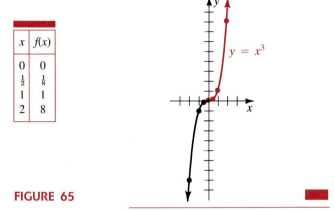

x	$f(x)$
0	0
$\frac{1}{2}$	$\frac{1}{8}$
1	1
2	8

FIGURE 65

EXAMPLE 9

Graph the function $f(x) = x^{2/3}$.

Solution We note that the function can also be written in the form

$$f(x) = (x^{1/3})^2 = (\sqrt[3]{x})^2.$$

Since it is possible to find the cube root of any real number, the domain of f is the set R of real numbers. As in Example 8, the x- and y-intercepts are 0. Now from

$$\begin{aligned} f(-x) &= (\sqrt[3]{-x})^2 \\ &= (-\sqrt[3]{x})^2 \\ &= (\sqrt[3]{x})^2 \\ &= f(x), \end{aligned}$$

we see that f is an even function. Plotting the points from the table and using the intercepts and symmetry with respect to the y-axis, we obtain the graph shown in Figure 66.

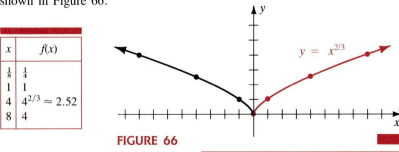

x	$f(x)$
$\frac{1}{8}$	$\frac{1}{4}$
1	1
4	$4^{2/3} \approx 2.52$
8	4

FIGURE 66

EXERCISE 1.5

In Problems 1–36, graph the given function and find the intercepts. Use symmetry when possible.

1. $f(x) = 3$
2. $f(x) = -1$
3. $f(x) = x$
4. $f(x) = x - 5$
5. $f(x) = 2x + 1$
6. $f(x) = -3x + 6$
7. $f(x) = \sqrt{x}$
8. $f(x) = 3 + \sqrt{x}$
9. $f(x) = \sqrt{x - 2}$
10. $f(x) = \sqrt{4 - x}$
11. $f(x) = x^{1/3}$
12. $f(x) = \sqrt[3]{x - 8}$
13. $f(x) = -x^2$
14. $f(x) = 10x^2$
15. $f(x) = x^2 + 1$
16. $f(x) = 4 - x^2$
17. $f(x) = x^2 - 2x$
18. $f(x) = x^2 + x$
19. $f(x) = x^2 - 4x + 4$
20. $f(x) = x^2 - 5x + 4$
21. $f(x) = -x^3$
22. $f(x) = x^3 - 1$
23. $f(x) = 1 - x^3$
24. $f(x) = x^3 - x$
25. $f(x) = |x| - 3$
26. $f(x) = |x - 3|$
27. $f(x) = |x + 2|$
28. $f(x) = |x| + 2$
29. $f(x) = \dfrac{1}{|x|}$
30. $f(x) = \dfrac{|x|}{x}$
31. $f(x) = \dfrac{1}{x}$
32. $f(x) = \dfrac{1}{x^2}$
33. $f(x) = x^4$
34. $f(x) = (x - 1)^4$
35. $f(x) = \sqrt{x^2 - 1}$
36. $f(x) = \sqrt{9 - x^2}$

In Problems 37–42, find all intercepts of the graph of the given function.

37. $f(x) = (x + 2)^2 - 49$
38. $f(x) = x^4 - 16$
39. $f(x) = \sqrt{\dfrac{1 - x^2}{2 - x}}$
40. $f(x) = \dfrac{x^2 + 5x - 14}{2x - 8}$
41. $f(x) = \dfrac{6x - 4}{x}$
42. $f(x) = \dfrac{30}{x^2 + 5}$

In Problems 43–50, determine whether the given graph is the graph of a function.

43.

44.

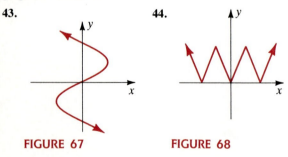

FIGURE 67   **FIGURE 68**

45.

46.

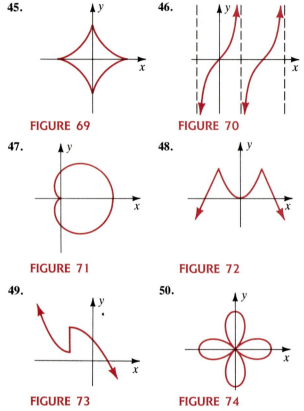

FIGURE 69   **FIGURE 70**

47.   **48.**

FIGURE 71   **FIGURE 72**

49.   **50.**

FIGURE 73   **FIGURE 74**

In Problems 51–54, the given graph is the graph of a function f. From the figure, determine the domain and the range of f.

51.

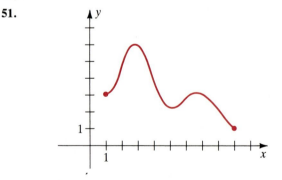

FIGURE 75

52.

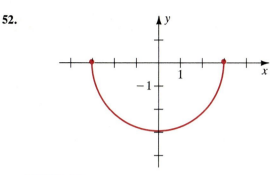

FIGURE 76

53.

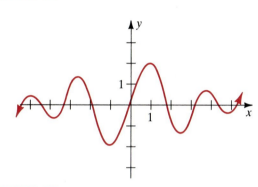

FIGURE 77

54.

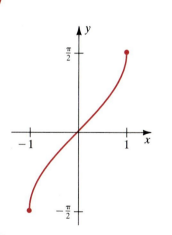

FIGURE 78

55. The **greatest integer function** $f(x) = [x]$ is defined to be the largest integer n for which $n \leq x$. Find $f(-3.5)$, $f(-3)$, $f(-1.4)$, $f(0)$, $f(1.7)$, $f(2.3)$, $f(2.8)$, and $f(3.1)$. Sketch the graph of the greatest integer function.

56. Let $f(x) = [x]$ be the greatest integer function defined in Problem 55. Sketch the graph of $y = [x + 1]$.

In Problems 57–62, the given graph is the graph of a function f. From the figure, determine whether f is one-to-one.

57.

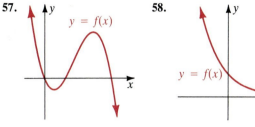

FIGURE 79

58.

FIGURE 80

59.

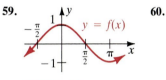

FIGURE 81

60.

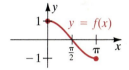

FIGURE 82

61.

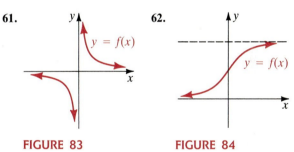

FIGURE 83

62.

FIGURE 84

In Problems 63 and 64, complete the graph (a) if f is an even function and (b) if f is an odd function.

63.

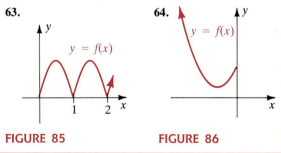

FIGURE 85

64.

FIGURE 86

1.6

Combining Functions

Functions can be combined through the arithmetic operations of addition, subtraction, multiplication, and division to yield new functions. For two functions f and g, the **sum** $f + g$, the **difference** $f - g$, the **product** fg, and the **quotient** f/g are defined as follows.

$$(f + g)(x) = f(x) + g(x)$$
$$(f - g)(x) = f(x) - g(x)$$
$$(fg)(x) = f(x)g(x)$$
$$(f/g)(x) = \frac{f(x)}{g(x)}, \quad (g(x) \neq 0).$$

DOMAINS

The domain of each of the functions $f + g$, $f - g$, fg, and f/g is the *intersection* of the domain of f with the domain of g. In the case of the quotient f/g, we must also exclude values of x for which the denominator $g(x)$ is 0.

■ EXAMPLE 1

For $f(x) = x^2 + 4x$ and $g(x) = x^2 - 9$, we have

$$(f + g)(x) = (x^2 + 4x) + (x^2 - 9) = 2x^2 + 4x - 9,$$
$$(f - g)(x) = (x^2 + 4x) - (x^2 - 9) = 4x + 9,$$
$$(fg)(x) = (x^2 + 4x)(x^2 - 9) = x^4 + 4x^3 - 9x^2 - 36x,$$

and $\qquad (f/g)(x) = \dfrac{x^2 + 4x}{x^2 - 9}.$ ■

In Example 1 we note that $x = 3$ and $x = -3$ are not in the domain of $(f/g)(x)$.

■ EXAMPLE 2

If we add

$$f(x) = \sqrt{1 - x}, \quad x \leq 1 \qquad \text{and} \qquad g(x) = \sqrt{x + 2}, \quad x \geq -2,$$

we obtain

$$(f + g)(x) = \sqrt{1 - x} + \sqrt{x + 2}.$$

The domain of this new function is the interval $[-2, 1]$, which is the set of numbers common to both domains.

ADDITION OF *y*-COORDINATES

The graph of the sum of two functions f and g that have the same domain can be obtained by the **addition of y-coordinates.** Since $(f + g)(x) = f(x) + g(x)$, we see from Figure 87 that the y-coordinate of a point (x_1, y_1) on the graph of $f + g$ is simply the sum of the y-coordinates of the points $(x_1, f(x_1))$ and $(x_1, g(x_1))$ on the graphs of f and g, respectively.

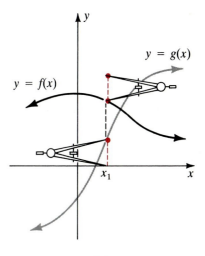

(a)

■ EXAMPLE 3

If $f(x) = x$ and $g(x) = \sqrt{x}$, graph $(f + g)(x)$.

Solution First, we note that the domain of

$$(f + g)(x) = x + \sqrt{x}$$

is $[0, \infty)$. Figure 88(a) shows the graphs of f and g for $x \geq 0$. As indicated, at any given value of x, we can obtain the y-coordinate of the corresponding point on the graph of $f + g$ by adding y_1 and y_2. The graph of the sum is shown in Figure 88(b).

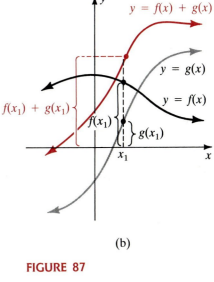

(b)

FIGURE 87

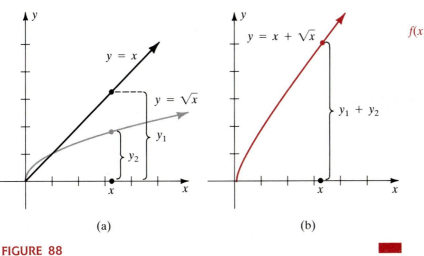

(a)   (b)

FIGURE 88

FUNCTION COMPOSITION

Another method of combining two functions f and g is called **function composition.** We define the composition of f and g, denoted by $f \circ g$, to be the function

$$(f \circ g)(x) = f(g(x)),$$

where it is understood that the functional values $g(x)$ (that is, elements in the range of g) are in the domain of f. Similarly, the composition of g and f is defined by

$$(g \circ f)(x) = g(f(x)).$$

Of course, the functional values $f(x)$ must be in the domain of g. The functions $f \circ g$ and $g \circ f$ are called **composite functions.**

■ EXAMPLE 4

If $f(x) = x^2 + 3x - 1$ and $g(x) = 2x^2 + 1$, find $(f \circ g)(x)$.

Solution   For emphasis we write f in the form

$$f(\quad) = (\quad)^2 + 3(\quad) - 1.$$

Thus to evaluate $(f \circ g)(x) = f(g(x))$, we can substitute $g(x)$ into each set of parentheses. We find

$$
\begin{aligned}
(f \circ g)(x) = f(g(x)) &= f(2x^2 + 1) \\
&= (2x^2 + 1)^2 + 3(2x^2 + 1) - 1 \\
&= 4x^4 + 4x^2 + 1 + 6x^2 + 3 - 1 \\
&= 4x^4 + 10x^2 + 3.
\end{aligned}
$$

■ EXAMPLE 5

Find $(g \circ f)(x)$ for the functions given in Example 4.

Solution   In this case,

$$g(\quad) = 2(\quad)^2 + 1.$$

Thus,
$$
\begin{aligned}
(g \circ f)(x) = g(f(x)) &= g(x^2 + 3x - 1) \\
&= 2(x^2 + 3x - 1)^2 + 1 \\
&= 2(x^4 + 6x^3 + 7x^2 - 6x + 1) + 1 \\
&= 2x^4 + 12x^3 + 14x^2 - 12x + 3.
\end{aligned}
$$

Note of Caution: Examples 4 and 5 illustrate that, in general, $f \circ g \neq g \circ f$.

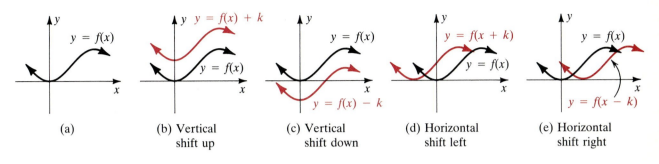

(a)   (b) Vertical shift up   (c) Vertical shift down   (d) Horizontal shift left   (e) Horizontal shift right

SHIFTED GRAPHS

FIGURE 89

If f is a function and k is a positive constant, then the graphs of the sum $f(x) + k$, the difference $f(x) - k$, and the compositions $f(x + k)$ and $f(x - k)$ can be obtained from the graph of f by either a vertical or a horizontal **shift,** or **translation,** by an amount k. Figure 89 illustrates the four cases that are summarized in the following table.

FUNCTION, $k > 0$	GRAPH OF $y = f(x)$
$y = f(x) + k$	Shifted *up* k units
$y = f(x) - k$	Shifted *down* k units
$y = f(x + k)$	Shifted to the *left* k units
$y = f(x - k)$	Shifted to the *right* k units

EXAMPLE 6

The graph of the function $f(x) = \sqrt{x}$ is shown in Figure 90(a). The graphs of $y = \sqrt{x} + 1$, $y = \sqrt{x} - 1$, $y = \sqrt{x + 1}$, and $y = \sqrt{x - 1}$, shown in Figures 90(b), (c), (d), and (e) were obtained by shifting the graph of $f(x) = \sqrt{x}$, in turn, up one unit, down one unit, to the left one unit, and to the right one unit.

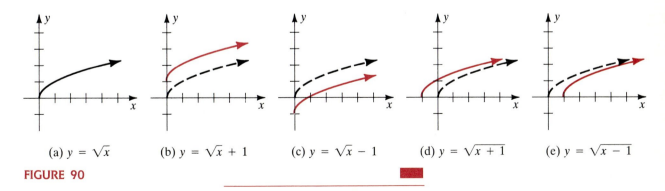

(a) $y = \sqrt{x}$ (b) $y = \sqrt{x} + 1$ (c) $y = \sqrt{x} - 1$ (d) $y = \sqrt{x + 1}$ (e) $y = \sqrt{x - 1}$

FIGURE 90

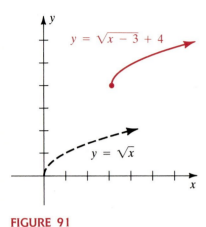

FIGURE 91

In general, the graph of

$$y = f(x \pm k_1) \pm k_2, \quad k_1 > 0, \quad k_2 > 0,$$

can be found from the graph of the function $y = f(x)$ by combining a horizontal and a vertical shift. For example, the graph of $y = f(x + k_1) - k_2$ is the graph of $y = f(x)$ shifted horizontally k_1 units to the left and then shifted vertically k_2 units down.

■ EXAMPLE 7

Graph $y = \sqrt{x - 3} + 4$.

Solution The graph of $y = \sqrt{x - 3} + 4$ is the graph of $f(x) = \sqrt{x}$ shifted 3 units to the right and 4 units upward. The graph is given in Figure 91. ■

REFLECTIONS

The graph of $y = -f(x)$ is a **reflection** of the graph of $y = f(x)$ through the x-axis. In other words, to graph $y = -f(x)$ we simply turn the graph of $y = f(x)$ upside down.

■ EXAMPLE 8

Graph $y = -\sqrt{x}$.

Solution The graph of $f(x) = \sqrt{x}$, shown as a dashed curve in Figure 92, is reflected in the x-axis, resulting in the curve shown in color. ■

INVERSE FUNCTIONS

If f is a one-to-one function, then there exists a function g for which $f(g(x)) = x$ and $g(f(x)) = x$. The function g is called the **inverse of f.**

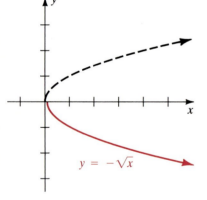

FIGURE 92

DEFINITION 6

Let f be a one-to-one function with domain X and range Y. The **inverse of f** is a function g with domain Y and range X for which

$$f(g(x)) = x \quad \text{for every } x \text{ in the domain of } g$$

and $$g(f(x)) = x \quad \text{for every } x \text{ in the domain of } f.$$

(7)

■ EXAMPLE 9

The function $g(x) = \frac{1}{5}(x + 7)$ is the inverse of $f(x) = 5x - 7$. To verify this, we form

$$f(g(x)) = f(\tfrac{1}{5}(x + 7)) = 5(\tfrac{1}{5}(x + 7)) - 7 = x + 7 - 7 = x;$$
$$g(f(x)) = g(5x - 7) = \tfrac{1}{5}((5x - 7) + 7) = \tfrac{1}{5}(5x) = x.$$

We note that the domain and the range of f consist of the set R of real numbers. Consequently, the set R of real numbers is also the domain and the range of the inverse function g. ■

The inverse of a one-to-one function f is denoted by the symbol f^{-1}. It is important to note that "-1" in f^{-1} is *not* an exponent; that is, $f^{-1} \neq 1/f$. In Example 9, the inverse of the function $f(x) = 5x - 7$ is written $f^{-1}(x) = \frac{1}{5}(x + 7)$.

THE GRAPH OF f^{-1}

Suppose that f is a one-to-one function and that (a, b) denotes any point on the graph of f. Then $f(a) = b$, and, from (17),

$$f^{-1}(b) = f^{-1}(f(a)) = a.$$

This means that the point (b, a) is on the graph of f^{-1}. But since the points (a, b) and (b, a) are symmetric with respect to the line $y = x$, we conclude that the graph of f^{-1} is a reflection of the graph of $y = f(x)$ in the line $y = x$ (see Figure 93).

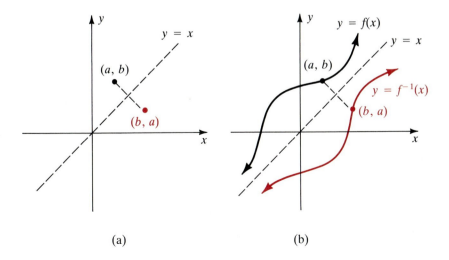

(a) (b) **FIGURE 93**

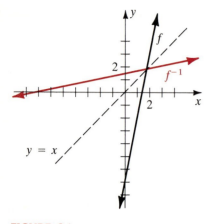

FIGURE 94

■ EXAMPLE 10

In Example 9 we saw that the inverse of

$$f(x) = 5x - 7 \quad \text{is} \quad f^{-1}(x) = \tfrac{1}{5}x + \tfrac{7}{5}.$$

Since both f and f^{-1} are linear functions, their graphs are straight lines. The two graphs of f and f^{-1} are compared in Figure 94.   ■

A METHOD FOR FINDING f^{-1}

Since (a, b) is on the graph of a one-to-one function if and only if (b, a) is on the graph of f^{-1}, we can find the inverse of a function by simply interchanging, or relabeling, the variables x and y. The procedure is summarized as follows.

A Method for Finding f^{-1}

To find f^{-1} for a one-to-one function f:

1. Interchange the variables x and y in the equation $y = f(x)$, and
2. Solve the resulting equation $x = f(y)$ for the symbol y.

■ EXAMPLE 11

Find the inverse of the function f in Example 9.

Solution   From Example 9, we write the given function as

$$y = 5x - 7.$$

Interchanging the variables x and y in the last equation gives

$$x = 5y - 7.$$

Solving for y then yields

$$y = \tfrac{1}{5}x + \tfrac{7}{5}, \quad \text{or} \quad f^{-1}(x) = \tfrac{1}{5}x + \tfrac{7}{5}.$$ ■

■ EXAMPLE 12

Find the inverse of $f(x) = \sqrt{3x - 6}$.

Solution   We first write the function as

$$y = \sqrt{3x - 6}$$

and then interchange the variables and solve for y:

$$x = \sqrt{3y - 6}$$
$$x^2 = 3y - 6$$
$$3y = x^2 + 6$$
$$y = \tfrac{1}{3}(x^2 + 6).$$

Since the domain and the range of f are $[2, \infty)$ and $[0, \infty)$, respectively, the domain and the range of f^{-1} are, in turn, $[0, \infty)$ and $[2, \infty)$. In other words,

$$f^{-1}(x) = \tfrac{1}{3}(x^2 + 6), \quad x \geq 0.$$

The graph of f^{-1} is shown in color in Figure 95.

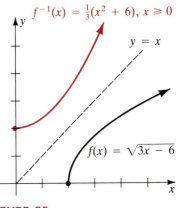

$f^{-1}(x) = \tfrac{1}{3}(x^2 + 6), \, x \geq 0$

$y = x$

$f(x) = \sqrt{3x - 6}$

FIGURE 95

In Problems 1–8, find the indicated functions and give their domains.

1. $f(x) = 2x^2 - x + 3$, $g(x) = x^2 + 1$; $f + g$, fg
2. $f(x) = x$, $g(x) = \sqrt{x - 1}$; fg, f/g
3. $f(x) = 2x - (1/\sqrt{x})$, $g(x) = 2x + (1/\sqrt{x})$; $f + g$, fg
4. $f(x) = 3x^3 - 4x^2 + 5x - 6$, $g(x) = (1 - x)^3$, $f + g$, $f - g$
5. $f(x) = x^2 - 4$, $g(x) = x + 2$; fg, f/g
6. $f(x) = (2x + 3)^{1/2}$, $g(x) = 2x + 3 + (2x + 3)^{1/2}$; $f - g$, fg
7. $f(x) = \sqrt{1 - x}$, $g(x) = \sqrt{x + 2}$; $f + g$, fg
8. $f(x) = 2 + \sqrt{x + 2}$, $g(x) = \sqrt{5x + 5}$; $f - g$, f/g

In Problems 9 and 10, use addition of y-coordinates to graph the function $f + g$.

9. $f(x) = x$, $g(x) = |x|$
10. $f(x) = x^2$, $g(x) = 2x$

In Problems 11 and 12, use the given graphs of $y = f(x)$ and $y = g(x)$ to graph $y = f(x) + g(x)$.

11.

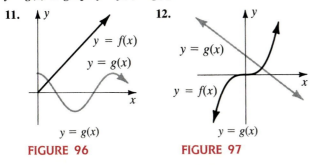

FIGURE 96

12.

FIGURE 97

In Problems 13–22, find $f \circ g$ and $g \circ f$.

13. $f(x) = 1 + x^2$, $g(x) = \sqrt{x - 1}$
14. $f(x) = x^2 - x + 5$, $g(x) = -x + 4$
15. $f(x) = \dfrac{1}{2x - 1}$, $g(x) = x^2 + 1$
16. $f(x) = \dfrac{x + 1}{x}$, $g(x) = \dfrac{1}{x}$
17. $f(x) = 2x - 3$, $g(x) = \dfrac{x + 3}{2}$
18. $f(x) = x - 1$, $g(x) = x^3$
19. $f(x) = x + \dfrac{1}{x^2}$, $g(x) = \dfrac{1}{x}$
20. $f(x) = \sqrt{x - 4}$, $g(x) = x^2$
21. $f(x) = x + 1$, $g(x) = x + \sqrt{x - 1}$
22. $f(x) = x^3 - 4$, $g(x) = \sqrt[3]{x + 4}$

In Problems 23–28, find the domain of $f \circ g$.

23. $f(x) = \sqrt{x + 1}$, $g(x) = 2x + 1$
24. $f(x) = \sqrt{x - 4}$, $g(x) = x^2 - 5$
25. $f(x) = \dfrac{1}{x}$, $g(x) = \dfrac{x}{x - 6}$
26. $f(x) = \dfrac{x}{x^2 - 1}$, $g(x) = 4x$
27. $f(x) = x^2 + 1$, $g(x) = \sqrt{x}$
28. $f(x) = \dfrac{x + 1}{x}$, $g(x) = \sqrt{x + 3}$

In Problems 29 and 30, find the graphs of the indicated functions by shifting the graph of the given function.

29.   **30.**

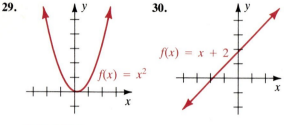

FIGURE 98   **FIGURE 99**

(a) $y = f(x) + 1$ **(a)** $y = f(x) + 3$
(b) $y = f(x) - 1$ **(b)** $y = f(x) - 2$
(c) $y = f(x + 1)$ **(c)** $y = f(x + 2)$
(d) $y = f(x - 1)$ **(d)** $y = f(x - 5)$

In Problems 31–36, the given graph is a shifted graph of the indicated function. Find an equation of the graph.

31. $f(x) = |x|$ **32.** $f(x) = x^2 + 1$

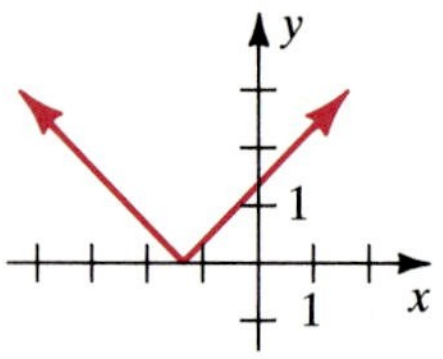

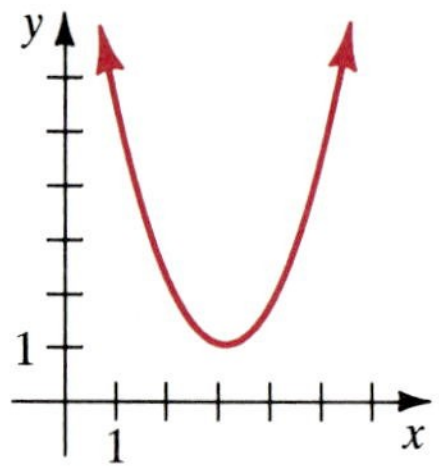

FIGURE 100   **FIGURE 101**

33. $f(x) = -x^2$ **34.** $f(x) = \sqrt{1 - x}$

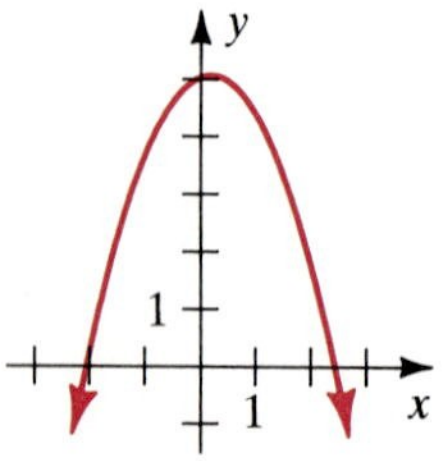

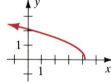

FIGURE 102   **FIGURE 103**

35. $f(x) = x^2$ **36.** $f(x) = -|x|$

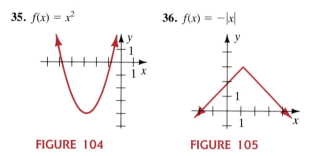

FIGURE 104   **FIGURE 105**

In Problems 37–42, find the graph of the indicated function from the graph of f.

37. $f(x) = x^2$; $y = (x - 2)^2 + 3$
38. $f(x) = \sqrt{x}$; $y = \sqrt{x + 4} - 1$
39. $f(x) = |x|$; $y = |x - 1| + 4$
40. $f(x) = x^3$; $y = (x - 2)^3 - 2$
41. $f(x) = x^2 - 4$; $y = -(x^2 - 4)$
42. $f(x) = \sqrt{x}$; $y = -\sqrt{x + 4}$

In Problems 43–50, the given function is one-to-one. Find f^{-1}.

43. $f(x) = 3x - 9$ **44.** $f(x) = \frac{1}{2}x - 2$
45. $f(x) = x^3 + 2$ **46.** $f(x) = 1 - x^3$
47. $f(x) = \sqrt{x}$ **48.** $f(x) = 6 - 9\sqrt{x}$
49. $f(x) = \dfrac{1}{x} + 4$ **50.** $f(x) = \dfrac{7x}{2x - 3}$

In Problems 51–54, verify that the given functions are inverse functions of each other.

51. $f(x) = \frac{1}{4}x + 2$, $g(x) = 4x - 8$
52. $f(x) = x^5 + 6$, $g(x) = \sqrt[5]{x - 6}$
53. $f(x) = \dfrac{x - 2}{x + 2}$, $g(x) = \dfrac{2(1 + x)}{1 - x}$
54. $f(x) = \dfrac{x}{4x + 3}$, $g(x) = \dfrac{3x}{1 - 4x}$

In Problems 55 and 56, determine the domain and the range of f^{-1} without finding the inverse.

55. $f(x) = \sqrt{x - 3}$ **56.** $f(x) = 2 + \sqrt{x}$

In Problems 57–60, the given function is one-to-one. Without finding f^{-1}, find, at the indicated value of x, the corresponding point on the graph of f^{-1}.

57. $f(x) = 2x^3 + 2x$; $x = 2$ **58.** $f(x) = 8x - 3$; $x = 5$
59. $f(x) = x + \sqrt{x}$; $x = 9$ **60.** $f(x) = \dfrac{4x}{x + 1}$; $x = \dfrac{1}{2}$

In Problems 61–64, sketch the graphs of f and f^{-1} using the same coordinate axes.

61. $f(x) = 2x + 2$ **62.** $f(x) = -2x + 3$
63. $f(x) = x^3$ **64.** $f(x) = 2 + \sqrt{x}$

In Problems 65 and 66, sketch the graph of f^{-1} from the graph of f.

65.

66.

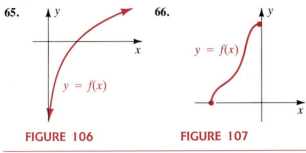

FIGURE 106   FIGURE 107

In Problems 67 and 68, sketch the graph of f from the graph of f^{-1}.

67.   **68.**

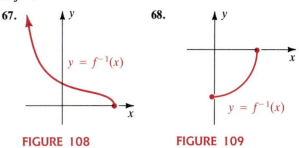

FIGURE 108   FIGURE 109

IMPORTANT CONCEPTS

Set	Domain of variable	Symmetry
element	Equation	Distance formula
subset	identity	Circle
Real numbers	conditional	Function
natural numbers	equivalent	Domain
integers	Solution	Dependent variable
rational numbers	Quadratic formula	Independent variable
irrational numbers	Cartesian coordinate system	Range
decimal numbers	Coordinate axes	Vertical line test
Real number line	Origin	Linear function
origin	Quadrants	One-to-one function
coordinate	Point	Horizontal line test
Less than	x-coordinate	Even function
Greater than	y-coordinate	Odd function
Simultaneous inequality	Relation	Addition of y-coordinates
Interval	Graphs	Function composition
Distance	of a relation	Shifted graphs
Absolute value	of an equation	Reflections
Triangle inequality	Intercepts	Inverse functions
Variable		

CHAPTER 1 REVIEW EXERCISE

In Problems 1–20, fill in the blank or answer true or false.

1. $\frac{9}{0}$ is a real number. ___

2. π is a rational number. ___

3. Every real number can be written as a quotient of two integers. ___

4. If x is negative, then $|x| =$ _____.

5. The distance from a to b on the number line is given by _____.

6. The domain of the variable x in $(3x + 1)/(x^2 - 1)$ is _____.

7. If (a, b) is a point in the second quadrant, then $(a, -b)$ is a point in the _____ quadrant.

8. The distance between the points $(5, 1)$ and $(-1, 9)$ is _____.

9. If the graph of an equation contains the point $(2, 3)$ and is symmetric with respect to the x-axis, then the graph also contains the point _____.

10. The center and the radius of the circle $(x - 2)^2 + (y + 7)^2 = 8$ are _____.

11. If f is a function such that $f(a) = f(b)$, then $a = b$. ___

12. $f(x) = (x^3 + x)^5$ is an odd function. ___

13. The x- and y-intercepts of the graph of the function $f(x) = x^2 - 4x + 4$ are _____.

14. The domain of the function $f(x) = 1/\sqrt{3 - x}$ is _____.

15. The graph of a nonzero function cannot be symmetric with respect to the x-axis. ___

16. The graph of a linear function is a straight line. ___

17. The graph of a function can possess only one y-intercept. ___

18. The function $f(x) = 5$ has an inverse. ___

19. If $(5, -7)$ is on the graph of f, then $(\underline{\quad}, \underline{\quad})$ is on the graph of f^{-1}.

20. The inverse of the one-to-one function $f(x) = x + 1$ is $f^{-1}(x) = x - 1$. ___

In Problems 21–24, insert the appropriate sign: $<$, $>$, or $=$.

21. -1.4, $\quad -\sqrt{2}$

22. 0.50, $\quad \frac{1}{2}$

23. $\frac{2}{3}$, $\quad 0.67$

24. -0.9, $\quad -0.8$

In Problems 25–30, find the indicated absolute value.

25. $|\sqrt{8} - 3|$

26. $|-(\sqrt{15} - 4)|$

27. $|x^2 + 5|$

28. $\dfrac{|x|}{|-x|}$, $\quad x \neq 0$

29. $|t + 5|$, if $t < -5$

30. $|r - s|$, if $r > s$

In Problems 31–36, solve the given equation.

31. $2x^2 - 6x - 3 = 0$

32. $4x^2 + 20x + 25 = 0$

33. $x^3 = 8x$

34. $x^3 - 4x^2 = 0$

35. $x^4 - 8x^2 + 12 = 0$

36. $4x^4 - 4x^2 + 1 = 0$

37. Determine whether the points $A(1, 1)$, $B(3, 3)$, and $C(5, 1)$ are vertices of a right triangle.

38. Find an equation of a circle with center $(3, 3)$ that passes through the origin.

39. Find all numbers in the domain of $f(x) = x^2 + 2x$ that correspond to the number 15 in the range.

40. From the graph of $y = f(x)$ shown in Figure 110, sketch the graphs of each of the following.
  (a) $y = f(x) + 1$
  (b) $y = f(x + 1)$
  (c) $y = f(x) - 2$
  (d) $y = f(x - 2)$
  (e) $y = f\left(x - \dfrac{1}{2}\right)$
  (f) $y = -f(x)$

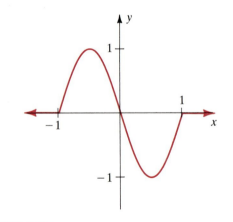

FIGURE 110

In Problems 41 and 42, graph the given relation.

41. $|y| \leq x$

42. $y > x^2 + 1$

In Problems 43–46, graph the given equation.

43. $|y| = |x|$ **44.** $|x| + |y| = 1$

45. $y^2 - x^4 = 0$ **46.** $x^2 + y^2 - 25 = 0$

In Problems 47–52, graph the given function.

47. $f(x) = 3x - 9$ **48.** $f(x) = -5x$

49. $f(x) = 2x^2 + 1$ **50.** $f(x) = x(x - 3)$

51. $f(x) = |x| + |x - 3|$ **52.** $f(x) = x|x|$

In Problems 53 and 54, find $f \circ g$, $g \circ f$, $f \circ f$, $g \circ g$, and $f \circ (1/g)$ for each of the given functions.

53. $f(x) = x^3$, $g(x) = 2/x^2$

54. $f(x) = x^2 + 2x$, $g(x) = 3x + 1$

In Problems 55 and 56, find the inverse of the given one-to-one function.

55. $f(x) = 5/(x - 2)$ **56.** $f(x) = 3 + 4\sqrt{x}$

57. Express the radius r of a circle as a function of its area A.

58. A rectangular box, open at the top, has a square base. Let x denote the length of one side of the base. If the volume of the box is 1000 in³, express the total surface area S of the box as a function of x.

59. Consider the rectangle inscribed in the circle $x^2 + y^2 = 9$, shown in Figure 111. Express the area A as a function of x.

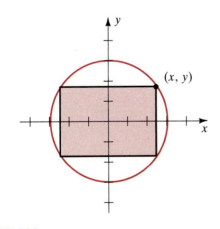

FIGURE 111

60. Consider the four circles of radius h shown in Figure 112. Express the area A of the shaded region as a function of h.

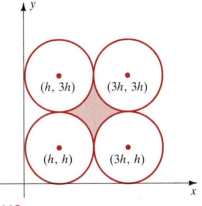

FIGURE 112

61. If
$$f(x) = \begin{cases} 1, & x \text{ a rational number,} \\ 0, & x \text{ an irrational number,} \end{cases}$$
find each of the following.

(a) $f(\frac{1}{3})$ **(b)** $f(-1)$ **(c)** $f(\sqrt{2})$

(d) $f(\sqrt{3}/2)$ **(e)** $f(\pi)$ **(f)** $f(5.72)$

62. Fill in the blanks by referring to the graph of the function $y = f(x)$ shown in Figure 113.

$f(-2) = $ _____ $f(2.5) = $ _____

$f(-1) = $ _____ $f(3) = $ _____

$f(0) = $ _____ $f(4) = $ _____

$f(1) = $ _____ $f(5) = $ _____

$f(2) = $ _____ $f(5.5) = $ _____

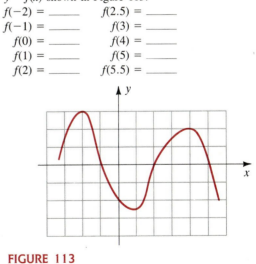

FIGURE 113

Triangle Trigonometry

2

Hipparchus

Trigonometry was developed from early efforts to advance the study of astronomy by predicting the paths and positions of the heavenly bodies and to improve accuracy in navigation and the reckoning of time and calendars.

The Greek astronomer and mathematician Hipparchus, who lived in the second century B.C., was a principal developer of trigonometry. The tables of ''chords'' that he constructed were precursors to our tables of the trigonometric functions. The word trigonometry is derived from two Greek words: *trigon* meaning triangle and *metra* meaning measurement. Thus the name trigonometry refers to the various relationships between the angles of a triangle and its sides.

We begin this chapter with a discussion of angles and then define the trigonometric functions for acute angles in a right triangle. We expand these definitions to general angles and examine a wide variety of applications of triangle trigonometry.

2.1

Angles and Their Measurement

We begin our study of trigonometry by discussing angles and two methods of measuring them.

ANGLE

An **angle** is formed by two rays that have a common endpoint, called the **vertex.** We designate one ray the **initial side** of the angle and the other the **terminal side.** It is convenient to consider the angle as having been formed by a rotation from the initial side to the terminal side, as shown in Figure 1(a).

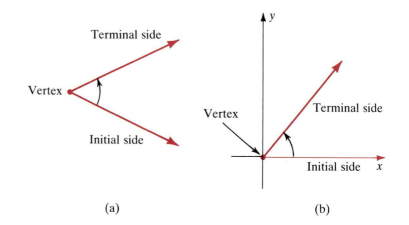

FIGURE 1   (a)   (b)

STANDARD POSITION

As shown in Figure 1(b), we can place the angle in a Cartesian coordinate plane with its vertex at the origin and its initial side coinciding with the positive x-axis. Such an angle is said to be in **standard position.** For the remainder of this section, all angles are considered to be in standard position.

DEGREE MEASURE

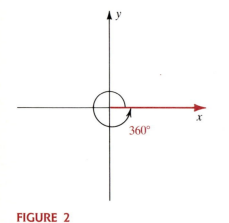

FIGURE 2

Two units of measure commonly used for angles are **degrees** and **radians.** The first of these is based on the assignment of 360 degrees (written 360°) to the angle formed by one complete counterclockwise rotation, as shown in Figure 2. Other angles are then measured in terms of a 360° angle. If the rotation is counterclockwise, the measure will be *positive;* if clockwise, the measure will be *negative.* For example, two counterclockwise rotations would result in a 720° angle; three clockwise rotations would yield an angle of −1080°. See Figures 3(a) and (b). An angle obtained by one fourth of a complete *counterclockwise* rotation will be

$$\tfrac{1}{4}(360°) = 90°,$$

but if the angle is obtained by two thirds of a complete *clockwise* rotation, it will be −240°. See Figures 3(c) and (d). An angle of 1° is formed by $\frac{1}{360}$ of a complete counterclockwise rotation.

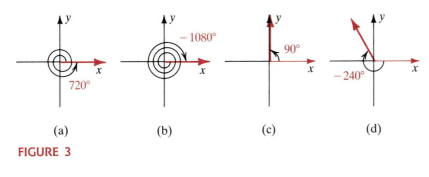

    (a)   (b)   (c)   (d)

FIGURE 3

■ EXAMPLE 1

Sketch an angle of 120°.

Solution Since $120° = \tfrac{1}{3}(360°)$, this angle corresponds to one third of a complete counterclockwise rotation, as shown in Figure 4. ■

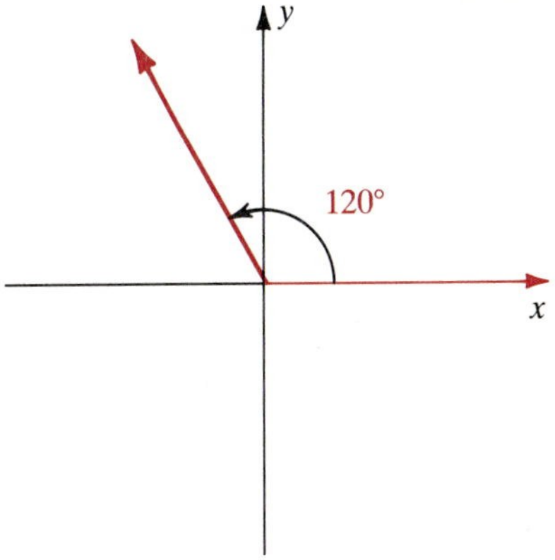

■ EXAMPLE 2

Locate the terminal side of an angle of 960°. Sketch the angle.

Solution We first determine how many full rotations are made in forming this angle. Dividing 960 by 360, we obtain a quotient of 2 and a remainder of 240; that is,

$$960 = 2(360) + 240.$$

Thus this angle is formed by making two counterclockwise rotations before completing

$$\tfrac{240}{360} = \tfrac{2}{3}$$

of another rotation. As illustrated in Figure 5, the terminal side lies in the third quadrant. ■

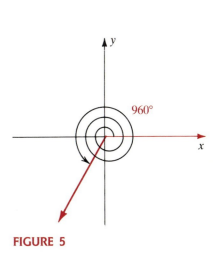

FIGURE 4

FIGURE 5

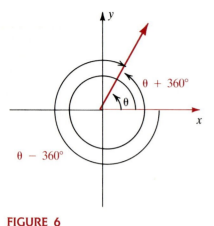

FIGURE 6

COTERMINAL ANGLES

From Example 2 we can see that the terminal side of a 960° angle coincides with the terminal side of a 240° angle. When two angles in standard position have the same terminal sides, we say that they are **coterminal.** For example, the angles θ, $\theta + 360°$, and $\theta - 360°$ shown in Figure 6 are coterminal. In fact, the addition of any integer multiple of 360° to a given angle results in a coterminal angle. Conversely, any two coterminal angles have degree measures that differ by an integer multiple of 360°.

◼ EXAMPLE 3

Find the angle between 0° and 360° that is coterminal with an angle of −690°.

Solution   We can repeatedly add 360° to −690° until we obtain an angle between 0° and 360°:

$$-690° + 360° + 360° = 30°.$$

Thus a 30° angle is coterminal with a −690° angle.   ◼

MINUTES AND SECONDS

With calculators it is convenient to represent fractions of degrees by decimals, such as 42.23°. Traditionally, however, fractions of degrees were expressed in **minutes** and **seconds,** where

$$1° = 60 \text{ minutes (written } 60')*$$

and $$1' = 60 \text{ seconds (written } 60'').$$

For example, an angle of 7 degrees, 30 minutes, and 5 seconds is expressed as 7°30′5″. Some scientific calculators have a special $\boxed{\text{D M S}}$ key for converting an angle given in decimal degrees to *d*egrees, *m*inutes, and *s*econds (DMS notation), and vice versa. The following examples show how to perform these conversions manually.

---

*The use of the number 60 as a **base** dates back to the Babylonians. Another example of the use of this base in our culture is the measurement of time (1 hour = 60 minutes).

■■■■ **EXAMPLE 4**

Convert 86.23° to degrees, minutes, and seconds.

Solution   Since 0.23° represents $\frac{23}{100}$ of 1° and 1° = 60 minutes, we have

$$86.23° = 86° + 0.23°$$
$$= 86° + (0.23)(60')$$
$$= 86° + 13.8'.$$

Now $13.8' = 13' + 0.8'$, so we must convert $0.8'$ to seconds. Since $0.8'$ represents $\frac{8}{10}$ of $1'$ and $1' = 60$ seconds, we have

$$86° + 13' + 0.8' = 86° + 13' + (0.8)(60'')$$
$$= 86° + 13' + 48''.$$

Hence, $$86.23° = 86°13'48''.$$ ■■■■

■■■■ **EXAMPLE 5**

Convert $17°47'13''$ to decimal notation.

Solution   Since $1° = 60'$, it follows that $1' = (\frac{1}{60})°$. Similarly, $1'' = (\frac{1}{60})' = (\frac{1}{3600})°$. Thus we have

$$17°47'13'' = 17° + 47' + 13''$$
$$= 17° + 47(\tfrac{1}{60})° + 13(\tfrac{1}{3600})°$$
$$\approx 17° + 0.7833° + 0.0036°$$
$$= 17.7869°.$$ ■■■■

You may recall from your study of geometry that a 90° angle is called a **right angle** and a 180° angle is called a **straight angle.** An **acute angle** has measure between 0° and 90° and an **obtuse angle** has measure between 90° and 180°. Two acute angles are said to be **complementary** if their sum is 90°. Two positive angles are **supplementary** if their sum is 180°. A **quadrantal angle** is an angle whose terminal side coincides with a coordinate axis when the angle is placed in standard position (for example, 0°, 90°, 180°, 270°, or 360°).

■■■■ **EXAMPLE 6**

Find the angle that is complementary to the given angle.

(a) $\alpha = 74.23°$ **(b)** $\beta = 34°15'$

Solution

(a) Since two angles are complementary if their sum is 90°, we must find 90° − α:

$$90° − α = 90° − 74.23° = 15.77°.$$

(b) To perform the subtraction 90° − β, we arrange the work vertically and convert 90° to 89°60′:

$$\begin{array}{r} 89°60' \\ -\ 34°15' \\ \hline 90° − β = 55°45'. \end{array}$$

⟵ 90°

RADIAN MEASURE

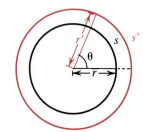

FIGURE 7

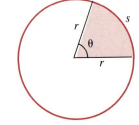

FIGURE 8

Another measure for angles is **radian measure,** which is generally used in almost all applications of trigonometry that involve calculus.

The radian measure of an angle θ is based on the length of an arc on a circle. If we place the vertex of the angle θ at the center of a circle of radius r, then θ is called a **central angle.** The region inside the circle contained within a central angle is called a **sector.** As shown in Figure 7, let s denote the length of the arc subtended or cut off by θ. Then the radian measure of θ is defined by

$$θ = s/r.$$

This definition does not depend on the size of the circle. To see this, we take another circle centered at the vertex of θ with radius r' and subtended arclength s'. As we can see in Figure 8, the two circular sectors are similar and so the ratios s/r and s'/r' are equal. Therefore, regardless of which circle we use, we obtain the same radian measure for θ.

Figure 9(a) illustrates an angle of one radian, since it subtends an arc of

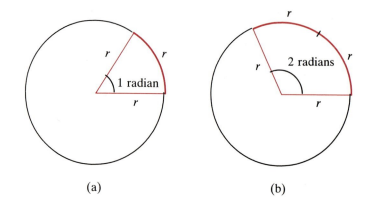

FIGURE 9

(a)

(b)

length s equal to r, the radius of the circle. The angle shown in Figure 9(b) has radian measure 2, since it subtends an arc of length $2r$ on the circle of radius r centered at the vertex.

One complete rotation will subtend an arc equal in length to the circumference of the circle $2\pi r$ (see Figure 10). It follows that

$$\text{one rotation} = s/r = 2\pi r/r = 2\pi \text{ radians.}$$

We have the same convention as before: An angle formed by a counterclockwise rotation is considered positive, whereas an angle formed by a clockwise rotation is negative. In Figure 11 we illustrate angles in standard position of $\pi/2$, $-\pi/2$, π, and 3π radians, respectively. Note that the angle of $\pi/2$ radians shown in (a) is obtained by one fourth of a complete counterclockwise rotation; that is,

$$\frac{1}{4}(2\pi \text{ radians}) = \frac{\pi}{2} \text{ radians.}$$

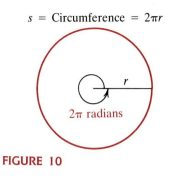

$s = \text{Circumference} = 2\pi r$

2π radians

FIGURE 10

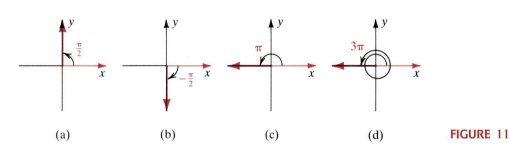

(a) (b) (c) (d)

FIGURE 11

The angle shown in (b), obtained by one fourth of a complete clockwise rotation, is $-\pi/2$ radians. The angle shown in (c) is coterminal with the angle shown in (d). In general, two angles measured in radians are coterminal if and only if they differ by a multiple of 2π radians.

CONVERSION FORMULAS

Since there are two forms of measurement for angles, we wish to be able to convert from one to the other. The degree measure of the angle corresponding to one complete counterclockwise rotation is $360°$, while the radian measure of the same angle is 2π radians. Thus we have $360° = 2\pi$ radians,

or $$180° = \pi \text{ radians.*} \tag{1}$$

___

*This does not mean that $180 = \pi$, just as 12 inches = 1 foot does not mean that $12 = 1$. Rather, it says that two measures of the one angle are equivalent.

From (1) we obtain the following formulas for converting between degrees and radians.

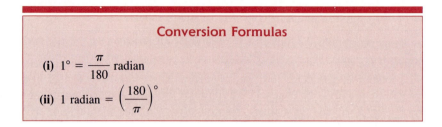

Conversion Formulas

(i) $1° = \dfrac{\pi}{180}$ radian

(ii) 1 radian $= \left(\dfrac{180}{\pi}\right)°$

Using a calculator to carry out the division in (i) and (ii), we find that

$$1° \approx 0.0174533 \text{ radian}$$

and

$$1 \text{ radian} \approx 57.29578°,$$

respectively.

Formula (i) enables us to convert from degrees to radians and (ii) allows us to convert from radians to degrees. As an aid for recalling these formulas, you can work with the ratio

$$\frac{\text{radian measure of } \theta}{\text{degree measure of } \theta} = \frac{\pi \text{ radians}}{180°}. \tag{2}$$

■ **EXAMPLE 7**

Convert 20° to radians.

Solution From (2) with $\theta = 20°$, we have

$$\frac{\text{radian measure of } \theta}{20°} = \frac{\pi \text{ radians}}{180°}$$

or radian measure of $\theta = 20°\left(\dfrac{\pi \text{ radians}}{180°}\right) = \dfrac{\pi}{9}$ radian. ■

Since the radian measure of an angle is the ratio of two lengths, it is dimensionless. Consequently, the word "radian" is often omitted when angles are measured in radians. It is understood that radian measure is intended when no unit of measure is indicated.

EXAMPLE 8

Convert the given angle to degrees.

(a) $\alpha = \dfrac{7\pi}{6}$ **(b)** $\beta = 2$

Solution

(a) Since no unit of angular measure is given, we know that α is measured in radians. From formula (ii) we know that the number of degrees in 1 radian is $180/\pi$. Thus the conversion is

$$\frac{7\pi}{6} = \frac{7\pi}{6}\left(\frac{180}{\pi}\right)^{\circ} = 210^{\circ}.$$

(b) Similarly, from (ii) we have

$$2 = 2\left(\frac{180}{\pi}\right)^{\circ} = \left(\frac{360}{\pi}\right)^{\circ} \approx 114.59^{\circ}.$$

EXAMPLE 9

Convert $153°40'$ to radians.

Solution We first write $153°40'$ in decimal form:

$$153°40' = (153 + \tfrac{40}{60})^{\circ}$$
$$= (153 + \tfrac{2}{3})^{\circ}$$
$$\approx 153.67^{\circ}.$$

Then we convert from degrees to radians using formula (i):

$$153.67^{\circ} = (153.67)\left(\frac{\pi}{180}\right) \text{radians}$$
$$\approx 0.85\pi \text{ radians}$$
$$\approx 2.68 \text{ radians}.$$

EXAMPLE 10

Find an angle between 0 and 2π that is coterminal with $\theta = 11\pi/4$. Sketch the angle.

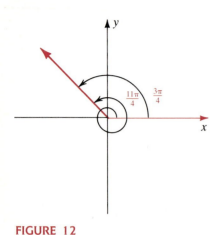

FIGURE 12

FIGURE 13

Solution   Since $2\pi < 11\pi/4 < 3\pi$, we subtract one revolution, or 2π radians, to obtain

$$\frac{11\pi}{4} - 2\pi = \frac{11\pi}{4} - \frac{8\pi}{4} = \frac{3\pi}{4}.$$

Thus, $3\pi/4$ is coterminal with $11\pi/4$, as shown in Figure 12.

ARC LENGTH

In many applications it is necessary to find the length s of the arc subtended by a central angle θ in a circle of radius r (see Figure 13). From the definition of radian measure,

$$\theta \text{ (in radians)} = s/r,$$

we obtain the **arc length formula:**

$$s = r\theta, \quad \text{where } \theta \text{ is measured in radians.} \tag{3}$$

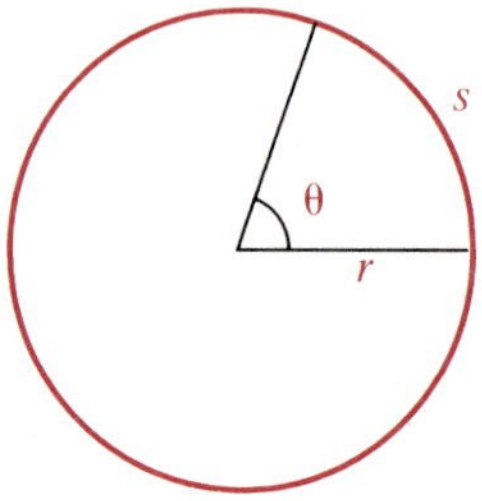

■ EXAMPLE 11

Find the arc length subtended by a central angle of 2 radians in a circle of **(a)** radius 4 and **(b)** radius 6.

Solution

(a) From the arc length formula (3) with $\theta = 2$ and $r = 4$, we have

$$s = r\theta = (4)(2) = 8.$$

(b) From the arc length formula (3) with $\theta = 2$ and $r = 6$, we find

$$s = r\theta = (6)(2) = 12.$$

Note of Caution: Students often apply the arc length formula $s = r\theta$ incorrectly by using degree measure for θ. Remember that $s = r\theta$ is valid *only if* θ *is measured in radians*.

EXERCISE 2.1 ▬▬▬▬▬▬▬▬▬▬▬▬▬▬▬▬▬▬▬

In Problems 1–18, draw the given angle in standard position.

1. 60°

2. −120°

3. 135°

4. 150°

5. 1140°

6. −315°

7. −240°

8. −210°

9. $\dfrac{\pi}{3}$

10. $\dfrac{5\pi}{4}$ **11.** $\dfrac{7\pi}{6}$ **12.** $-\dfrac{2\pi}{3}$

13. $-\dfrac{\pi}{6}$ **14.** -3π **15.** $\dfrac{5\pi}{2}$

16. $\dfrac{3\pi}{4}$ **17.** 3 **18.** 4

In Problems 19–22, express the given angle in decimal notation.

19. $10°39'17''$ **20.** $143°7'2''$
21. $5°10'$ **22.** $10°25'$

In Problems 23–26, express the given angle in terms of degrees, minutes, and seconds.

23. $210.78°$ **24.** $15.45°$
25. $30.81°$ **26.** $110.5°$

In Problems 27–38, convert from degrees to radians.

27. $45°$ **28.** $30°$ **29.** $270°$
30. $60°$ **31.** $1°$ **32.** $0°$
33. $131°40'$ **34.** $-120°$ **35.** $-230°$
36. $52°$ **37.** $540°$ **38.** $-47.2°$

In Problems 39–50, convert from radians to degrees.

39. $\dfrac{2\pi}{3}$ **40.** $\dfrac{\pi}{12}$ **41.** $\dfrac{\pi}{6}$

42. 7π **43.** 3.1 **44.** $\dfrac{19\pi}{2}$

45. 1.5 **46.** 0.76 **47.** 12

48. 17π **49.** $\dfrac{5\pi}{4}$ **50.** $\dfrac{3\pi}{8}$

In Problems 51–54, find the angle between 0° and 360° that is coterminal with the given angle.

51. $875°$ **52.** $400°$ **53.** $-610°$ **54.** $-150°$

In Problems 55–60, find the angle between 0 and 2π radians that is coterminal with the given angle.

55. $-\dfrac{\pi}{4}$ **56.** $\dfrac{17\pi}{2}$ **57.** 5.3π

58. $-\dfrac{9\pi}{5}$ **59.** -4 **60.** 7.5

In Problems 61–64, find an angle that is (a) complementary, and (b) supplementary to the given angle, or state why no such angle can be found.

61. $48°15'$ **62.** $92°55'$
63. $98.4°$ **64.** $63.08°$
65. Find both the degree and the radian measures of the angle formed by (a) three-fifths of a counterclockwise rotation and (b) five and one-eighth clockwise rotations.
66. Find both the degree and the radian measures of the obtuse angle formed by the hands of a clock (a) at 8:00; (b) at 1:00; and (c) at 7:30.
67. Find both the degree and the radian measures of the angle through which the hour hand on a clock rotates in 2 hours.
68. Answer the question in Problem 67 for the minute hand.
69. The earth rotates once every 24 hours. How long does it take the earth to rotate through an angle of (a) 240° and (b) $\pi/6$ radians?
70. The planet Mercury completes one rotation every 59 days. Through what angle (measured in degrees) does it rotate in (a) 1 day; (b) 1 hour; and (c) 1 minute?
71. Find the arc length subtended by a central angle of 3 radians in a circle of (a) radius 3 and (b) radius 5.
72. Find the arc length subtended by a central angle of 30° in a circle of (a) radius 2 and (b) radius 4.
73. Find the measure of a central angle θ in a circle of radius 5 if θ subtends an arc of length 7.5. Give θ in (a) radians and (b) degrees.
74. Find the measure of a central angle θ in a circle of radius 1 if θ subtends an arc of length $\pi/3$. Give θ in (a) radians and (b) degrees.
75. A yo-yo is whirled around in a circle at the end of its 100-cm string.
   (a) If it makes 6 revolutions in 4 seconds, find its rate of turning (**angular speed**) in radians per second.
   (b) Find the speed at which the yo-yo travels in centimeters per second. (This is called its **linear speed.***)
76. If there is a knot in the yo-yo string described in Problem 75 at a point 40 cm from the yo-yo, find (a) the angular speed of the knot and (b) the linear speed. Which of these speeds is independent of the radius?

---

*For further discussion of linear and angular speed, see Appendix A.

77. An automobile is traveling at a rate of 55 mph and the diameter of its tires is 26 in.

(a) Find the number of revolutions per minute that its wheels are making.

(b) Find the angular speed of its wheels in radians per minute.

78. Prove that the area A of a sector formed by a central angle of θ radians in a circle of radius r is given by $A = \frac{1}{2}r^2\theta$. [*Hint*: Use the proportionality property from geometry that the ratio of the area A of a circular sector to the total area πr^2 of the circle equals the ratio of the central angle θ to one complete revolution 2π.]

79. What is the area of the shaded circular band shown in Figure 14 if θ is measured (a) in radians and (b) in degrees? [*Hint*: Use the result of Problem 78.]

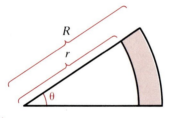

FIGURE 14

80. A clock pendulum is 1.3 m long and swings back and forth along a 15-cm arc. Find (a) the central angle and (b) the area of the sector through which the pendulum sweeps in one swing. [*Hint*: To answer part (b), use the result of Problem 78.]

81. A **nautical mile** is defined as the arc length subtended on the surface of the earth by an angle of measure 1 minute. If the diameter of the earth is 7927 miles, find how many **statute** (land) **miles** there are in a nautical mile.

82. Around 230 B.C. Eratosthenes calculated the circumference of the earth from the following observations. At noon on the longest day of the year, the sun was directly overhead in Syene, while it was inclined 7.2° from the vertical in Alexandria. He believed the two cities to be on the same longitudinal line and assumed that the rays of the sun are parallel. Thus he concluded that the arc from Syene to Alexandria was subtended by a central angle of 7.2° at the center of the earth (see Figure 15). At that time the distance from Syene to Alexandria was measured as 5000 stades. If one stade = 559 feet, find the circumference of the earth in (a) stades and (b) miles. Show that Eratosthenes' data gives a result that is within 7% of the correct value if the polar diameter of the earth is 7900 miles (to the nearest mile).

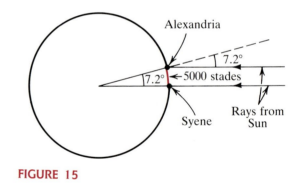

FIGURE 15

2.2

Trigonometric Functions of Acute Angles in Right Triangles

As we said in the introduction to this chapter, the word trigonometry refers to the measurement of triangles. In this section we define the six trigonometric functions—**sine, cosine, tangent, cosecant, secant,** and **cotangent**—as ratios of the lengths of the sides of a right triangle. The names of these functions are abbreviated as **sin, cos, tan, csc, sec,** and **cot,** respectively. In Section 2.4 we will extend the definitions to general angles.

As Figure 16 shows, if $\triangle OAB$ is any right triangle, then the side AB is said to be the **side opposite** the angle θ. The side OA is called the **side adjacent** to the angle θ. The **hypotenuse,** OB, is the side opposite the right angle. The lengths of these sides are denoted by **opp, adj,** and **hyp,** respectively.

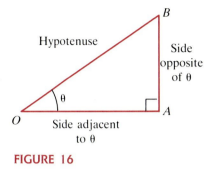

FIGURE 16

DEFINITION 1

The six trigonometric functions of an acute angle θ in a right triangle are defined as follows:

$$\sin \theta = \frac{\text{opp}}{\text{hyp}}, \qquad \csc \theta = \frac{\text{hyp}}{\text{opp}},$$

$$\cos \theta = \frac{\text{adj}}{\text{hyp}}, \qquad \sec \theta = \frac{\text{hyp}}{\text{adj}},$$

$$\tan \theta = \frac{\text{opp}}{\text{adj}}, \qquad \cot \theta = \frac{\text{adj}}{\text{opp}}.$$

The domain of each of these trigonometric functions is the set of all acute angles. In Section 2.4 we will extend these domains to include angles other than acute angles. Then in Chapter 3 we will see how the trigonometric functions can be defined with domains consisting of real numbers rather than angles.

The values of the six trigonometric functions depend only on the size of the angle θ and not on the size of the right triangle. This can be seen as follows. As Figure 17 shows, two right triangles with the same acute angle θ are similar, and thus the ratios of the corresponding sides are equal. For example, from the smaller triangle we have

$$\sin \theta = \frac{\text{opp}}{\text{hyp}} = \frac{\overline{AB}}{\overline{OB}},$$

whereas in the larger triangle we find that

$$\sin \theta = \frac{\overline{A'B'}}{\overline{OB'}}.$$

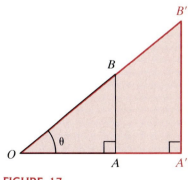

FIGURE 17

But since triangle AOB is similar to triangle $A'OB'$, we have

$$\frac{\overline{AB}}{\overline{OB}} = \frac{\overline{A'B'}}{\overline{OB'}}.$$

Thus we get the same value for sin θ regardless of which right triangle we use to compute it. A similar argument can be made for the other trigonometric functions.

▬▬ EXAMPLE 1

Find the values of the six trigonometric functions of the angle θ shown in Figure 18.

Solution   From Figure 18 we see that for the angle θ, opp = 8 and adj = 15. The value of hyp can be found from the Pythagorean theorem as follows:

$$(\text{hyp})^2 = 8^2 + 15^2 = 64 + 225 = 289$$
$$\text{hyp} = \sqrt{289} = 17.$$

Thus the values of the six trigonometric functions are

$$\sin \theta = \frac{\text{opp}}{\text{hyp}} = \frac{8}{17}, \qquad \csc \theta = \frac{\text{hyp}}{\text{opp}} = \frac{17}{8},$$

$$\cos \theta = \frac{\text{adj}}{\text{hyp}} = \frac{15}{17}, \qquad \sec \theta = \frac{\text{hyp}}{\text{adj}} = \frac{17}{15},$$

$$\tan \theta = \frac{\text{opp}}{\text{adj}} = \frac{8}{15}, \qquad \cot \theta = \frac{\text{adj}}{\text{opp}} = \frac{15}{8}.$$ ▬▬

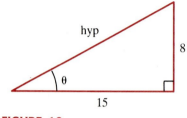

FIGURE 18

hyp

8

θ

15

FUNDAMENTAL IDENTITIES

There are many important relationships among the trigonometric functions. The basic ones are referred to as the **fundamental identities** and are worth memorizing. The following identities are easily derived from the definitions of the trigonometric functions.

Quotient Identities

$$\tan \theta = \frac{\sin \theta}{\cos \theta}, \qquad \cot \theta = \frac{\cos \theta}{\sin \theta}$$

Reciprocal Identities

$$\sec \theta = \frac{1}{\cos \theta}, \qquad \csc \theta = \frac{1}{\sin \theta}, \qquad \cot \theta = \frac{1}{\tan \theta}$$

For example, the first of the quotient identities is verified as follows:

$$\frac{\sin \theta}{\cos \theta} = \frac{\text{opp/hyp}}{\text{adj/hyp}} = \frac{\text{opp}}{\text{adj}} = \tan \theta.$$

The others can be verified in a similar manner (see Problems 51–54). Using these identities, we can find the values of all six trigonometric functions once we know the values of $\sin \theta$ and $\cos \theta$.

■ EXAMPLE 2

Given $\sin \theta = \frac{4}{5}$ and $\cos \theta = \frac{3}{5}$, find the values of the remaining trigonometric functions.

Solution From the fundamental identities, we have

$$\tan \theta = \frac{\sin \theta}{\cos \theta} = \frac{4/5}{3/5} = \frac{4}{3},$$

$$\sec \theta = \frac{1}{\cos \theta} = \frac{1}{3/5} = \frac{5}{3},$$

$$\csc \theta = \frac{1}{\sin \theta} = \frac{1}{4/5} = \frac{5}{4},$$

$$\cot \theta = \frac{1}{\tan \theta} = \frac{1}{4/3} = \frac{3}{4}.$$

The following example illustrates that if just one trigonometric function value of an acute angle is known, it is possible to find the other five function values by drawing an appropriate triangle.

■ EXAMPLE 3

If θ is an acute angle and $\sin \theta = \frac{2}{7}$, find the values of the other trigonometric functions at θ.

Solution We will sketch a right triangle with an acute angle θ satisfying $\sin \theta = \frac{2}{7}$. This is accomplished by making opp $= 2$ and hyp $= 7$, as shown in Figure 19. We find adj from the Pythagorean theorem:

$$2^2 + (\text{adj})^2 = 7^2$$

so that

$$(\text{adj})^2 = 7^2 - 2^2 = 49 - 4 = 45.$$

Thus,

$$\text{adj} = \sqrt{45} = 3\sqrt{5}.$$

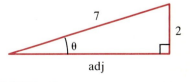

FIGURE 19

The remaining trigonometric function values are

$$\cos \theta = \frac{3\sqrt{5}}{7},$$

$$\tan \theta = \frac{2}{3\sqrt{5}} = \frac{2\sqrt{5}}{15},$$

$$\csc \theta = \frac{7}{2},$$

$$\sec \theta = \frac{7}{3\sqrt{5}} = \frac{7\sqrt{5}}{15},$$

$$\cot \theta = \frac{3\sqrt{5}}{2}.$$

SPECIAL ANGLES

The angles with measures 30° ($\pi/6$ radian), 45° ($\pi/4$ radian), and 60° ($\pi/3$ radians) occur frequently in trigonometry. We can find the values of the six trigonometric functions of these special angles using some results from plane geometry.

To find the values of the trigonometric functions of a 45° angle, we consider the isosceles right triangle with two equal sides of length 1 shown in Figure 20. From plane geometry we know that the acute angles in this triangle are equal; therefore, each acute angle measures 45°. From the Pythagorean theorem, we can find the length of the hypotenuse:

$$(\text{hyp})^2 = (1)^2 + (1)^2 = 2$$
$$\text{hyp} = \sqrt{2}.$$

Thus we have

$$\sin 45° = \frac{1}{\sqrt{2}} = \frac{\sqrt{2}}{2}, \qquad \csc 45° = \sqrt{2},$$

$$\cos 45° = \frac{1}{\sqrt{2}} = \frac{\sqrt{2}}{2}, \qquad \sec 45° = \sqrt{2},$$

$$\tan 45° = \frac{\sqrt{2}/2}{\sqrt{2}/2} = 1, \qquad \cot 45° = 1.$$

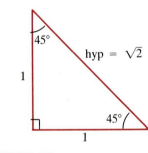

FIGURE 20

To find the values of the trigonometric functions of the 30° and 60° angles, we consider the equilateral triangle *AOB* with sides of length 2 shown in Figure 21(a). From plane geometry we know that the three angles of an equilateral triangle each measure 60°. As shown in Figure 21(b), if we bisect

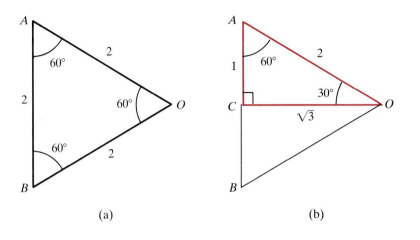

(a)   (b)   **FIGURE 21**

the angle at O, then CO is the perpendicular bisector of AB. It follows that

$$\angle AOC = \tfrac{1}{2} \angle AOB = \tfrac{1}{2}(60°) = 30°,$$
$$\overline{AC} = \tfrac{1}{2} \overline{AB} = \tfrac{1}{2}(2) = 1,$$

and
$$\angle ACO = 90°.$$

If we apply the Pythagorean theorem to the right triangle ACO, we get

$$(\overline{CO})^2 + 1^2 = 2^2,$$

or
$$(\overline{CO})^2 = 2^2 - 1^2 = 3$$

so that
$$\overline{CO} = \sqrt{3}.$$

Therefore, from triangle ACO in Figure 21(b), we can obtain the following values:

$$\theta = 30°: \quad \sin 30° = \frac{1}{2}, \quad \cos 30° = \frac{\sqrt{3}}{2}, \quad \tan 30° = \frac{1}{\sqrt{3}} = \frac{\sqrt{3}}{3};$$

$$\theta = 60°: \quad \sin 60° = \frac{\sqrt{3}}{2}, \quad \cos 60° = \frac{1}{2}, \quad \tan 60° = \sqrt{3}.$$

Using the same figure or the fundamental identities, we can compute the remaining values for $30°$ and $60°$ angles:

$$\theta = 30°: \quad \csc 30° = 2, \qquad \sec 30° = \frac{2}{\sqrt{3}} = \frac{2\sqrt{3}}{3}, \qquad \cot 30° = \sqrt{3};$$

$$\theta = 60°: \quad \csc 60° = \frac{2}{\sqrt{3}} \qquad \sec 60° = 2, \qquad \cot 60° = \frac{1}{\sqrt{3}}.$$
$$= \frac{2\sqrt{3}}{3}, \qquad\qquad\qquad\qquad\qquad\qquad = \frac{\sqrt{3}}{3}$$

The following table summarizes the values of the sine and cosine functions that we have just determined for the special angles 30°, 45°, and 60°. These are used so frequently that they should be memorized. Knowing these values and the fundamental identities we discussed earlier will enable you to determine any of the trigonometric functions for these special angles.

θ (degrees)	θ (radians)	$\sin \theta$	$\cos \theta$
30°	$\dfrac{\pi}{6}$	$\dfrac{1}{2}$	$\dfrac{\sqrt{3}}{2}$
45°	$\dfrac{\pi}{4}$	$\dfrac{\sqrt{2}}{2}$	$\dfrac{\sqrt{2}}{2}$
60°	$\dfrac{\pi}{3}$	$\dfrac{\sqrt{3}}{2}$	$\dfrac{1}{2}$

You can obtain the values in the table above by first constructing the 45°–45°–90° and 30°–60°–90° triangles as we did in Figures 20 and 21. Then you can determine the functional values from the triangles. This approach is summarized in the following table.

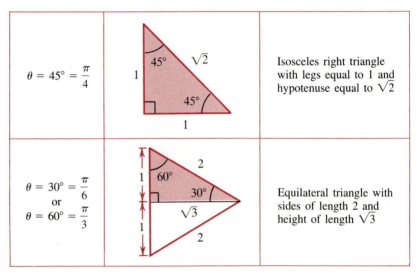

$\theta = 45° = \dfrac{\pi}{4}$		Isosceles right triangle with legs equal to 1 and hypotenuse equal to $\sqrt{2}$
$\theta = 30° = \dfrac{\pi}{6}$ or $\theta = 60° = \dfrac{\pi}{3}$		Equilateral triangle with sides of length 2 and height of length $\sqrt{3}$

See Problem 56 for another memory aid for the sines and cosines of these special angles.

USE OF A CALCULATOR

Approximations to the values of the trigonometric functions can be obtained using a scientific calculator.

Note of Caution: Before using a calculator to find trigonometric function values of angles measured in radians, you must set the calculator in radian mode. If the angles are measured in degrees, then you must select the degree mode before making your calculations. Also, if the angles are given in degrees, minutes, and seconds, they must be converted to decimal form first.

Most scientific calculators have keys labeled $\boxed{\text{sin}}$, $\boxed{\text{cos}}$, and $\boxed{\text{tan}}$ for computing the values of these functions. To obtain the values of csc, sec, or cot, we can use the $\boxed{\text{sin}}$, $\boxed{\text{cos}}$, or $\boxed{\text{tan}}$ keys with the reciprocal key $\boxed{1/x}$. The following example illustrates the process.

■ EXAMPLE 4

Use a calculator to approximate each of the following.

(a) $\sin 45°$ (b) $\cos 8°15'$ (c) $\tan 1.4$ (d) $\sec 0.23$ (e) $\cot \dfrac{\pi}{7}$

Solution

(a) First we make sure that the calculator is set in degree mode. Then we enter 45 and press the $\boxed{\text{sin}}$ key to obtain

$$\sin 45° \approx 0.7071068,$$

which is a seven-decimal-place approximation to the exact value $\sqrt{2}/2$.

(b) Since the angle is given in degrees and minutes, we must first convert it to decimal form:

$$8°15' = 8° + (\tfrac{15}{60})° = 8.25°.$$

Now with the calculator set in *degree mode,* we enter 8.25 and press the $\boxed{\text{cos}}$ key to obtain

$$\cos 8°15' = \cos 8.25° \approx 0.9896514.$$

(c) Since degrees are not indicated, we recognize that this angle is measured in radians. Therefore, we first set the calculator in *radian mode,* enter 1.4, and press the $\boxed{\text{tan}}$ key to get

$$\tan 1.4 \approx 5.7978837.$$

(d) To evaluate $\sec 0.23$, we will use the fundamental identity

$$\sec \theta = \frac{1}{\cos \theta}.$$

With the calculator set in radian mode, we enter 0.23, press the $\boxed{\text{cos}}$ key, and then take the reciprocal of the result by pressing the $\boxed{1/x}$ key.

Thus we have

$$\sec 0.23 = \frac{1}{\cos 0.23} \approx 1.0270458.$$

(e) We observe that this angle is measured in radians and set the calculator accordingly. We press the $\boxed{\pi}$ key (or enter a decimal approximation such as 3.141592654), divide by 7, and press the $\boxed{\texttt{t a n}}$ key and the $\boxed{\texttt{1/x}}$ key to obtain

$$\cot \frac{\pi}{7} = \frac{1}{\tan \dfrac{\pi}{7}} \approx 2.0765214.$$

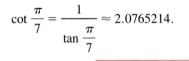

In many applications we are able to determine the value of one of the trigonometric functions, say $\sin \theta$, and we wish to find θ. We can do so by using the inverse trigonometric functions, which are discussed in detail in Section 3.4. For now we are able to find an acute angle θ given $\sin \theta$ (or one of the other trigonometric functions of θ) by using the inverse key $\boxed{\texttt{I N V}}$ on the calculator. For example, if we are given $\sin \theta = 0.625$ and we wish to find θ, we enter 0.625, press the inverse key $\boxed{\texttt{I N V}}$, and then press the sine key $\boxed{\texttt{s i n}}$. If the calculator is set in *degree mode,* we obtain $\theta \approx 38.6821875°$. However if the calculator is set in radian mode, we would get $\theta \approx 0.6751315$.

The inverse trigonometric functions are denoted by $\sin^{-1}$ or arcsin, $\cos^{-1}$ or arccos, $\tan^{-1}$ or arctan, and so on. Some calculators have keys labeled $\boxed{\texttt{s i n}^{-1}}$, $\boxed{\texttt{c o s}^{-1}}$, and $\boxed{\texttt{t a n}^{-1}}$ and do not require the use of the $\boxed{\texttt{I N V}}$ key. (Read the manual for your calculator if you need more explanation.)

■ EXAMPLE 5

Use a calculator to find an acute angle θ measured in (a) degrees and (b) radians for which $\cos \theta = 0.5$.

Solution

(a) We must first set the calculator in degree mode. We enter 0.5 and then press the $\boxed{\texttt{I N V}}$ key and then the $\boxed{\texttt{c o s}}$ key. The resulting display is 60. Thus, $\theta = 60°$.

(b) With the calculator set in radian mode, we enter 0.5 and then press $\boxed{\texttt{I N V}}$ followed by $\boxed{\texttt{c o s}}$. The display reads 1.0471976. Thus for $\cos \theta = 0.5$, $\theta \approx 1.0471976$ radians. Note that 1.0471976 is a decimal approximation of $\pi/3$. ■

TRIGONOMETRIC TABLES

Before the development of the scientific calculator, tables of numerical values for the trigonometric functions were essential in solving trigonometry problems. Tables III and IV at the back of the book contain approximations to four decimal places for sin θ, cos θ, tan θ, and cot θ for any acute angle θ measured in degrees and radians, respectively. See Appendix C for instructions on how to use these tables if an electronic calculator is not available to you.

COFUNCTIONS

The use of the terminology sine and *co*sine, tangent and *co*tangent, secant and *co*secant is a result of the following observation. As shown in Figure 22, if the two acute angles of a right triangle ABC are labeled α and β and a is the length of the side opposite α, b is the length of the side opposite β, and c is the length of the side opposite the right angle, then

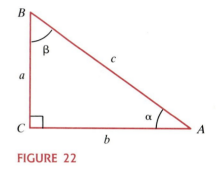

FIGURE 22

$$\sin \alpha = \frac{a}{c} = \cos \beta, \qquad \csc \alpha = \frac{c}{a} = \sec \beta,$$

$$\cos \alpha = \frac{b}{c} = \sin \beta, \qquad \sec \alpha = \frac{c}{b} = \csc \beta,$$

$$\tan \alpha = \frac{a}{b} = \cot \beta, \qquad \cot \alpha = \frac{b}{a} = \tan \beta.$$

Thus the cosine of an acute angle equals the sine of the complementary angle; the cotangent of an acute angle equals the tangent of the complementary angle; the cosecant of an acute angle equals the secant of the complementary angle; and conversely. For this reason we say that sine and cosine, tangent and cotangent, and secant and cosecant are **cofunctions** of one another. As a result, we have verified the following identities for any acute angle θ.

θ (radians)	θ (degrees)
$\sin\left(\dfrac{\pi}{2} - \theta\right) = \cos \theta$	$\sin (90° - \theta) = \cos \theta$
$\tan\left(\dfrac{\pi}{2} - \theta\right) = \cot \theta$	$\tan (90° - \theta) = \cot \theta$
$\sec\left(\dfrac{\pi}{2} - \theta\right) = \csc \theta$	$\sec (90° - \theta) = \csc \theta$

As we will see in Section 4.2, these identities hold for any angle θ (not just acute angles).

EXERCISE 2.2

In Problems 1–10, find the values of the six trigonometric functions of the angle θ in the given triangle.

1.
2.

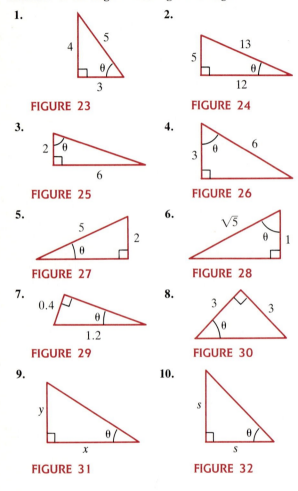

FIGURE 23

FIGURE 24

3.
4.

FIGURE 25

FIGURE 26

5.
6.

FIGURE 27

FIGURE 28

7.
8.

FIGURE 29

FIGURE 30

9.
10.

FIGURE 31

FIGURE 32

In Problems 11–18, use the fundamental identities to find the values of the remaining trigonometric functions at θ.

11. $\sin \theta = \dfrac{2}{\sqrt{13}}$, $\cos \theta = \dfrac{3}{\sqrt{13}}$

12. $\sin \theta = \dfrac{1}{\sqrt{10}}$, $\cos \theta = \dfrac{3}{\sqrt{10}}$

13. $\sin \theta = \dfrac{2}{7}$, $\cos \theta = \dfrac{3\sqrt{5}}{7}$

14. $\sin \theta = \dfrac{5}{\sqrt{26}}$, $\cos \theta = \dfrac{1}{\sqrt{26}}$

15. $\sin \theta = \dfrac{1}{\sqrt{65}}$, $\tan \theta = \dfrac{1}{8}$

16. $\cos \theta = \dfrac{5}{\sqrt{29}}$, $\cot \theta = \dfrac{5}{2}$

17. $\csc \theta = \dfrac{5}{3}$, $\sec \theta = \dfrac{5}{4}$

18. $\sin \theta = \dfrac{1}{\sqrt{50}}$, $\cot \theta = 7$

In Problems 19–26, find the values of the remaining trigonometric functions by drawing an appropriate triangle.

19. $\sin \theta = \dfrac{12}{13}$
20. $\cos \theta = \dfrac{2}{\sqrt{5}}$

21. $\sec \theta = \dfrac{2}{\sqrt{3}}$
22. $\csc \theta = \sqrt{10}$

23. $\tan \theta = \dfrac{2}{5}$
24. $\cot \theta = \dfrac{1}{7}$

25. $\cos \theta = \dfrac{a}{\sqrt{a^2 + b^2}}$, $a^2 + b^2 \neq 0$

26. $\tan \theta = \dfrac{b}{a}$, $a \neq 0$

In Problems 27–34, use a calculator to find the approximate values of the six trigonometric functions of the given angle.

27. $17°$
28. $82°$
29. $14.3°$
30. $46°15'8''$
31. $\dfrac{\pi}{5}$
32. $\dfrac{\pi}{10}$
33. 0.6725
34. 1.24

In Problems 35–44, use a calculator to approximate the acute angle θ, measured in (a) radians and (b) degrees, satisfying the given condition.

35. $\sin \theta = 0.5260$
36. $\cos \theta = 0.8964$
37. $\tan \theta = 2.4$
38. $\sin \theta = 0.752$
39. $\cos \theta = 0.2$
40. $\tan \theta = 3.15$

41. $\sin \theta = \frac{1}{3}$

42. $\cos \theta = \frac{1}{4}$

43. $\sec \theta = 3.81$

44. $\csc \theta = 1.05$

In Problems 45–50, answer true or false.

45. $\cot 47° = \dfrac{\cos 47°}{\sin 47°}$ ____

46. If θ is measured in radians, $\cos\left(\dfrac{\pi}{2} - \theta\right) = \sin \theta.$ ____

47. $\sin 61° = \cos 29°$ ____

48. $\sin \frac{1}{2} = 30°$ ____

49. $\cos \dfrac{\pi}{6} = \sin \dfrac{\pi}{3}$ ____

50. $\cos \dfrac{\pi}{3} = \sec \dfrac{3}{\pi}$ ____

In Problems 51–54, use the definitions of the trigonometric functions listed on page 71 to verify the given identity.

51. $\cot \theta = \dfrac{\cos \theta}{\sin \theta}$

52. $\csc \theta = \dfrac{1}{\sin \theta}$

53. $\sec \theta = \dfrac{1}{\cos \theta}$

54. $\cot \theta = \dfrac{1}{\tan \theta}$

55. Using Figure 33, show that if θ is an acute angle in a right triangle, then

$$(\cos \theta)^2 + (\sin \theta)^2 = 1.$$

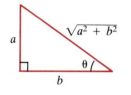

FIGURE 33

56. The pattern in the following table can be helpful in memorizing the sines and cosines of special angles. The pattern utilizes the fact that $\sqrt{1} = 1$. Verify the entries in the table.

θ (degrees)	θ (radians)	$\sin \theta$	$\cos \theta$
30°	$\dfrac{\pi}{6}$	$\dfrac{\sqrt{1}}{2}$	$\dfrac{\sqrt{3}}{2}$
45°	$\dfrac{\pi}{4}$	$\dfrac{\sqrt{2}}{2}$	$\dfrac{\sqrt{2}}{2}$
60°	$\dfrac{\pi}{3}$	$\dfrac{\sqrt{3}}{2}$	$\dfrac{\sqrt{1}}{2}$

57. As shown in Figure 34, two tracking stations S_1 and S_2 sight a weather balloon between them at elevation angles α and β, respectively. Show that the height h of the balloon is given by

$$h = \dfrac{c}{\cot \alpha + \cot \beta},$$

where c is the distance between the tracking stations. [*Hint:* Find $\cot \alpha + \cot \beta$.]

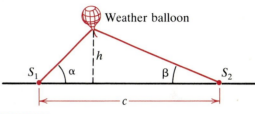

FIGURE 34

58. Use the cofunction identity $\tan (90° - \theta) = \cot \theta$ to prove that the altitude to the hypotenuse of a right triangle is the mean proportional between the segments of the hypotenuse; that is (referring to Figure 35),

$$\dfrac{m}{h} = \dfrac{h}{n}.$$

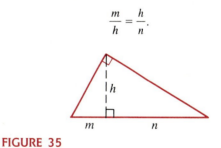

FIGURE 35

2.3

Applications of Trigonometry to Right Triangles

SOLVING RIGHT TRIANGLES

Applications of trigonometry in fields such as surveying and navigation involves **solving right triangles.** The expression "to solve a triangle" means that we wish to find the length of each side and the measure of each angle in the triangle. We will see in this section that we can solve any right triangle given either

(i) the lengths of any two sides, or
(ii) the length of one side and the measure of one acute angle.

As the following examples show, sketching and labeling the triangle is an essential part of the solution process. It will be our general practice to label a right triangle as shown in Figure 36. The three vertices of the triangle are denoted by A, B, and C with C at the vertex of the right angle. We denote the angles at A and B by α and β, and the lengths of the sides opposite these angles by a and b, respectively. The length of the side opposite the right angle at C is denoted by c.

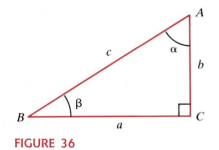

FIGURE 36

▰▰ EXAMPLE 1

Solve the right triangle having a hypotenuse of length $4\sqrt{3}$ and one $60°$ angle.

Solution First we make a sketch of the triangle and label it as shown in Figure 37. We wish to find a, b, and β. Since α and β are complementary,

$$\beta = 90° - \alpha = 90° - 60° = 30°.$$

We know the length of the *hypotenuse*. To find a, the side *opposite* the $60°$ angle, we select the sine function. From $\sin \alpha = \text{opp/hyp}$, we obtain

$$\sin 60° = \frac{a}{4\sqrt{3}}, \quad \text{or} \quad a = 4\sqrt{3}\, \sin 60°.$$

Recall from Section 2.2 that $\sin 60° = \sqrt{3}/2$, so that

$$a = 4\sqrt{3}\, \sin 60° = 4\sqrt{3}\left(\frac{\sqrt{3}}{2}\right) = 6.$$

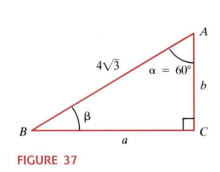

FIGURE 37

To find b, the side *adjacent* to the 60° angle, we select the cosine function. From $\cos \alpha = \text{adj/hyp}$, we obtain

$$\cos 60° = \frac{b}{4\sqrt{3}}, \quad \text{or} \quad b = 4\sqrt{3} \cos 60°.$$

Thus, using $\cos 60° = \frac{1}{2}$, we have

$$b = 4\sqrt{3} \cos 60° = 4\sqrt{3}(\tfrac{1}{2}) = 2\sqrt{3}.$$

(Alternatively, once we determined a, we could have found b by using either the Pythagorean theorem or the tangent function. In general, there can be many ways to approach solving a triangle.)

If angles other than the special angles 30°, 45°, or 60° are involved in the problem, either a calculator or trigonometric tables must be used to obtain *approximations* of the desired trigonometric function values. For the remainder of this chapter, whenever an approximation is used, we will round the final results to the nearest hundredth unless the problem specifies otherwise.

USE OF A CALCULATOR

In the following examples, the computations have been done with a scientific calculator. However, if Table IV is used to obtain the values of the trigonometric functions, the results may differ somewhat from those shown. This is due to the fact that the calculator is working internally with eight or more significant digits, although the table provides only four significant digits. To take full advantage of the calculator's greater accuracy, the computed values of the trigonometric functions should be retained or stored in the calculator for subsequent use. If, instead, a displayed value is written down and then later a rounded version of that value is keyed into the calculator, the accuracy of the final result may be diminished.

EXAMPLE 2

Solve the right triangle that has a 57.5° angle with the opposite side of length 10.

Solution First we sketch and label the triangle as shown in Figure 38. From the sketch we can see that we need to find β, b, and c. Since α and β are complementary angles,

$$\beta = 90° - \alpha = 90° - 57.5° = 32.5°.$$

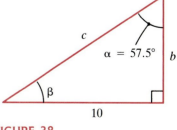

FIGURE 38

We know the length of the side *opposite* α. To find the length of the *adjacent* side, we use the tangent function. From $\tan \alpha = \text{opp/adj}$, we have

$$\tan 57.5° = \frac{10}{b}, \quad \text{or} \quad b = \frac{10}{\tan 57.5°}.$$

Using a calculator, we find that $\tan 57.5° \approx 1.5696856$ so that

$$b \approx \frac{10}{1.5696856^*} \approx 6.37.$$

(For the preceding computation, one possible keying sequence on most calculators is: Set the calculator in degree mode, enter 57.5, press $\boxed{\tan}$, then press $\boxed{1/x}$, and finally multiply by 10.)

To find the *hypotenuse c,* we use $\sin \alpha = \text{opp/hyp}$ to obtain

$$\sin 57.5° = \frac{10}{c}, \quad \text{or} \quad c = \frac{10}{\sin 57.5°}.$$

Thus,

$$c \approx \frac{10}{0.8433914} \approx 11.86.$$

■

■ EXAMPLE 3

Solve the right triangle with legs of length 4 and 5.

Solution   After sketching and labeling the triangle as shown in Figure 39, we see that we need to find c, α, and β. From the Pythagorean theorem, the hypotenuse c is given by

$$c = \sqrt{5^2 + 4^2} = \sqrt{41} \approx 6.40.$$

To find β, we use $\tan \beta = \text{opp/adj}$. (By choosing to work with the given quantities, we avoid error due to previous approximations.) Thus we have

$$\tan \beta = \tfrac{4}{5} = 0.8.$$

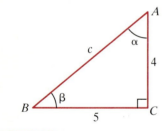

FIGURE 39

To find β with a calculator set in degree mode, we enter 0.8, press $\boxed{INV}$, and then press $\boxed{\tan}$. The result is

$$\beta \approx 38.6598083° \approx 38.66°.$$

Then

$$\alpha = 90° - \beta \approx 90° - 38.66° = 51.34°.$$

■

---

*At this point we are showing the value of the trigonometric function as a check for you. When working with a calculator this step would normally not be written down. We will continue to provide this check in subsequent examples in this section.

■■■ EXAMPLE 4

A kite is caught in the top branches of a tree. If the 90-ft kite string makes an angle of 22° with the ground, estimate the height of the tree by finding the distance from the kite to the ground.

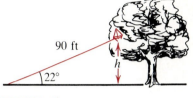

90 ft

Let h denote the height of the kite. From Figure 40 we see that

$$\frac{h}{90} = \sin 22°, \quad \text{or} \quad h = 90 \sin 22°.$$

Thus, $\qquad h \approx 90(0.3746066) \approx 33.71 \text{ ft.}$ ■■■

FIGURE 40

■■■ EXAMPLE 5

The distance between the earth and the moon varies as the moon revolves around the earth. At a particular time an amateur astronomer measures the angle of 1° shown in Figure 41.* Calculate to the nearest hundred miles the distance between the center of the earth and the center of the moon at this instant. Assume that the radius of the earth is 3963 mi.

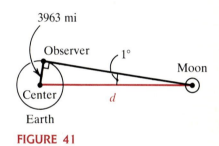

FIGURE 41

Solution Let d represent the distance between the center of the earth and the center of the moon. From the definition of the sine function, we have

$$\sin 1° = \frac{3963}{d}, \quad \text{or} \quad d = \frac{3963}{\sin 1°}.$$

With a calculator, we find that

$$d \approx \frac{3963}{0.0174524} \approx 227,100 \text{ mi}$$

rounded to the nearest hundred miles. ■■■

■■■ EXAMPLE 6

A carpenter cuts the end of a 4-in.-wide board on a 25° bevel from the vertical, starting at a point $1\frac{1}{2}$ in. from the end of the board. Find the lengths of the diagonal cut and the remaining side. See Figure 42.

---

*This angle is known as the **geocentric parallax.** The determination of this angle actually depends on two observations made from earth.

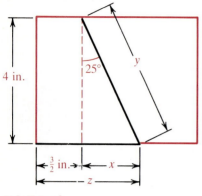

FIGURE 42

Solution   Let x, y, and z be the (unknown) dimensions, as labeled in Figure 42. It follows from the definition of the tangent function that

$$\tan 25° = \frac{x}{4}.$$

Thus, $x = 4 \tan 25° \approx 4(0.4663077) \approx 1.87$ in.

To find y we observe that

$$\frac{y}{4} = \sec 25°, \quad \text{or} \quad y = 4 \sec 25°.$$

With the aid of a calculator (using the $\boxed{\cos}$ and $\boxed{1/x}$ keys), we obtain

$$y \approx 4(1.1033779) \approx 4.41 \text{ in.}$$

Since $z = \frac{3}{2} + x$ and $x \approx 1.87$ in., we see that

$$z \approx 1.5 + 1.87 \approx 3.37 \text{ in.}$$

ANGLES OF ELEVATION AND DEPRESSION

The angle between an observer's line of sight to an object and the horizontal is given a special name. As Figure 43 illustrates, if the line of sight is to an object above the horizontal, the angle is called an **angle of elevation,** whereas if the line of sight is to an object below the horizontal, the angle is called an **angle of depression.**

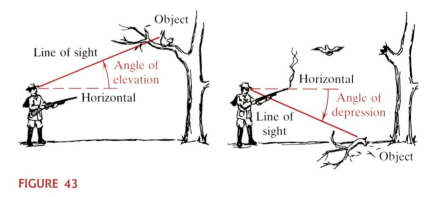

FIGURE 43

▬ EXAMPLE 7

A surveyor uses an instrument called a theodolite to measure the angle of elevation between the top of a mountain and ground level. At one point the angle of elevation is measured to be 41°. A half kilometer farther from the base of the mountain, the angle of elevation is measured to be 37°. How high is the mountain?

Solution Let h represent the height. Figure 44 shows that there are two right triangles sharing the common side h, so we obtain two equations in two unknowns z and h:

$$\frac{h}{z + 0.5} = \tan 37° \quad \text{and} \quad \frac{h}{z} = \tan 41°.$$

We can solve each of these for h, obtaining

$$h = (z + 0.5) \tan 37° \quad \text{and} \quad h = z \tan 41°.$$

Equating the last two results gives an equation from which we can determine z:

$$(z + 0.5) \tan 37° = z \tan 41°.$$

Solving for z gives us

$$z = \frac{-0.5 \tan 37°}{\tan 37° - \tan 41°}$$
$$\approx 3.2556 \text{ km}.$$

Now h can be found:

$$h = z \tan 41°$$
$$\approx 3.2556 \tan 41°$$
$$\approx 2.83 \text{ km}.$$

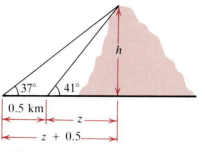

FIGURE 44

■ **EXAMPLE 8**

Most airplanes approach San Francisco International Airport (SFO) on a straight 3° glide path starting at a point 5.5 mi from the field. Using an experimental computerized technique, called the *two-segment approach,* a plane approaches the field on a 6° glide path starting at a point 5.5 mi out and then switches to a 3° glide path 1.5 mi from the point of touchdown. The purpose of this new approach is, of course, noise reduction. What is the height in feet of a plane P using the experimental glide path when it switches to the 3° glide path? Compare the height of this plane with a plane P' using the standard 3° approach, when both planes are 5.5 mi from the field.

Solution For purposes of illustration, the angles and distances shown in Figure 45 are slightly exaggerated.

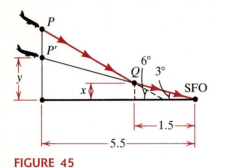

FIGURE 45

Let x be the height of plane P at the point Q 1.5 mi out from the field when the plane switches to the 3° glide path. From the figure we see that

$$\frac{x}{1.5} = \tan 3°, \quad \text{or} \quad x = 1.5 \tan 3°.$$

Thus,
$$x \approx 1.5(0.0524078)$$
$$\approx 0.0786 \text{ mi}$$
$$= 0.0786(5280) \text{ ft}$$
$$\approx 415 \text{ ft}.$$

If y is the height of the plane P' on the standard 3° approach when it is 5.5 mi out from the field, then

$$\frac{y}{5.5} = \tan 3°, \quad \text{or} \quad y = 5.5 \tan 3°.$$

Thus,
$$y \approx 5.5(0.0524078)$$
$$\approx 0.2882 \text{ mi}$$
$$= 0.2882(5280) \text{ ft}$$
$$\approx 1522 \text{ ft}.$$

Now, as Figure 46 shows, the height of plane P at 5.5 mi out is given by

$$z = w + x,$$

where
$$\frac{w}{4} = \tan 6°, \quad \text{or} \quad w = 4 \tan 6°.$$

Thus,
$$w \approx 4(0.1051042)$$
$$\approx 0.4204 \text{ mi}$$
$$= 0.4204(5280) \text{ ft}$$
$$\approx 2220 \text{ ft}.$$

Therefore, the approximate height of plane P at a point 5.5 mi out is

$$z \approx 2220 + 415 = 2635 \text{ ft}.$$

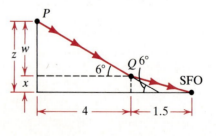

FIGURE 46

EXERCISE 2.3

In Problems 1–14, find the indicated unknowns. Each problem refers to the triangle shown in Figure 47.

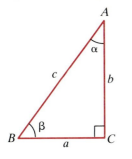

FIGURE 47

1. $a = 4$, $\beta = 27°$; b, c **2.** $c = 10$, $\beta = 49°$; a, b
3. $b = 8$, $\beta = 34°20'$; a, c **4.** $c = 25$, $\alpha = 50°$; a, b
5. $a = 6$, $\alpha = 61°10'$; b, c **6.** $a = 5$, $b = 2$; α, β, c
7. $b = 1.5$, $c = 3$; α, β, a **8.** $b = 4$, $\alpha = 58°$; a, c
9. $a = 4$, $b = 10$; α, β, c **10.** $b = 3$, $c = 6$; α, β, a
11. $a = 9$, $c = 12$; α, β, b **12.** $a = 11$, $\alpha = 33.5°$; b, c
13. $b = 20$, $\alpha = 23°$; a, c **14.** $c = 15$, $\beta = 31°40'$; a, b

15. A building casts a shadow 20 m long. If the angle from the tip of the shadow to a point on top of the building is 69°, how high is the building?

16. Two trees are on opposite sides of a river, as shown in Figure 48. A base line of 100 ft is measured from tree T_1, and from that position the angle β to T_2 is measured to be 29.7°.

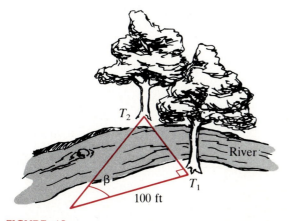

FIGURE 48

If the base line is perpendicular to the line segment between T_1 and T_2, find the distance between the two trees.

17. A 50-ft tower is located on the edge of a river. The angle of elevation between the opposite bank and the top of the tower is 37°. How wide is the river?

18. A surveyor uses a geodometer to measure the straight-line distance from a point on the ground to a point on top of a mountain. Use the information given in Figure 49 to find the height of the mountain.

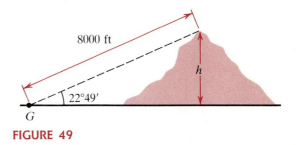

FIGURE 49

19. An observer on the roof of building A measures a 27° angle of depression between the horizontal and the base of building B. The angle of elevation from the same point to the roof of the second building is 41°25'. What is the height of building B if the height of building A is 150 ft?

20. Find the height h of a mountain using the information given in Figure 50.

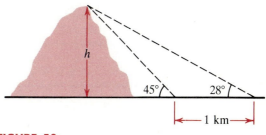

FIGURE 50

21. The top of a 20-ft ladder is leaning against the edge of the roof of a house. If the angle of inclination of the ladder from the horizontal is 51°, what is the approximate height of the house and how far is the bottom of the ladder from the base of the house?

22. An airplane at an altitude of 25,000 ft approaches a radar station located on a 2000-ft-high hill. At one instant in time, the angle between the radar dish pointed at the plane and the horizontal is 57°. What is the straight-line distance in miles between the airplane and the radar station at that particular instant?

23. A 5-mi straight segment of a road climbs a 4000-ft hill. Determine the angle that the road makes with the horizontal.

24. A box has dimensions as shown in Figure 51. Find the length of the diagonal between the corners P and Q. What is the angle θ formed between the diagonal and the bottom edge of the box?

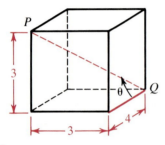

FIGURE 51

25. Observers in two towns A and B on either side of a 12,000-ft mountain measure the angles of elevation between the ground and the top of the mountain (see Figure 52). Assuming that the towns lie in the same vertical plane, find the horizontal distance between them.

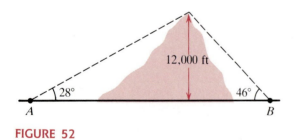

FIGURE 52

26. A drawbridge measures 7.5 m from shore to shore, and when completely open it makes an angle of 43° with the horizontal (see Figure 53(b)). When the bridge is closed, the angle of depression from the shore to a point on the surface of the water below the opposite end is 27° (see

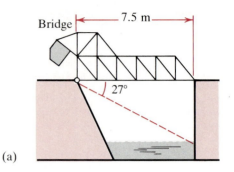

(a)

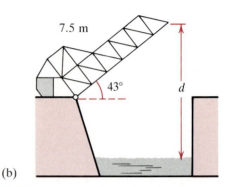

(b)

FIGURE 53

Figure 53(a)). When the bridge is fully open, what is the distance d between the highest point of the bridge and the water below?

27. A man standing 50 ft from a 20-ft tall house looks up at a TV antenna located on the edge of the roof (see Figure 54). If the angle between his line of sight to the edge of the roof and his line of sight to the top of the antenna is 12°, how tall is the antenna?

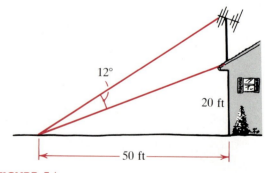

FIGURE 54

28. A flagpole is located at the edge of a sheer 50-ft cliff at the bank of a river of width 40 feet (see Figure 55). An observer on the opposite side of the river measures an angle of 9° between her line of sight to the top of the flagpole and her line of sight to the top of the cliff. Find the height of the flagpole.

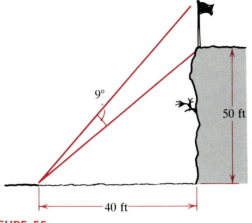

9°

50 ft

←— 40 ft —→

FIGURE 55

29. The length of a Boeing 727 airplane is 131.1 ft. What is the plane's altitude if it subtends an angle of 2° when it is directly above an observer on the ground? See Figure 56.

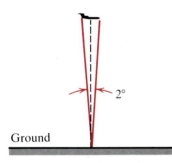

2°

Ground

FIGURE 56

30. The gnomon (pin) of a sundial is 4 in. in height. If it casts a 6-in. shadow, what is the angle of elevation of the sun?

31. Weather radar is capable of measuring both the angle of elevation to the top of a thunderstorm and its range (the horizontal distance to the storm). If the range of a storm is 90 km and the angle of elevation is 4°, can a passenger plane that is able to climb to 10 km fly over the storm?

32. Cloud ceiling is the lowest altitude at which solid cloud is

present. The cloud ceiling at airports must be sufficiently high for safe takeoffs and landings. At night the cloud ceiling can be determined by illuminating the base of the clouds with a searchlight pointed vertically upward. If an observer is 1 km from the searchlight and the angle of elevation to the base of the illuminated cloud is 8°, find the cloud ceiling (see Figure 57). (During the day cloud ceilings are generally estimated by sight. However, if an accurate reading is required, a balloon is inflated so that it will rise at a known constant rate. Then it is released and timed until it disappears into the cloud. The cloud ceiling is determined by multiplying the rate by the time of the ascent; trigonometry is not required for this calculation.)

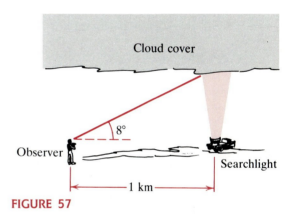

Cloud cover

8°

Observer

Searchlight

←— 1 km —→

FIGURE 57

33. Assuming the earth is a sphere, show that

$$C_\theta = C_e \cos \theta,$$

where C_θ is the circumference of the parallel of latitude at the latitude angle θ and C_e is the earth's circumference at the equator. See Figure 58. [*Hint:* $R \cos \theta = r$.]

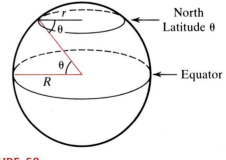

r

North
Latitude θ

θ

θ

R

Equator

FIGURE 58

In Problems 34 and 35, use the result of Problem 33. (Take 6400 km as the radius R of the earth.)

34. Find the circumference of the Arctic Circle, which lies at 66°33′N latitude.

35. Find the distance "around the world" at the 58°40′N latitude.

36. Derive the formula $A = \frac{1}{2}ac \sin \beta$ for the area of the triangle shown in Figure 47.

37. The final length of a volcanic lava flow seems to decrease as the elevation of the lava vent from which it originates increases. An empirical study of Mt. Etna gives the final lava flow length L in terms of elevation h by the formula

$$L = 23 - 0.0053h,$$

where L is measured in kilometers and h is measured in meters.

Suppose that a Sicilian village at elevation 750 m is on a

10° slope directly below a lava vent at 2500 m (see Figure 59). According to the formula, how close will the lava flow get to the village?

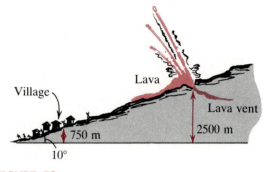

FIGURE 59

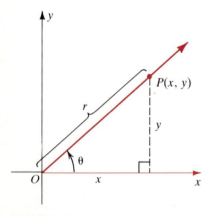

FIGURE 60

2.4

Trigonometric Functions of General Angles

Until now we have defined the trigonometric functions only for acute angles. However, many applications of trigonometry involve angles that are not acute. Consequently it is necessary to extend the definition of the six trigonometric functions to general angles. Naturally, we want the extended definition to agree with the earlier definition whenever the angle is acute. To accomplish this we proceed as follows.

Let θ be an acute angle in standard position and choose a point $P(x, y)$ on the terminal side of θ, as shown in Figure 60. If we let $r = d(O, P) = \sqrt{x^2 + y^2}$, then $y = \text{opp}$, $x = \text{adj}$, and $r = \text{hyp}$, and we have

$$\sin \theta = \frac{y}{r}, \qquad \cos \theta = \frac{x}{r}, \quad \text{and} \quad \tan \theta = \frac{y}{x}. \qquad (4)$$

The expressions in (4) provide us with a model on which to base our extended definition for *any* angle θ in standard position, such as those illustrated in Figure 61.

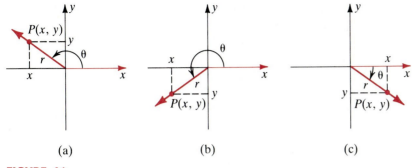

(a)   (b)   (c)

FIGURE 61

We now have the following definition of the **trigonometric functions of a general angle.**

DEFINITION 2

Let θ be any angle in standard position, and let $P(x, y)$ be any point other than $(0, 0)$ on the terminal side of θ. If $r = \sqrt{x^2 + y^2}$ is the distance between $(0, 0)$ and (x, y), then the six trigonometric functions of θ are defined to be

$$\sin \theta = \frac{y}{r}, \qquad \csc \theta = \frac{r}{y}, \; y \neq 0,$$

$$\cos \theta = \frac{x}{r}, \qquad \sec \theta = \frac{r}{x}, \; x \neq 0,$$

$$\tan \theta = \frac{y}{x}, \; x \neq 0, \qquad \cot \theta = \frac{x}{y}, \; y \neq 0.$$

It can be shown by using similar triangles that the values of the six trigonometric functions depend only on the angle θ and not on which point $P(x, y)$ is chosen on the terminal side of θ. (The justification is like the one made for acute angles on page 71. See Problem 65.)

The trigonometric functions will be undefined if the denominators of their formulas are zero. Since $P(x, y) \neq (0, 0)$, then $r = \sqrt{x^2 + y^2}$ is never zero. Thus the domains of the sine and the cosine functions consist of all angles θ. However, the tangent and the secant functions are undefined if the terminal side of θ lies on the y-axis because then $x = 0$. Therefore, the domains of $\tan \theta$ and $\sec \theta$ consist of all angles θ *except* those having radian measure $\pm \pi/2, \pm 3\pi/2, \pm 5\pi/2$, and so on. Using set notation, we can write the domains of the tangent and the secant functions as $\{\theta | \theta \neq (2n + 1)\pi/2, n = 0, \pm 1, \pm 2, \ldots\}$ or $\{\theta | \theta \neq (2n + 1)90°, n = 0, \pm 1, \pm 2, \ldots\}$. The cotangent and the cosecant functions are not defined for angles with their terminal sides on the x-axis because then $y = 0$. Thus the domains of cot θ and csc θ consist of all angles θ except those having radian measure 0, $\pm\pi$, $\pm 2\pi$, $\pm 3\pi$, and so on; that is, $\{\theta | \theta \neq n\pi, \ n = 0, \pm 1, \pm 2, \ldots\}$ or $\{\theta | \theta \neq 180°n, \ n = 0, \pm 1, \pm 2, \ldots\}$.

Since $r = \sqrt{x^2 + y^2}$, it follows that $|x| \leq r$ and $|y| \leq r$. Therefore,

$$|\sin \theta| \leq 1, \qquad |\cos \theta| \leq 1, \qquad |\csc \theta| \geq 1, \quad \text{and} \quad |\sec \theta| \geq 1$$

for every θ in the domain of each of these functions.

■ EXAMPLE 1

Find the exact values of the six trigonometric functions of the angle θ if θ is in standard position and the terminal side of θ contains the point $P(-3, 1)$.

Solution The terminal side of θ is sketched in Figure 62. We have $x = -3$ and $y = 1$, so

$$r = \sqrt{x^2 + y^2} = \sqrt{(-3)^2 + (1)^2} = \sqrt{10}.$$

Hence,

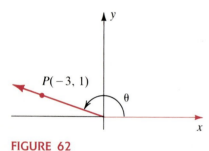

FIGURE 62

$$\sin \theta = \frac{y}{r} = \frac{1}{\sqrt{10}} = \frac{\sqrt{10}}{10}, \qquad \csc \theta = \frac{r}{y} = \frac{\sqrt{10}}{1} = \sqrt{10},$$

$$\cos \theta = \frac{x}{r} = \frac{-3}{\sqrt{10}} = -\frac{3\sqrt{10}}{10}, \qquad \sec \theta = \frac{\sqrt{10}}{-3} = -\frac{\sqrt{10}}{3},$$

$$\tan \theta = \frac{y}{x} = \frac{1}{-3} = -\frac{1}{3}, \qquad \cot \theta = \frac{-3}{1} = -3. \qquad \blacksquare$$

■ EXAMPLE 2

Find the values of the six trigonometric functions of θ if $\theta = -\pi/2$.

Solution First we place θ in standard position as shown in Figure 63.

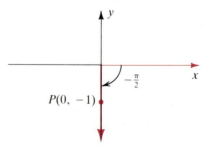

FIGURE 63

According to the definition, we can choose any point $P(x, y)$ on the terminal side of θ. For convenience, we select $P(0, -1)$, so that $x = 0$, $y = -1$, and $r = \sqrt{x^2 + y^2} = 1$. Thus,

$$\sin\left(-\frac{\pi}{2}\right) = \frac{-1}{1} = -1, \qquad \csc\left(-\frac{\pi}{2}\right) = \frac{1}{-1} = -1,$$

$$\cos\left(-\frac{\pi}{2}\right) = \frac{0}{1} = 0, \qquad \cot\left(-\frac{\pi}{2}\right) = \frac{0}{-1} = 0.$$

However, the expressions $\tan\theta = y/x$ and $\sec\theta = r/x$ are undefined for $\theta = -\pi/2$ since $x = 0$.

ALGEBRAIC SIGNS OF THE TRIGONOMETRIC FUNCTIONS

Depending on the quadrant in which the terminal side of θ lies, one or both coordinates of $P(x, y)$ may be negative. Since $r = \sqrt{x^2 + y^2}$ is *always positive,* each of the six trigonometric functions of θ has negative as well as positive values. For example, as Figure 64 shows, $\sin\theta = y/r$ is positive if the terminal side of θ lies in quadrants I or II (where y is positive), and it is negative if the terminal side of θ lies in quadrants III or IV (where y is negative).

The following table summarizes the algebraic signs of the six trigonometric functions. For convenience, if the terminal side of θ lies in quadrant II, we will refer to θ as a quadrant II angle or say that θ is in quadrant II. We will use similar terminology when we refer to angles with terminal sides in quadrants I, III, or IV.

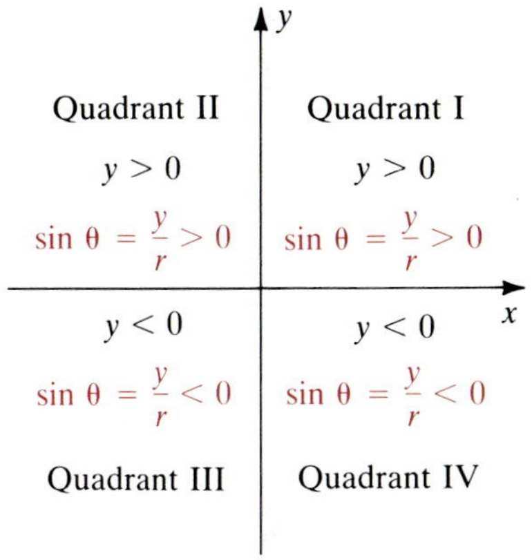

FIGURE 64

Signs of the Trigonometric Functions						
QUADRANT CONTAINING TERMINAL SIDE OF θ	$\sin\theta$	$\cos\theta$	$\tan\theta$	$\csc\theta$	$\sec\theta$	$\cot\theta$
I $(0 < \theta < \pi/2)$	+	+	+	+	+	+
II $(\pi/2 < \theta < \pi)$	+	−	−	+	−	−
III $(\pi < \theta < 3\pi/2)$	−	−	+	−	−	+
IV $(3\pi/2 < \theta < 2\pi)$	−	+	−	−	+	−

■ EXAMPLE 3

In which quadrant does the terminal side of θ lie if $\sin\theta > 0$ and $\tan\theta < 0$?

Solution   Since the sine function is positive for angles in quadrants I and II and the tangent function is negative in quadrants II and IV, the terminal side of θ must lie in quadrant II.

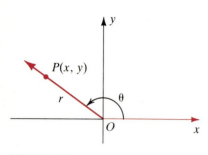

FIGURE 65

PYTHAGOREAN IDENTITIES

The reciprocal and quotient identities for acute angles given in Section 2.2 also hold for general angles. (See Problems 66 and 67.) To our collection of fundamental identities we now add three very useful identities called the **Pythagorean identities.** To derive the first of these, let θ be any angle in standard position. As shown in Figure 65, we let $P(x, y)$ be any point other than the origin on the terminal side of θ. If we again let $r = d(O, P) = \sqrt{x^2 + y^2}$, then we have

$$x^2 + y^2 = r^2.$$

Dividing both sides by r^2, we can write

$$\frac{x^2}{r^2} + \frac{y^2}{r^2} = 1, \quad \text{or} \quad \left(\frac{x}{r}\right)^2 + \left(\frac{y}{r}\right)^2 = 1.$$

Recognizing that $x/r = \cos\theta$ and $y/r = \sin\theta$, we obtain the basic Pythagorean identity

$$(\cos\theta)^2 + (\sin\theta)^2 = 1. \qquad\qquad \textbf{(5)}$$

It is standard practice to write $\cos^2\theta$ instead of $(\cos\theta)^2$ and $\sin^2\theta$ instead of $(\sin\theta)^2$. Similar notation is used for the other trigonometric functions and for all powers *except* -1. (As we remarked in Section 2.2, $\sin^{-1}\theta$, $\cos^{-1}\theta$, and so on, refer to the corresponding inverse trigonometric functions, which will be discussed in Chapter 3.)

With this new notation (5) becomes

$$\cos^2\theta + \sin^2\theta = 1. \qquad\qquad \textbf{(6)}$$

Dividing both sides of this equation by $\cos^2\theta$, we obtain

$$\frac{\cos^2\theta}{\cos^2\theta} + \frac{\sin^2\theta}{\cos^2\theta} = \frac{1}{\cos^2\theta},$$

or

$$1 + \left(\frac{\sin\theta}{\cos\theta}\right)^2 = \left(\frac{1}{\cos\theta}\right)^2.$$

Since $\tan\theta = \sin\theta/\cos\theta$ and $\sec\theta = 1/\cos\theta$, this simplifies to

$$1 + \tan^2\theta = \sec^2\theta.$$

The final Pythagorean identity

$$\cot^2\theta + 1 = \csc^2\theta$$

is obtained by dividing both sides of (6) by $\sin^2\theta$. (See Problem 68.)

For your convenience a summary of basic trigonometry including all the fundamental identities can be found inside the cover of this text.

■■■■ **EXAMPLE 4**

Given that cos $\theta = \frac{1}{3}$ and that θ is a quadrant IV angle, find the exact values of the remaining five trigonometric functions of θ.

Solution Substituting cos $\theta = \frac{1}{3}$ into $\cos^2 \theta + \sin^2 \theta = 1$ gives

$$(\tfrac{1}{3})^2 + \sin^2 \theta = 1$$
$$\sin^2 \theta = 1 - \tfrac{1}{9} = \tfrac{8}{9}.$$

Since the terminal side of θ is in quadrant IV, sin θ is negative. Therefore, we *must* select the negative square root of $\frac{8}{9}$:

$$\sin \theta = -\sqrt{\frac{8}{9}} = -\frac{2\sqrt{2}}{3}.$$

It follows that

$$\tan \theta = \frac{\sin \theta}{\cos \theta} = \frac{-2\sqrt{2}/3}{1/3} = -2\sqrt{2},$$

$$\cot \theta = \frac{1}{\tan \theta} = \frac{1}{-2\sqrt{2}} = -\frac{\sqrt{2}}{4},$$

$$\sec \theta = \frac{1}{\cos \theta} = \frac{1}{1/3} = 3,$$

$$\csc \theta = \frac{1}{\sin \theta} = \frac{1}{-2\sqrt{2}/3} = -\frac{3\sqrt{2}}{4}.$$ ■■■■

■■■■ **EXAMPLE 5**

Given tan $\theta = -2$ and sin $\theta > 0$, find the exact values of the remaining five trigonometric functions of θ.

Solution Letting tan $\theta = -2$ in the identity $1 + \tan^2 \theta = \sec^2 \theta$, we find

$$\sec^2 \theta = 1 + (-2)^2 = 5.$$

Since tan θ is negative in quadrants II and IV and sin θ is positive in quadrants I and II, the terminal side of θ must lie in quadrant II. Thus we must take

$$\sec \theta = -\sqrt{5}.$$

From sec $\theta = 1/\cos \theta$, it follows that

$$\cos \theta = \frac{1}{\sec \theta} = \frac{1}{-\sqrt{5}} = -\frac{\sqrt{5}}{5}.$$

Using $\tan \theta = \sin \theta / \cos \theta$, we obtain

$$\sin \theta = \cos \theta \tan \theta = \left(-\frac{\sqrt{5}}{5}\right)(-2) = \frac{2\sqrt{5}}{5}.$$

Then

$$\csc \theta = \frac{1}{\sin \theta} = \frac{1}{2\sqrt{5}/5} = \frac{\sqrt{5}}{2}$$

and

$$\cot \theta = \frac{1}{\tan \theta} = \frac{1}{-2} = -\frac{1}{2}.$$

REFERENCE ANGLE

In Section 2.2 we found exact values for the six trigonometric functions of the special angles of 30°, 45°, and 60° (or $\pi/6$, $\pi/4$, and $\pi/3$ in radian measure). These values can be used to determine the exact trigonometric function values of certain nonacute angles by means of a **reference angle.**

DEFINITION 3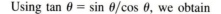

Let θ be an angle in standard position such that its terminal side does not lie on a coordinate axis. The **reference angle** θ' for θ is defined to be the acute angle formed by the terminal side of θ and the x-axis.

Figure 66 illustrates this definition for angles with terminal sides in each of the four quadrants.

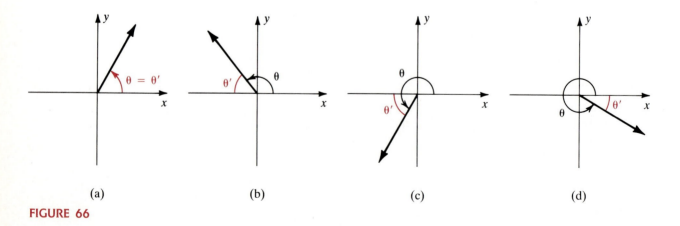

(a)   (b)   (c)   (d)

FIGURE 66

EXAMPLE 6

Find the reference angle θ' for each angle θ.

(a) $\theta = 40°$ **(b)** $\theta = \dfrac{2\pi}{3}$ **(c)** $\theta = 210°$ **(d)** $\theta = -\dfrac{9\pi}{4}$

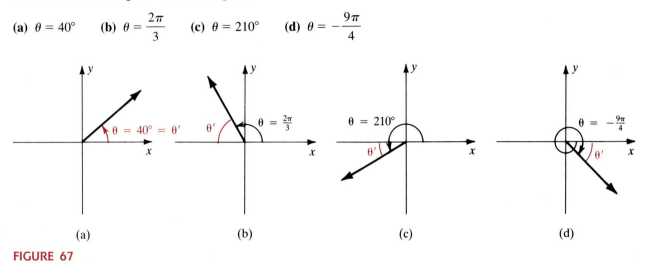

(a)   (b)   (c)   (d)

FIGURE 67

Solution

(a) From Figure 67(a) we see that $\theta' = 40°$.
(b) From Figure 67(b), $\theta' = \pi - \theta = \pi - 2\pi/3 = \pi/3$.
(c) From Figure 67(c), $\theta' = \theta - 180° = 210° - 180° = 30°$.
(d) Since $\theta = -9\pi/4$ is coterminal with

$$-\frac{9\pi}{4} + 2\pi = -\frac{\pi}{4},$$

we find that $\theta' = \pi/4$ (see Figure 67(d)).

The usefulness of reference angles in evaluating trigonometric functions is a result of the following property.

Property of Reference Angles

The absolute value of any trigonometric function of an angle θ equals the value of that function for the reference angle θ'.

For instance,

$$|\sin \theta| = \sin \theta', \qquad |\cos \theta| = \cos \theta',$$

and so on.

We verify the property for the sine function and leave the remaining cases as an exercise (see Problem 69). If the terminal side of θ lies in quadrant I, then $\theta = \theta'$ and $\sin \theta$ is positive, so

$$\sin \theta' = \sin \theta = |\sin \theta|.$$

From Figure 68, we see that if θ is a quadrant II, III, or IV angle, then we have

$$\sin \theta' = \frac{|y|}{r} = \left|\frac{y}{r}\right| = |\sin \theta|,$$

where $P(x, y)$ is any point on the terminal side of θ and $r = \sqrt{x^2 + y^2}$.

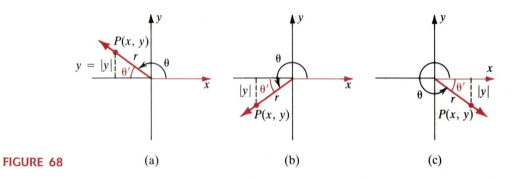

FIGURE 68   (a)   (b)   (c)

We can now describe a step-by-step procedure for determining a trigonometric function value of any angle θ.

Finding the Value of a Trigonometric Function of an Angle θ

1. Find the reference angle θ'.
2. Determine the trigonometric function value for θ'.
3. Select the correct algebraic sign by considering in which quadrant the terminal side of θ lies.

■ EXAMPLE 7

Find the exact values of $\sin \theta$, $\cos \theta$, and $\tan \theta$ for each of the following.

(a) $\theta = \dfrac{2\pi}{3}$ **(b)** $\theta = 210°$ **(c)** $\theta = -\dfrac{9\pi}{4}$

Solution

(a) We follow the procedure just discussed.

1. In Example 6(b) we found the reference angle for $\theta = 2\pi/3$ to be $\theta' = \pi/3$.
2. From Section 2.2 we know that $\sin (\pi/3) = \sqrt{3}/2$, $\cos (\pi/3) = 1/2$, and $\tan (\pi/3) = \sqrt{3}$.
3. Since $\theta = 2\pi/3$ is a quadrant II angle, where the sine is positive but the cosine and the tangent are negative, we have

$$\sin \frac{2\pi}{3} = \frac{\sqrt{3}}{2}, \qquad \cos \frac{2\pi}{3} = -\frac{1}{2}, \quad \text{and} \quad \tan \frac{2\pi}{3} = -\sqrt{3}.$$

(b) Referring to Example 6(c), we see that the reference angle is $\theta' = 30°$. Using the property of reference angles and the fact that the terminal side of $\theta = 210°$ lies in quadrant III, we obtain

$$\sin 210° = -\sin 30° = -\frac{1}{2},$$

$$\cos 210° = -\cos 30° = -\frac{\sqrt{3}}{2},$$

$$\tan 210° = \tan 30° = \frac{\sqrt{3}}{3}.$$

(c) From Example 6(d) we know that the reference angle $\theta' = \pi/4$. Since $\theta = -9\pi/4$ is a quadrant IV angle, we find

$$\sin \left(-\frac{9\pi}{4} \right) = -\sin \frac{\pi}{4} = -\frac{\sqrt{2}}{2},$$

$$\cos \left(-\frac{9\pi}{4} \right) = \cos \frac{\pi}{4} = \frac{\sqrt{2}}{2},$$

$$\tan \left(-\frac{9\pi}{4} \right) = -\tan \frac{\pi}{4} = -1.$$

◼ EXAMPLE 8

Find all angles θ satisfying $0° \le \theta < 360°$ such that $\sin \theta = \frac{1}{2}$.

Solution   From what we know about the special angles 30°, 60°, and 90°, we have that $\theta = 30°$ is one solution. Using 30° as a reference angle in the second quadrant, as shown in Figure 69, we find $\theta = 150°$ as a second solution. Since the sine function is negative for angles in quadrants III and IV, there are no additional solutions satisfying $0 \le \theta < 360°$. ◼

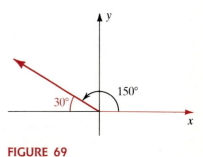

FIGURE 69

EXAMPLE 9

Find all angles θ satisfying $0 \le \theta < 2\pi$ such that $\cos \theta = -\sqrt{2}/2$.

Solution   Since the given value of the cosine function is negative, we first determine the reference angle θ' such that $\cos \theta' = \sqrt{2}/2$. From Section 2.2 we know that $\theta' = \pi/4$. Since the cosine function is negative for angles in quadrants II and III, we position the reference angle $\theta' = \pi/4$ as shown in Figure 70. We then obtain $\theta = 3\pi/4$ and $\theta = 5\pi/4$ as solutions.

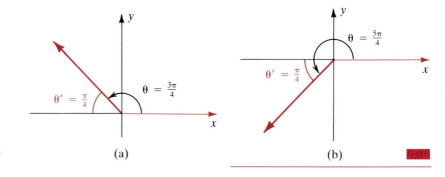

FIGURE 70   (a)   (b)

Note of Caution: In this section we have purposely avoided using calculators. For a complete understanding of trigonometry, it is essential that you master the concepts and be able to perform *without the aid of a calculator* the types of calculations and simplifications we have discussed.

EXERCISE 2.4

(Do not use a calculator or tables in solving Problems 1–73.)

In Problems 1–10, evaluate the six trigonometric functions of the angle θ if θ is in standard position and the terminal side of θ contains the given point.

1. (6, 8)   **2.** (−1, 2)   **3.** (5, −12)   **4.** (−8, −15)
5. (0, 2)   **6.** (−3, 0)   **7.** (−2, 3)   **8.** (5, −1)
9. $(-\sqrt{2}, -1)$ **10.** $(\sqrt{3}, \sqrt{2})$

In Problems 11–18, find the quadrant in which the terminal side of θ lies if θ satisfies the given conditions.

11. $\sin \theta < 0$ and $\tan \theta > 0$ **12.** $\cos \theta > 0$ and $\sin \theta < 0$
13. $\tan \theta < 0$ and $\sec \theta < 0$ **14.** $\sec \theta < 0$ and $\csc \theta < 0$
15. $\cot \theta > 0$ and $\sin \theta > 0$ **16.** $\csc \theta > 0$ and $\cot \theta < 0$
17. $\sin \theta > 0$ and $\cos \theta < 0$ **18.** $\tan \theta < 0$ and $\csc \theta > 0$

In Problems 19–28, the value of one of the trigonometric functions of an angle θ is given. From the given value and the additional information, determine the values of the five remaining trigonometric functions of θ.

19. $\sin \theta = \dfrac{1}{4}$, θ is in quadrant II

20. $\cos \theta = -\dfrac{2}{5}$, θ is in quadrant II

21. $\tan \theta = 3$, θ is in quadrant III

22. $\cot \theta = 2$, θ is in quadrant III

23. $\csc \theta = -10$, θ is in quadrant IV
24. $\sec \theta = 3$, θ is in quadrant IV
25. $\sin \theta = -\dfrac{1}{5}$, $\cos \theta > 0$ **26.** $\cos \theta = -\dfrac{2}{3}$, $\sin \theta < 0$
27. $\tan \theta = 8$, $\sec \theta > 0$
28. $\sec \theta = -4$, $\csc \theta < 0$
29. If $\cos \theta = \dfrac{3}{10}$, find all possible values of $\sin \theta$.

30. If $\sin \theta = -\dfrac{2}{7}$, find all possible values of $\cos \theta$.

31. If $2 \sin \theta - \cos \theta = 0$, find all possible values of $\sin \theta$ and $\cos \theta$.

32. If $\cot \theta = \dfrac{3}{4}$, find all possible values of $\csc \theta$.

33. If $\sec \theta = -5$, find all possible values of $\sin \theta$ and $\cos \theta$.
34. If $3 \cos \theta = \sin \theta$, find all possible values of $\tan \theta$, $\cot \theta$, $\sec \theta$, and $\csc \theta$.

35. Complete the following table.

θ (degrees)	θ (radians)	$\sin \theta$	$\cos \theta$	$\tan \theta$
0°	0	0	1	0
30°	$\dfrac{\pi}{6}$	$\dfrac{1}{2}$	$\dfrac{\sqrt{3}}{2}$	$\dfrac{\sqrt{3}}{3}$
45°	$\dfrac{\pi}{4}$	$\dfrac{\sqrt{2}}{2}$	$\dfrac{\sqrt{2}}{2}$	1
60°	$\dfrac{\pi}{3}$	$\dfrac{\sqrt{3}}{2}$	$\dfrac{1}{2}$	$\sqrt{3}$
90°	$\dfrac{\pi}{2}$	1	0	—
120°	$\dfrac{2\pi}{3}$	$\dfrac{\sqrt{3}}{2}$	$-\dfrac{1}{2}$	$-\sqrt{3}$
135°	$\dfrac{3\pi}{4}$			
150°	$\dfrac{5\pi}{6}$			
180°	π			
210°	$\dfrac{7\pi}{6}$	$-\dfrac{1}{2}$	$-\dfrac{\sqrt{3}}{2}$	$\dfrac{\sqrt{3}}{3}$
225°	$\dfrac{5\pi}{4}$			
240°	$\dfrac{4\pi}{3}$			
270°	$\dfrac{3\pi}{2}$			
300°	$\dfrac{5\pi}{3}$			
315°	$\dfrac{7\pi}{4}$			
330°	$\dfrac{11\pi}{6}$			
360°	2π			

36. Complete the following table.

θ (degrees)	θ (radians)	$\csc \theta$	$\sec \theta$	$\cot \theta$
0°	0	—	1	—
30°	$\dfrac{\pi}{6}$	2	$\dfrac{2\sqrt{3}}{3}$	$\sqrt{3}$
45°	$\dfrac{\pi}{4}$	$\sqrt{2}$	$\sqrt{2}$	1
60°	$\dfrac{\pi}{3}$	$\dfrac{2\sqrt{3}}{3}$	2	$\dfrac{\sqrt{3}}{3}$
90°	$\dfrac{\pi}{2}$	1	—	0
120°	$\dfrac{2\pi}{3}$			
135°	$\dfrac{3\pi}{4}$			
150°	$\dfrac{5\pi}{6}$			
180°	π			
210°	$\dfrac{7\pi}{6}$			
225°	$\dfrac{5\pi}{4}$			
240°	$\dfrac{4\pi}{3}$			
270°	$\dfrac{3\pi}{2}$			
300°	$\dfrac{5\pi}{3}$			
315°	$\dfrac{7\pi}{4}$			
330°	$\dfrac{11\pi}{6}$			
360°	2π			

In Problems 37–52, find the exact value of the given expression.

37. $\cos 5\pi$

38. $\sin\left(-\dfrac{7\pi}{6}\right)$

39. $\cot \dfrac{13\pi}{6}$

40. $\tan \dfrac{9\pi}{2}$

41. $\sin\left(-\dfrac{4\pi}{3}\right)$

42. $\cos \dfrac{23\pi}{4}$

43. $\csc\left(-\dfrac{\pi}{6}\right)$

44. $\tan \dfrac{23\pi}{4}$

45. $\sec (-120°)$

46. $\csc 495°$

47. $\sin 150°$

48. $\cos (-45°)$

49. $\tan 405°$

50. $\sin 315°$

51. $\cot (-720°)$

52. $\sec (-300°)$

In Problems 53–58, find all angles θ, where $0 \le \theta < 360°$, satisfying the given condition.

53. $\tan \theta = \sqrt{3}$

54. $\sin \theta = -\dfrac{1}{2}$

55. $\cos \theta = -\dfrac{\sqrt{2}}{2}$

56. $\sec \theta = \dfrac{2\sqrt{3}}{3}$

57. $\csc \theta = -1$

58. $\cot \theta = -\dfrac{1}{\sqrt{3}}$

In Problems 59–64, find all angles θ, where $0 \le \theta < 2\pi$, satisfying the given condition.

59. $\sin \theta = 0$

60. $\cos \theta = -1$

61. $\sec \theta = -\sqrt{2}$

62. $\csc \theta = 2$

63. $\cot \theta = -\sqrt{3}$

64. $\tan \theta = 1$

65. Using similar triangles, show that the values of the six trigonometric functions of a general angle θ defined on page 93 depend only on the angle θ and not on the choice of the point $P(x, y)$ on the terminal side of θ.

66. If θ is any angle for which the functions are defined, prove that

$$\csc \theta = \frac{1}{\sin \theta}, \quad \sec \theta = \frac{1}{\cos \theta}, \quad \text{and} \quad \cot \theta = \frac{1}{\tan \theta}.$$

67. If θ is any angle for which the functions are defined, prove that

$$\tan \theta = \frac{\sin \theta}{\cos \theta} \quad \text{and} \quad \cot \theta = \frac{\cos \theta}{\sin \theta}.$$

68. If θ is any angle for which the functions are defined, prove that

$$\cot^2 \theta + 1 = \csc^2 \theta.$$

69. Verify the reference angle property for the cosine, tangent, cosecant, secant, and cotangent functions.

70. Let l be a nonvertical line that passes through the origin and makes an angle θ measured counterclockwise from the positive x-axis (see Figure 71). Prove that the slope of l equals $\tan \theta$.

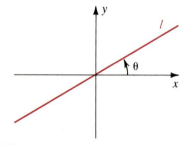

FIGURE 71

71. Is there an angle θ such that $\cos \theta = \frac{4}{3}$? Explain.

72. Is there an angle θ such that $2 \csc t = 1$? Explain.

73. The memory aid given in Problem 56 of Exercise 2.2 can now be expanded to include angles of $0°$ (0 radians) and $90°$ ($\pi/2$ radians). Verify the entries in the following table.

θ (degrees)	θ (radians)	$\sin \theta$	$\cos \theta$
$0°$	0	$\dfrac{\sqrt{0}}{2}$	$\dfrac{\sqrt{4}}{2}$
$30°$	$\dfrac{\pi}{6}$	$\dfrac{\sqrt{1}}{2}$	$\dfrac{\sqrt{3}}{2}$
$45°$	$\dfrac{\pi}{4}$	$\dfrac{\sqrt{2}}{2}$	$\dfrac{\sqrt{2}}{2}$
$60°$	$\dfrac{\pi}{3}$	$\dfrac{\sqrt{3}}{2}$	$\dfrac{\sqrt{1}}{2}$
$90°$	$\dfrac{\pi}{2}$	$\dfrac{\sqrt{4}}{2}$	$\dfrac{\sqrt{0}}{2}$

74. Because of its rotation, the earth bulges at the equator and is flattened at the poles. As a result, the acceleration due to gravity actually varies with latitude θ. Satellite studies have shown that the acceleration due to gravity g_{sat} is approximated by the function

$$g_{sat} = 978.0309 + 5.18552 \sin^2 \theta - 0.00570 \sin^2 2\theta.$$

(a) Find g_{sat} at the equator ($\theta = 0°$), (b) at the north pole, and (c) at 45° north latitude.

75. Under certain conditions the maximum height y attained by a basketball released from a height h at an angle α measured from the horizontal with an initial velocity v is given by

$$y = h + \frac{v^2 \sin^2 \alpha}{2g},$$

where g is the acceleration due to gravity. Compute the maximum height reached by a free throw if $h = 2.15$ m, $v = 8$ m/sec, $\alpha = 64°28'$, and $g = 9.81$ m/sec².

76. The range of a shot put released from a height h above the ground with an initial velocity v at an angle α to the horizontal can be approximated by

$$R = \frac{v \cos^2 \alpha}{g} [v \sin \alpha + \sqrt{v^2 \sin^2 \alpha + 2gh}],$$

where g is the acceleration due to gravity. Compare the ranges achieved for the release heights (a) $h = 2.0$ m and

(b) $h = 2.4$ m if $v = 13.7$ m/sec and $\alpha = 40°$. Take $g = 9.81$ m/sec².
(c) Explain why an increase in h yields an increase in R if the other parameters are held fixed.
(d) What does this imply about the advantage that height gives a shot-putter?

The Law of Sines 2.5

In Section 2.3 we saw how to solve right triangles. In this section and the next we will consider techniques for solving general triangles.

Consider the triangle ABC, shown in Figure 72, with angles α, β, and γ, and opposite sides a, b, and c, respectively. If we know the length of one side and two other parts of the triangle, we can then find the remaining three parts. This can be accomplished by using either the **law of sines,**

$$\frac{\sin \alpha}{a} = \frac{\sin \beta}{b} = \frac{\sin \gamma}{c}, \tag{7}$$

or the law of cosines, which is developed in Section 2.6.

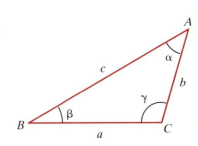

FIGURE 72

Although the law of sines is valid for any triangle, we will derive it only for acute triangles and leave the case of obtuse triangles as an exercise (see Problem 37).*

As shown in Figure 73, let h be the altitude from vertex A to side BC. It follows that

$$\frac{h}{c} = \sin \beta,$$

or

$$h = c \sin \beta. \tag{8}$$

Similarly,

$$\frac{h}{b} = \sin \gamma,$$

or

$$h = b \sin \gamma. \tag{9}$$

Equating the expressions in (8) and (9) gives

$$c \sin \beta = b \sin \gamma,$$

so that

$$\frac{\sin \beta}{b} = \frac{\sin \gamma}{c}. \tag{10}$$

In a similar manner, we can prove that

$$\frac{\sin \alpha}{a} = \frac{\sin \beta}{b} \tag{11}$$

(see Problem 36). Combining (10) and (11) gives us

$$\frac{\sin \alpha}{a} = \frac{\sin \beta}{b} = \frac{\sin \gamma}{c}.$$

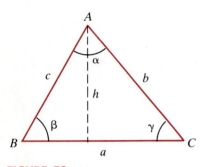

FIGURE 73

USE OF A CALCULATOR

For the remainder of this chapter we will rely on a scientific calculator to obtain approximations of trigonometric function values. To improve the accuracy of our answers all digits will be retained in the calculator for subsequent calculations, but for convenience only four decimal places will be displayed in the text from this point on. We will also round all final results to the nearest hundredth unless the problem specifies otherwise.

---

*An acute triangle is one in which all three angles are less than 90°. An obtuse triangle contains an angle with measure greater than 90°.

EXAMPLE 1

Determine the remaining parts of the triangle shown in Figure 74.

Solution Let $\beta = 20°$, $\alpha = 130°$, and $b = 6$. It follows immediately that

$$\gamma = 180° - 20° - 130°$$
$$= 30°.$$

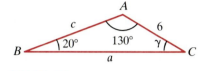

FIGURE 74

From the law of sines, we see that

$$\frac{\sin 130°}{a} = \frac{\sin 20°}{6} = \frac{\sin 30°}{c}.$$

Solving for a from

$$\frac{\sin 130°}{a} = \frac{\sin 20°}{6}$$

gives $\quad a = 6\left(\frac{\sin 130°}{\sin 20°}\right) \approx 6\left(\frac{0.7660}{0.3420}\right) \approx 13.44.$

To solve for c, we use

$$\frac{\sin 20°}{6} = \frac{\sin 30°}{c},$$

so that

$$c = 6\left(\frac{\sin 30°}{\sin 20°}\right) \approx 6\left(\frac{0.5000}{0.3420}\right) \approx 8.77.$$

Note of Caution: Results are generally more accurate if given values, rather than calculated values, are used to compute unknown values. Thus in Example 1 we used $(\sin 20°)/6 = (\sin 30°)/c$ to calculate c rather than $(\sin 130°)/a = (\sin 30°)/c$ and the calculated value $a = 13.44$.

SOLVING TRIANGLES: FOUR CASES

In general, we can use the law of sines to solve triangles for which we know (i) two angles and any side or (ii) two sides and an angle opposite one of these sides. Triangles for which we know either (iii) three sides or (iv) two sides and the included angle cannot be solved directly using the law of sines. In the next section we will consider a method for solving the latter two cases.

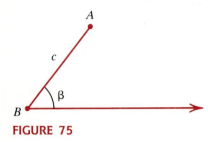

FIGURE 75

In Example 1, where we were given two angles and a side (case (i)), the triangle had a unique solution. However, this may not always be true for case (ii) triangles, that is, for triangles where we know two sides and an angle opposite one of these sides. For instance, suppose that the sides b and c and the angle β in triangle ABC are specified. As shown in Figure 75, we draw angle β and mark off side c to locate the vertices A and B. The third vertex C is located on the base by drawing an arc of a circle of radius b with center A. As illustrated in Figure 76, there are four possible outcomes for this construction:

(a) The arc does not intersect the base and no triangle is formed.
(b) The arc intersects the base in two distinct points C_1 and C_2 and two triangles are formed.
(c) The arc intersects the base in one point and one triangle is formed.
(d) The arc is tangent to the base and a single *right* triangle is formed.

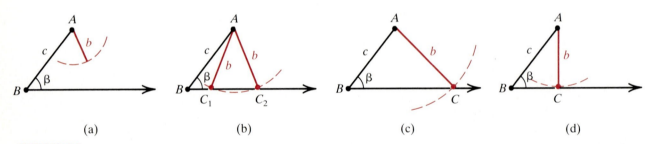

    (a)   (b)   (c)   (d)

FIGURE 76

Because these various possibilities exist, case (ii) is called the **ambiguous case.** The following three examples illustrate two-solution, single-solution, and no-solution outcomes for the ambiguous case.

■ EXAMPLE 2

Find the remaining parts of a triangle with $\beta = 50°$, $b = 5$, and $c = 6$.

Solution   From the law of sines, we have

$$\frac{\sin 50°}{5} = \frac{\sin \gamma}{6},$$

or

$$\sin \gamma = 6\left(\frac{\sin 50°}{5}\right) \approx 6\left(\frac{0.7660}{5}\right) \approx 0.9193.$$

From a calculator set in degree mode, we obtain $\gamma \approx 66.82°$. At this point in the solution it is essential to recall from Section 2.4 that the sine function is also positive for quadrant II angles. Hence there is another angle satisfying $0° \leq \gamma \leq 180°$ for which $\sin \gamma \approx 0.9193$. Using $66.82°$ as a reference angle, we find the quadrant II angle:

$$180° - 66.82° = 113.18°.$$

Therefore, the two possibilities for γ are

$$\gamma_1 \approx 66.82° \quad \text{and} \quad \gamma_2 \approx 113.18°.$$

Thus, as shown in Figure 77, there are two possible triangles ABC_1 and ABC_2 satisfying the given conditions.

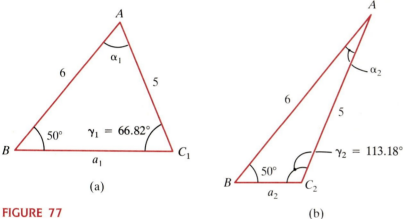

FIGURE 77

(a)

(b)

To complete the solution of triangle ABC_1 shown in Figure 77(a), we first find α_1:

$$\alpha_1 = 180° - \gamma_1 - \beta$$
$$\approx 180° - 66.82° - 50° = 63.18°.$$

To find a_1, we use

$$\frac{\sin 63.18°}{a_1} = \frac{\sin 50°}{5},$$

which gives

$$a_1 = 5\left(\frac{\sin 63.18°}{\sin 50°}\right) \approx 5\left(\frac{0.8925}{0.7660}\right) \approx 5.83.$$

We complete the solution of triangle ABC_2 shown in Figure 77(b) in a similar manner. Since $\gamma_2 \approx 113.18°$,

$$\alpha_2 \approx 180° - 113.18° - 50° = 16.82°.$$

EXERCISE 2.5

In Problems 1–26, solve the indicated triangle. The relative positions of α, β, γ, a, b, and c are shown in Figure 81.

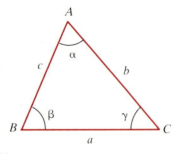

FIGURE 81

1. $\alpha = 80°$, $\beta = 20°$, $b = 7$
2. $\alpha = 60°$, $\beta = 15°$, $c = 30$
3. $\beta = 37°$, $\gamma = 51°$, $a = 5$
4. $\alpha = 30°$, $\gamma = 75°$, $a = 6$
5. $\beta = 72°$, $b = 12$, $c = 6$
6. $\alpha = 120°$, $a = 9$, $c = 4$
7. $\gamma = 62°$, $b = 7$, $c = 4$
8. $\beta = 110°$, $\gamma = 25°$, $a = 14$
9. $\gamma = 15°$, $a = 8$, $c = 5$
10. $\alpha = 55°$, $a = 20$, $c = 18$
11. $\gamma = 150°$, $b = 7$, $c = 5$
12. $\alpha = 140°$, $\gamma = 20°$, $c = 12$
13. $\beta = 13°20'$, $\gamma = 102°$, $b = 9$
14. $\alpha = 135°$, $a = 4$, $b = 5$
15. $\alpha = 20°$, $a = 8$, $c = 27$
16. $\beta = 47°10'$, $b = 20$, $c = 25$
17. $\beta = 30°$, $a = 10$, $b = 7$
18. $\alpha = 75°$, $\gamma = 45°$, $b = 8$
19. $\gamma = 80°$, $b = 4$, $c = 8$
20. $\alpha = 43°$, $\beta = 62°$, $c = 7$
21. $\beta = 100°$, $a = 9$, $b = 20$
22. $\alpha = 35°$, $a = 9$, $b = 12$
23. $\beta = 115°$, $b = 11$, $c = 15$ 24. $\alpha = 50°$, $a = 10$, $b = 15$
25. $\gamma = 95°$, $a = 20$, $c = 35$ 26. $\gamma = 27.3°$, $b = 3$, $c = 2$
27. A 10-ft rope that is available to measure the length between two points A and B at opposite ends of a kidney-shaped swimming pool is not long enough. A third point C is found such that the distance from A to C is 10 ft. It is determined that angle ACB is 115° and angle ABC is 35°. Find the distance from A to B. See Figure 82.

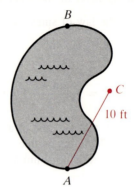

FIGURE 82

28. Two points A and B lie on opposite sides of a river. Another point C is located on the same side of the river as B at a distance of 230 ft from B. If angle ABC is 105° and angle ACB is 20°, find the distance across the river from A to B.

29. A telephone pole makes an angle of 82° with the level ground. The angle of elevation of the sun is 76° (see Figure 83). Find the length of the telephone pole if its shadow is 3.5 m. (Assume that the tilt of the pole is away from the sun and in the same plane as the pole and the sun.)

FIGURE 83

30. Suppose that a surveyor wants to find the straight-line distance between two points A and B at the same elevation on opposite sides of a mountain. The angle of elevation from A to the top of the mountain is $55°10'$ and the distance is 560 m. (There are instruments that will measure this distance provided they can be carried to the top of the mountain.) If the angle of elevation from B to the top of the mountain is $48°$, find the distance between A and B, assuming that they are in the same vertical plane as the top of the mountain.

31. The distance from the tee to the green on a particular golf hole is 370 yd. A golfer slices her drive and paces its distance off at 210 yd. From the point where the ball lies, she measures an angle of $160°$ from the tee to the green (see Figure 84). Find the angle of her slice.

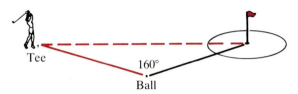

FIGURE 84

32. In Problem 31, what is the distance from the ball to the green?

33. A man 5 ft 9 in. tall stands on a sidewalk that slopes down at a constant angle. A vertical street lamp directly behind him causes his shadow to be 15 ft long. The angle of depression from the top of the man to the tip of his shadow is $31°$. Find the angle α, as shown in Figure 85, that the sidewalk makes with the horizontal.

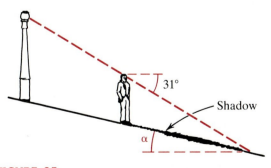

FIGURE 85

34. If the man in Problem 33 is 20 ft down the sidewalk from the lamppost, find the height of the light above the sidewalk.

35. Angles of elevation to an airplane are measured from the top and the base of a building that is 20 m tall. The angle from the top of the building is $38°$, and the angle from the base of the building is $40°$. Find the altitude of the airplane.

36. Derive equation (11) using Figure 86.

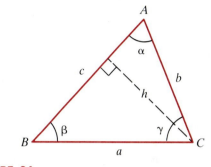

FIGURE 86

37. Derive the law of sines for a triangle with an obtuse angle as shown in Figure 87.

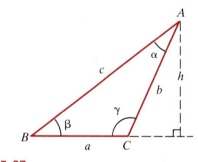

FIGURE 87

Problems 38–43 refer to the four cases described on page 107.

38. Explain why a unique triangle is determined if the sum of the two angles in case (i) is less than $180°$.

39. State a necessary and sufficient condition under which case (iii) will determine a triangle.

40. Explain why case (iv) will always determine a unique triangle provided that the included angle is less than $180°$.

41. Let $\beta = 45°$ and $c = 5$, as shown in Figure 88. Find all values of b for which:
   (a) the triangle is a right triangle;
   (b) the triangle has no solution;

(c) the triangle has two distinct solutions;
(d) the triangle has a unique solution.

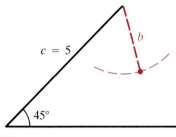

FIGURE 88

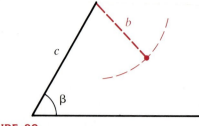

FIGURE 89

42. Let $0° < \beta < 90°$ and suppose that c is known (see Figure 89). Find all values of b for which:
(a) the triangle is a right triangle;
(b) the triangle has no solution;
(c) the triangle has two distinct solutions;
(d) the triangle has a unique solution.

43. Let $90° < \beta < 180°$ and suppose that c is known (see Figure 90). Find all values of b for which:
(a) the triangle has no solution;
(b) the triangle has a unique solution

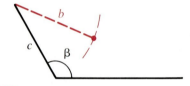

FIGURE 90

44. Find the perimeter of a regular pentagon inscribed in a circle of radius a.

2.6 The Law of Cosines

In a right triangle, such as the one shown in Figure 91, the lengths of the sides are related by the Pythagorean theorem,

$$c^2 = a^2 + b^2. \tag{13}$$

Actually this equation is a special case of a general formula that relates the lengths of the sides of *any* triangle. This generalization, called the **law of cosines,** enables us to solve triangles for which we know either three sides or two sides and the included angle.

FIGURE 91

A GENERALIZATION OF THE PYTHAGOREAN THEOREM

Suppose that the triangle in Figure 92(a) represents an arbitrary triangle, not necessarily a right triangle. If we introduce a Cartesian coordinate system with origin and x-axis as shown in Figure 92(b), then the coordinates of the vertices A, B, and C are as shown. Now, by the distance formula, the length of the side opposite the angle γ is

$$c = \sqrt{(b \cos \gamma - a)^2 + (b \sin \gamma)^2}.$$

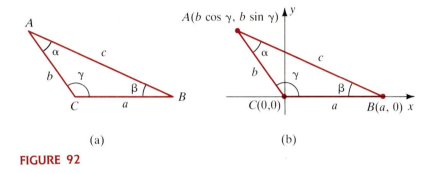

(a)   (b)

FIGURE 92

Thus,

$$c^2 = b^2 \cos^2 \gamma - 2ab \cos \gamma + a^2 + b^2 \sin^2 \gamma$$
$$= a^2 + b^2 \underbrace{(\cos^2 \gamma + \sin^2 \gamma)}_{1} - 2ab \cos \gamma$$

$$= a^2 + b^2 - 2ab \cos \gamma. \tag{14}$$

Note that equation (14) reduces to (13) when the angle γ is $90°$.

Since there is nothing special about placing the origin at the vertex C of the angle γ, we can repeat the argument above two more times. For example, had we chosen the origin at the vertex A and placed the x-axis along the side AC (see Figure 93), it would then follow, in exactly the same manner, that

$$a^2 = b^2 + c^2 - 2bc \cos \alpha. \tag{15}$$

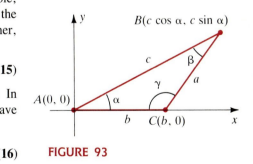

FIGURE 93

By a similar argument, we can express b in terms of a, c, and $\cos \beta$. In general, for any triangle such as the one shown in Figure 92(a), we have

$$a^2 = b^2 + c^2 - 2bc \cos \alpha,$$
$$b^2 = a^2 + c^2 - 2ac \cos \beta, \tag{16}$$
$$c^2 = a^2 + b^2 - 2ab \cos \gamma.$$

The equations above are known as the **law of cosines.**

We note that the results in (16) can be expressed as follows.

> ## Law of Cosines
>
> The square of the length of any side of a triangle equals the sum of the squares of the lengths of the other two sides minus twice the product of these lengths and the cosine of the included angle.

You are urged to learn and to work with the law of cosines as stated above instead of memorizing the three formulas given in (16).

■ EXAMPLE 1

Determine the remaining side of the triangle shown in Figure 94.

Solution If we call the unknown side b, then from the law of cosines we can write

$$b^2 = (10)^2 + (12)^2 - 2(10)(12)\cos 26°$$
$$\approx 100 + 144 - 240(0.8988)$$
$$\approx 244 - 215.7106 = 28.2894,$$

and therefore $b \approx \sqrt{28.2894} \approx 5.32.$ ■

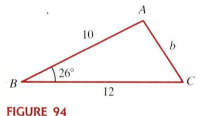

FIGURE 94

■ EXAMPLE 2

Determine the remaining angles in the triangle of the previous example (see Figure 95).

Solution First, we apply the law of cosines to the angle γ, which is included by the sides of length 5.32 and 12, and obtain

$$(10)^2 = (5.32)^2 + (12)^2 - 2(5.32)(12)\cos \gamma$$
$$100 = 28.3024 + 144 - 127.68 \cos \gamma$$
$$127.68 \cos \gamma = 72.3024$$
$$\cos \gamma = \frac{72.3024}{127.68} \approx 0.5663.$$

Using a calculator, we find that

$$\gamma \approx 55.51°.$$

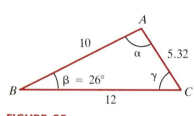

FIGURE 95

Since the cosine of an angle between 90° and 180° is negative, *there is no need to consider two possibilities* here as we did in the previous section when we used the sine function. Finally, from $\alpha + \beta + \gamma = 180°$, it follows that

$$\alpha \approx 180° - 55.51° - 26°$$
$$= 180° - 81.51°$$
$$= 98.49°.$$

Alternatively, since we know two sides and an angle opposite one of these sides, we could have used the law of sines to find γ. Furthermore, had we used the law of sines, there would have been no ambiguity about whether γ was acute or obtuse. Recall from geometry that a triangle can have at most one obtuse angle and, if there is one, it must be opposite the longest side. Since γ is not opposite the longest side in this triangle, it must be acute.

As we indicated in Section 2.5, a triangle for which we know either two sides and the included angle or three sides cannot be solved using the law of sines. The preceding examples illustrated how to solve the first type of triangle by the law of cosines. In the next example we consider the case in which three sides are given.

■ EXAMPLE 3

Determine the angles of triangle *ABC* shown in Figure 96 with sides of lengths 7, 6, and 9, respectively.

Solution   By using the law of cosines to find the angle opposite the longest side, we will immediately see whether the triangle contains an obtuse angle. Therefore, we solve for γ from

$$9^2 = 6^2 + 7^2 - 2(6)(7) \cos \gamma.$$

Combining and rearranging terms gives

$$84 \cos \gamma = 4,$$

or
$$\cos \gamma = \tfrac{4}{84}.$$

From a calculator we find that

$$\gamma \approx 87.27°.$$

Now we can use either the law of cosines or the law of sines to find another angle. We choose to find β by the law of sines:

$$\frac{\sin \beta}{b} = \frac{\sin \gamma}{c}.$$

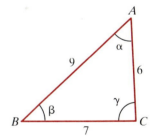

FIGURE 96

Substituting 6 for b, 9 for c, and 87.27° for γ (remember: for greater accuracy you should work with the actual value obtained by your calculator for γ, not the rounded value 87.27°), we find

$$\frac{\sin \beta}{6} \approx \frac{\sin 87.27°}{9},$$

or

$$\sin \beta \approx \frac{6}{9} \sin (87.27°) \approx 0.6659.$$

Since β is *not* the largest angle in the triangle, it must be acute. Therefore, $\beta \approx 41.75°$. Finally,

$$\alpha \approx 180° - 87.27° - 41.75° = 50.98°.$$

BEARING

In navigation directions are given using bearings. A **bearing** designates the acute angle that a line makes with the north–south line. For example, Figure 97(a) illustrates a bearing of S40°W, meaning *40 degrees west of south;* the bearings in Figures 97(b) and (c) are N65°E and S80°E, respectively.

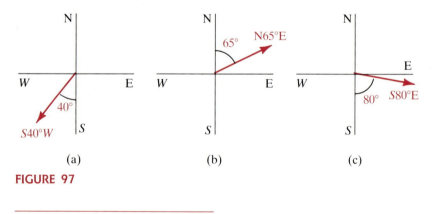

(a)   (b)   (c)

FIGURE 97

■ EXAMPLE 4

Two ships leave a port at 7:00 A.M., one traveling at 12 knots (nautical miles per hour) and the other at 10 knots. If the faster ship maintains a bearing of N47°W and the other ship maintains a bearing of S20°W, what is their separation (to the nearest nautical mile) at 11:00 A.M. that day?

Solution   Since the elapsed time is 4 hours, the faster ship has traveled $4 \cdot 12 = 48$ nautical miles from port and the slower ship $4 \cdot 10 = 40$ nautical miles. Using these distances and the given bearings, we can sketch the triangle shown in Figure 98. If we let c = the distance separating them at 11:00 A.M., we obtain from the law of cosines

$$c^2 = 48^2 + 40^2 - 2(48)(40) \cos \gamma.$$

Since $\gamma = 180° - 47° - 20° = 113°$, we find that

$$c^2 = 2304 + 1600 - 3840 \cos 113° \approx 5404.41,$$

or $$c \approx 73.51.$$

Thus the distance between them (to the nearest nautical mile) is 74 nautical miles.

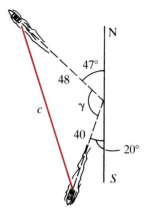

FIGURE 98

When a problem involves solving a triangle, it is important to determine which technique to use first. The following table summarizes the various types of problems and gives a correct approach for each. (The term **oblique triangle** refers to any triangle that is not a right triangle.)

SOLVING A TRIANGLE

TYPE OF TRIANGLE	INFORMATION GIVEN	TECHNIQUE
Right	Two sides or one angle and a side	Basic definitions for sine, cosine, or tangent; the Pythagorean theorem
Oblique	Three sides	The law of cosines
Oblique	Two sides and the included angle	The law of cosines
Oblique	Two angles and a side	The law of sines
Oblique	Two sides and an angle opposite one of the sides	The law of sines (If the given angle is acute, it is an ambiguous-case problem.)

Note of Caution:

(i) When three sides are given, if the length of the longest side is greater than or equal to the sum of the lengths of the other two sides, there is no solution. (This is because the shortest distance between two points is the length of the line segment joining them.)

(ii) In applying the law of sines, if you obtain a value greater than 1 for the sine of an angle, there is no solution.

(iii) In the ambiguous case, when solving for the first unknown angle, you must consider *both the acute angle found from a calculator and its supplement as possible solutions.* The supplement will be a solution if the sum of the supplement and the given angle is less than 180°.

EXERCISE 2.6

In Problems 1–16, solve the indicated triangle. The relative positions of α, β, γ, a, b, and c are shown in Figure 99.

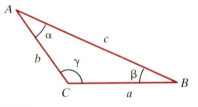

FIGURE 99

1. $\gamma = 65°$, $a = 5$, $b = 8$
2. $\beta = 48°$, $a = 7$, $c = 6$
3. $a = 8$, $b = 10$, $c = 7$
4. $\gamma = 31.5°$, $a = 4$, $b = 8$
5. $a = 3$, $b = 4$, $c = 5$
6. $a = 7$, $b = 9$, $c = 4$
7. $a = 11$, $b = 9.5$, $c = 8.2$
8. $\alpha = 162°$, $b = 11$, $c = 8$
9. $a = 5$, $b = 7$, $c = 10$
10. $a = 6$, $b = 5$, $c = 7$
11. $\gamma = 97°20'$, $a = 3$, $b = 6$
12. $\beta = 130°$, $a = 4$, $c = 7$
13. $a = 6$, $b = 8$, $c = 12$
14. $a = 5$, $b = 12$, $c = 13$
15. $\alpha = 22°$, $b = 3$, $c = 9$
16. $\beta = 100°$, $a = 22.3$, $b = 16.1$
17. A ship sails due west from a harbor for 22 nautical miles. It then sails S62°W for another 15 nautical miles. How far is the ship from the harbor?
18. Two hikers leave their camp simultaneously, taking bearings of N42°W and S20°E, respectively. If they each average a rate of 5 km/hr, how far apart are they after 1 hr?

19. On a hiker's map point A is 2.5 in. due west of point B and point C is 3.5 in. from B and 4.2 in. from A, respectively (see Figure 100). Find (a) the bearing of A from C and (b) the bearing of B from C.

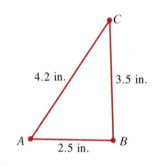

FIGURE 100

20. Two ships leave port simultaneously, one traveling at 15 knots and the other at 12 knots. They maintain bearings of S42°W and S10°E, respectively. After 3 hr the first ship runs aground and the second ship immediately goes to its aid.

    (a) How long will it take the second ship to reach the first ship if it travels at 14 knots?

    (b) What bearing should it take?

21. A two-dimensional robot arm "knows" where it is by keeping track of a "shoulder" angle α and an "elbow" angle β. As shown in Figure 101, this arm has a fixed point of rotation at the origin. The shoulder angle is measured counterclockwise from the x-axis, and the elbow angle is measured counterclockwise from the upper to the lower arm. Suppose that the upper and lower arms are both of length 2 and that the elbow angle β is prevented from "hyperextending" be-

yond 180°. Find the angles α and β that will position the robot's hand at the point (1, 2).

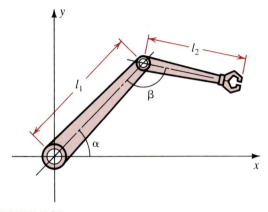

FIGURE 101

22. A rhombus has sides of length 10 cm. If the angle at one of the vertices is 50°, find the lengths of the diagonals.

23. A house measures 45 ft from front to back. The roof measures 32 ft from the front of the house to the peak and 18 ft from the peak to the back of the house (see Figure 102). Find the angles of elevation of the front and back parts of the roof.

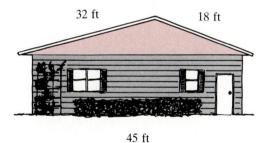

FIGURE 102

24. Two radar stations are situated 5000 m from each other. An airplane passes directly over the line between the two stations. At this instant the distances from the stations to the plane are 2300 and 4000 m. Find the altitude of the airplane.

25. A slanted roof makes an angle of 35° with the horizontal and measures 28 ft from base to peak. A television antenna 16 ft high is to be attached to the peak of the roof and secured by a wire from the top of the antenna to the nearest point at the base of the roof. Find the length of wire required.

26. Two lookout towers are situated on mountain tops A and B, 4 mi from each other. A helicopter firefighting team is located in a valley at point C, 3 mi from A and 2 mi from B. Using the line between A and B as a reference, lookouts spot a fire at an angle of 40° from tower A and 82° from tower B. (See Figure 103.) At what angle, measured from CB, should the helicopter fly in order to head directly for the fire?

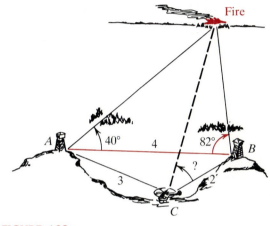

FIGURE 103

27. For the kite shown in Figure 104, find the length of doweling required for the diagonal supports.

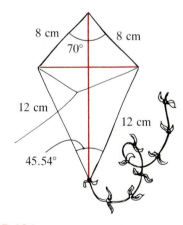

FIGURE 104

28. From the floor of a canyon it takes 62 ft of rope to reach the top of one canyon wall and 86 ft to reach the top of the opposite wall (see Figure 105). If the two ropes make an angle of 123°, what is the distance d from the top of one canyon wall to the other?

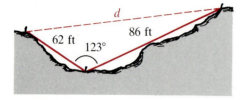

FIGURE 105

29. Use the law of cosines to derive **Heron's formula,** *

$$A = \sqrt{s(s-a)(s-b)(s-c)},$$

for the area of a triangle with sides a, b, and c, where $s = \frac{1}{2}(a + b + c)$.

In Problems 30–33, use Problem 29 to find the area of the given triangle.

30. $a = 5$, $b = 8$, $c = 4$ **31.** $a = 12$, $b = 5$, $c = 13$
32. $\gamma = 25°$, $a = 7$, $b = 10$ **33.** $\beta = 86.2°$, $a = 5.2$, $c = 7.3$
34. Use Heron's formula (Problem 29) to find the area of a triangular garden plot if the lengths of the three sides are 25, 32, and 41 m, respectively.

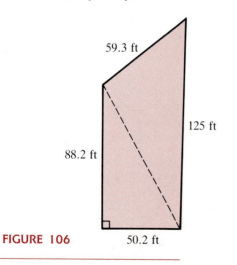

FIGURE 106

*This formula is named after the Greek mathematician Heron but should actually be credited to Archimedes.

35. Find the area of the irregular corner lot shown in Figure 106. [*Hint*: Divide the lot into two triangular lots as shown and then find the area of each triangle. Use Heron's formula (Problem 29) for the area of the acute triangle.]
36. Use Heron's formula (Problem 29) to find the area of a triangle with vertices located at (3, 2), (−3, −6), and (0, 6) in a rectangular coordinate system.
37. The effort in climbing a flight of stairs depends largely on the flexing angle of the leading knee. A simplified stick-figure model of a person walking up a staircase indicates that the maximum flexing of the knee occurs when the back leg is straight and the hips are directly over the heel of the front foot (see Figure 107). Show that

$$\cos\theta = \left(\frac{R}{a}\right)\sqrt{4 - \left(\frac{T}{a}\right)^2} + \frac{(T/a)^2 - (R/a)^2}{2} - 1,$$

where θ is the knee joint angle, $2a$ is the length of the leg, R is the rise of a single stairstep, and T is the width of a step. [*Hint*: Let h be the vertical distance from hip to heel of the leading leg, as shown in the figure. Set up two equations involving h: one by applying the Pythagorean theorem to the right triangle outlined in color and the other by using the law of cosines on the angle θ. Then eliminate h and solve for $\cos\theta$.]

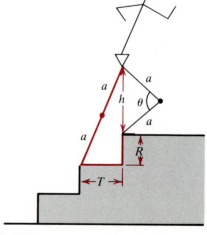

FIGURE 107

IMPORTANT CONCEPTS

Angle	Measure of an angle	secant
vertex	degree	cosecant
initial side	minutes	Fundamental identities
terminal side	seconds	quotient identities
standard position	radian	reciprocal identities
coterminal angles	Central angle	Pythagorean identities
right angle	Sector	Solving a triangle
straight angle	Arc length	Angle of elevation
acute angle	Trigonometric functions	Angle of depression
obtuse angle	sine	Reference angle
complementary angles	cosine	Law of sines
supplementary angles	tangent	ambiguous case
quadrantal angle	cotangent	Law of cosines

CHAPTER 2 REVIEW EXERCISE

In Problems 1–10, fill in the blank or answer true or false.

1. An angle that has negative measure was formed by a _____ rotation.

2. If $\alpha - \beta = 6\pi$, then α and β are coterminal. ___

3. $\tan\theta = \sin\theta/\cos\theta$ ___

4. $\sin(\pi/6) = \cos(\pi/3)$ ___

5. To solve a right triangle in which you know the side opposite θ and the side adjacent to θ, you would use the _____ function to find θ.

6. The reference angle for $4\pi/3$ is _____.

7. To solve a triangle in which you know two angles and a side opposite one of these angles, you would use the law of _____ first.

8. The ambiguous case refers to solving a triangle when _____ are given.

9. To solve a triangle in which you know two sides and the included angle, you would use the law of _____ first.

10. The _____ theorem is a special case of the law of cosines.

In Problems 11–14, draw the given angle in standard position.

11. $-5\pi/6$

12. $7\pi/3$

13. $225°$

14. $-450°$

In Problems 15–18, convert the given angle to radian measure.

15. $-120°$

16. $1°$

17. $48.3°$

18. $14°14'$

In Problems 19–22, convert the given angle to decimal degrees.

19. $\pi/9$

20. $78°15'$

21. 2.3

22. $7\pi/3$

In Problems 23–26, convert the given angle to degrees, minutes, and seconds.

23. $70.5°$

24. $170.15°$

25. 3.1

26. $\pi/10$

In Problems 27–28, find two positive and two negative angles that are coterminal with the given angle.

27. $85°$

28. $7\pi/6$

29. What is the length of arc of a circle of radius 16 in. subtended by a central angle of (a) 1 radian and (b) $\pi/10$ radian?

30. A fan of radius 10 in. revolves at a constant rate of 1000 revolutions/minute. Find its angular speed in radians/second. Determine the linear speed of the tip of one of its blades in inches/second. (See Problem 75 in Exercise 2.1.)

31. On one occasion the magnetic axis of the earth was measured at an inclination of 12° to the earth's geographic axis (see Figure 108). If the polar diameter of the earth is 7900 mi, find the distance (measured along the arc of a great circle) between the north pole and the "northern" magnetic pole.

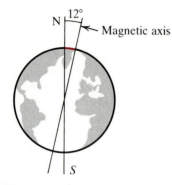

N 12°
Magnetic axis
S

FIGURE 108

32. A belt passing over a pulley moves at a speed of 6π ft/sec. If the pulley is revolving at a rate of 90 revolutions per minute, find the diameter of the pulley.

In Problems 33–38, solve the right triangle shown in Figure 109 using the given information.

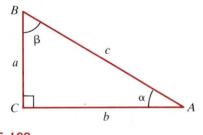

FIGURE 109

33. $a = 30$, $b = 40$
34. $a = 25$, $\alpha = 27.5°$
35. $b = 5$, $\alpha = 34°$
36. $b = 7$, $\beta = 45°$
37. $c = 10$, $\alpha = 41°40'$
38. $a = 2.3$, $\beta = 75.2°$
39. Determine the angles in the triangle with vertices (1, 1), (3, 1), and (3, 4).

40. A rocket is launched from ground level at an angle of elevation of 43°. If the rocket hits a drone target plane flying at 20,000 ft, find the horizontal distance between the launching site and the point directly below the plane. What is the straight-line distance between the rocket-launching site and the target plane?

41. A man 100 m from the base of an overhanging cliff measures a 28° angle of elevation from that point to the top of the cliff (see Figure 110). If the cliff makes an angle of 65° with the horizontal ground, determine its approximate height h.

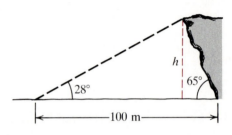

FIGURE 110

42. An escalator between the first and second floors of a department store is 58 ft long and makes an angle of 20° with the first floor (see Figure 111). Find the vertical distance between the two floors.

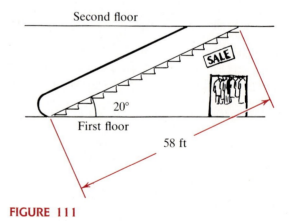

FIGURE 111

43. In 1972 a French helicopter attained a record height of 12,442 m. What would the angle of elevation to the helicopter have been from a point P on the ground 2000 m from the point directly beneath the helicopter? See Figure 112.

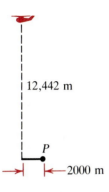

12,442 m

P

2000 m

FIGURE 112

44. A competition water skier leaves a ramp at point R and lands at S (see Figure 113). A judge located on shore at point J measures $\angle RJS$ as $47°$. If the distance from the ramp to the judge is 110 ft, find the length of the jump. (Assume that $\angle SRJ = 90°$.)

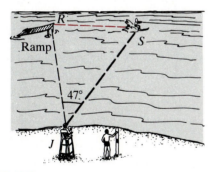

Ramp

R

S

$47°$

J

FIGURE 113

In Problems 45–48, evaluate the six trigonometric functions of the angle θ if θ is in standard position and the terminal side of θ contains the given point.

45. $(-1, 2)$ **46.** $(4, 7)$

47. $(-0.5, -0.3)$ **48.** $(\sqrt{2}, \sqrt{5})$

In Problems 49–54, the value of one of the trigonometric functions of an angle θ is given. From the given value and the additional information, determine the values of the five remaining trigonometric functions of θ.

49. $\cos \theta = -\dfrac{1}{7}$, θ is in quadrant III

50. $\sin \theta = \dfrac{2}{3}$, θ is in quadrant II

51. $\cot \theta = -5$, θ is in quadrant IV

52. $\sec \theta = 15$, $\sin \theta < 0$

53. $\csc \theta = -7$, $\tan \theta > 0$

54. $\tan \theta = \dfrac{1}{9}$, $\sec \theta < 0$

55. If $\cot \theta = -4$, find all possible values of $\sin \theta$, $\cos \theta$, $\tan \theta$, $\sec \theta$, and $\csc \theta$.

56. If $4 \sin \theta = 3 \cos \theta$, find all possible values of $\sin \theta$, $\cos \theta$, $\tan \theta$, $\sec \theta$, and $\csc \theta$.

In Problems 57–60, find the exact value of the given expression.

57. $\sin \left(-\dfrac{7\pi}{4} \right)$ **58.** $\csc \dfrac{13\pi}{6}$

59. $\tan 495°$ **60.** $\sin 330°$

In Problems 61–64, without using a calculator, find all angles θ, such that $0° \leq \theta < 360°$, satisfying the given condition.

61. $\sin \theta = \dfrac{\sqrt{2}}{2}$ **62.** $\tan \theta = -\dfrac{\sqrt{3}}{3}$

63. $\sec \theta = -2$ **64.** $\csc \theta = -\sqrt{2}$

In Problems 65–68, without using a calculator, find all angles θ, such that $0 \leq \theta < 2\pi$, satisfying the given condition.

65. $\sin \theta = \dfrac{\sqrt{3}}{2}$ **66.** $\csc \theta = -1$

67. $\cot \theta = -1$ **68.** $\cos \theta = \dfrac{1}{2}$

In Problems 69–72, solve the triangle satisfying the given conditions.

69. $\alpha = 30°$, $\beta = 70°$, $b = 10$

70. $\gamma = 145°$, $a = 25$, $c = 20$

71. $\beta = 45°$, $b = 7$, $c = 8$

72. $\beta = 100°$, $\gamma = 20°$, $b = 35$

73. An entry in a soapbox derby rolls down a hill. Using the information given in Figure 114, find the total distance $d_1 + d_2$ that the soap box travels.

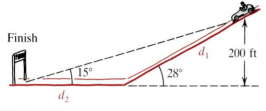

FIGURE 114

74. From two lifeguard towers, a swimmer is sighted on bearings of N46°E and N27°W, respectively. If the second tower is 250 ft due east of the first tower, what is the distance from each tower to the swimmer?

75. One Coast Guard vessel is located 4 nautical miles due south of a second Coast Guard vessel when they receive a distress signal from a sailboat. To offer assistance, the first vessel sails on a bearing of S50°E at 5 knots and the second vessel sails S10°E at 10 knots. Which one reaches the sailboat first?

76. The angle between two sides of a parallelogram is 40°. If the lengths of the sides are 5 and 10 cm, find the lengths of the two diagonals.

In Problems 77–80, solve the triangle satisfying the given conditions.

77. $\alpha = 51°$, $b = 20$, $c = 10$

78. $\gamma = 25°$, $a = 8$, $b = 5$

79. $a = 4$, $b = 6$, $c = 3$

80. $a = 1$, $b = 2$, $c = 3$

81. Determine the angles in the triangle with vertices at $(0, 0)$, $(5, 0)$, and $(4, 3)$.

82. A ship sails on a bearing of N85°E from a harbor for a distance of 10 nautical miles. At this point it changes its course to a bearing of N25°W and travels for 20 nautical miles. What is the straight-line distance from the harbor to this final point?

83. An airplane is supposed to fly 500 mi due west to a refueling rendezvous point. If a 5° error is made in the heading, how far is the plane from the rendezvous point after flying 400 mi? Through what angle must the plane turn in order to correct its course at that point?

84. Show that the area of a regular n-gon (a polygon with n sides) is given by

$$A = \frac{1}{4}s^2 n \cot\left(\frac{180°}{n}\right),$$

where s is the length of a side. (See Figure 115.)

Regular n-gon

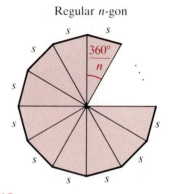

FIGURE 115

85. A weather satellite orbiting the equator at a height of $H = 36{,}000$ km spots a thunderstorm to the north at P at an angle of $\theta = 6.5°$ from its vertical (see Figure 116).

(a) Given that the earth's radius is approximately $R = 6370$ km, find the latitude angle ϕ of the thunderstorm.

(b) Show that θ and ϕ are related by the equation

$$\tan\theta = \frac{R\sin\phi}{H + R(1 - \cos\phi)}.$$

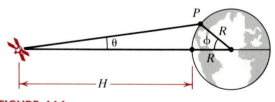

FIGURE 116

86. As shown in Figure 117, only a portion of the earth's surface can be observed from a spacecraft at an altitude H. The circle that bounds this region is called the **horizon circle.** Let C denote the center of the earth, S the satellite position, and A a point on the horizon circle.

(a) For the angles λ and θ illustrated in the figure, show that

$$\sin\theta = \cos\lambda = \frac{R}{R + H},$$

where R is the radius of the earth. [*Hint*: SA will be tangent to the earth.]

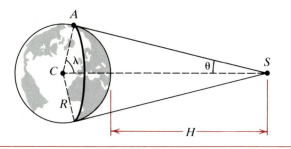

(b) Using $R = 6378$ km, find λ and θ corresponding to a point on the horizon circle of the Landsat 2 satellite when it is at its greatest distance (916 km) from the earth.

FIGURE 117

Analytic Trigonometry

3

Sophie Germaine

Much of the mathematics developed in the eighteenth century arose in an attempt to describe certain physical phenomena. What is the shape of a sail under the pressure of the wind? What is the shape of a vibrating elastic string (for example, a violin string) fixed at both ends? What is the shape of a vibrating plate? Answers to such questions frequently involve the trigonometric functions. Ernest Chladni, a German physicist, developed an interesting technique for studying the vibration of elastic plates. After sprinkling fine sand on drumlike surfaces, he drew a violin bow along the edges to start them vibrating. The French Academy of Sciences offered a prize for the best mathematical description of this phenomenon. In 1816 Sophie Germaine (1776–1831), a largely self-taught French mathematician, won the Paris Academy's prize with her paper on elasticity.

3.1

The Circular Functions

In Chapter 2 we considered trigonometric functions of *angles* measured either in degrees or in radians. For calculus and other advanced courses it is necessary to consider trigonometric functions with domains consisting of *real numbers* rather than angles. The transition from angles to real numbers is made by recognizing that to each real number t, there corresponds an angle of measure t radians.

We can visualize this correspondence using a circle centered at the origin with radius 1. This circle is called the **unit circle** (see Figure 1). From Section 1.3 it follows that the equation of the unit circle is $x^2 + y^2 = 1$.

We now consider an angle of t radians. From the definition of radian measure, $t = s/r$, the ratio of the subtended arc length s to the radius of the circle r. For the unit circle, $r = 1$, so that $t = s/1 = s$. Therefore, the angle of t radians shown in Figure 2 subtends an arc of length t units on the unit circle. It follows that for each real number t, the terminal side of an angle of t radians in standard position has traversed a distance of $|t|$ units along the circumference of the unit circle—counterclockwise if $t > 0$, clockwise if $t < 0$. This association of each real number t with an angle of t radians is illustrated in Figure 3.

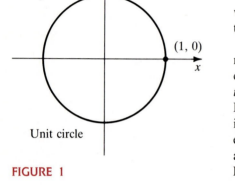

$x^2 + y^2 = 1$

$(1, 0)$

Unit circle

FIGURE 1

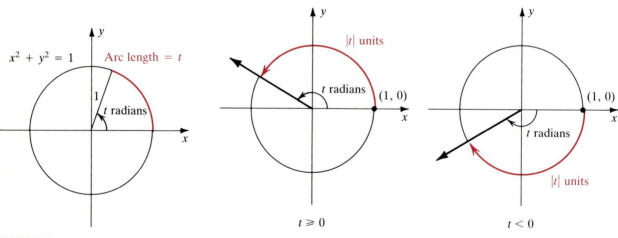

$x^2 + y^2 = 1$ Arc length $= t$

1

t radians

FIGURE 2

$|t|$ units

t radians

$(1, 0)$

$t \geq 0$

FIGURE 3

$(1, 0)$

t radians

$|t|$ units

$t < 0$

This association enables us to make the following definition.

DEFINITION 1

The value of each trigonometric function at a real number t is defined to be its value at an angle of t radians, provided that value exists.

For example, the sine of the real number $\pi/6$ is simply the sine of the angle of $\pi/6$ radian (which, as you know, is $\frac{1}{2}$). Thus there is really nothing new in evaluating the trigonometric function of a real number.

The unit circle is very helpful in describing the trigonometric functions of real numbers. For any real number t, consider the angle of t radians in standard position. As shown in Figure 4(a), let P_t be the point of intersection of the terminal side of the angle of t radians with the unit circle. Since P_t lies on the unit circle, the distance from P_t to the origin O is $r = 1$.

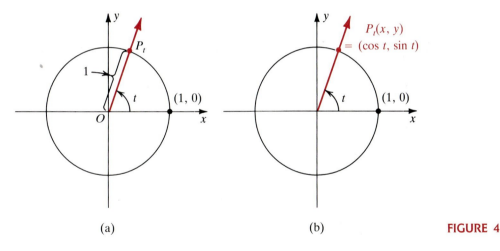

(a)   (b)   **FIGURE 4**

Let P_t have coordinates (x, y), as shown in Figure 4(b). From Section 2.4 we find that

$$\sin t = y/r = y/1 = y, \qquad \csc t = r/y = 1/y, \ y \neq 0,$$
$$\cos t = x/r = x/1 = x, \qquad \sec t = r/x = 1/x, \ x \neq 0,$$
$$\tan t = y/x, \ x \neq 0, \qquad \cot t = x/y, \ y \neq 0.$$

In particular, the coordinates of P_t are

$$P_t(x, y) = (\cos t, \sin t).$$

In other words:

For any real number t, $\cos t$ and $\sin t$ are the x- and y-coordinates, respectively, of the point of intersection of the terminal

As we will soon see, a number of important properties of the sine and the cosine functions can be obtained from this result. Because of the role played by the unit circle in this discussion, the trigonometric functions are sometimes referred to as the **circular functions.**

Because $P_t(x, y)$ lies on the unit circle, it follows that

$$-1 \le x \le 1 \quad \text{and} \quad -1 \le y \le 1.$$

Since $x = \cos t$ and $y = \sin t$, we obtain

$$|\cos t| \le 1 \quad \text{and} \quad |\sin t| \le 1.$$

DOMAIN AND RANGE

The observations above indicate that both $\cos t$ and $\sin t$ can be any number in the interval $[-1, 1]$. Thus we have the sine and cosine functions,

$$f(t) = \sin t \quad \text{and} \quad g(t) = \cos t,$$

each with domain the set R of all real numbers and range the interval $[-1, 1]$. The domains and the ranges of the other trigonometric functions will be discussed in Section 3.2.

■ EXAMPLE 1

Approximate $\cos 3$ and $\sin 3$ and give a geometric interpretation of these expressions.

Solution   From a calculator set in *radian mode,* we obtain

$$\cos 3 \approx -0.9899925 \quad \text{and} \quad \sin 3 \approx 0.1411200.$$

These values represent the x- and y-coordinates, respectively, of the point of intersection of the terminal side of the angle of 3 radians (in standard position) with the circle of radius 1 centered at the origin. As shown in Figure 5, this point lies in the second quadrant.   ■

PERIODICITY

In Section 2.1 we saw that for any real number t, the angles of t radians and $t \pm 2\pi$ radians are coterminal. Thus they determine the same point (x, y) on the unit circle. Therefore

$$\cos t = \cos (t \pm 2\pi) \quad \text{and} \quad \sin t = \sin (t \pm 2\pi).$$

In other words, the sine and the cosine functions repeat their values every 2π

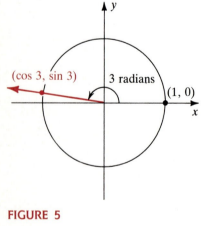

FIGURE 5

units. It follows that for any integer n:

$$\cos t = \cos (t + 2n\pi),$$
$$\sin t = \sin (t + 2n\pi).$$

In general, a nonconstant function f is said to be **periodic** if there is a positive number p such that

$$f(t) = f(t + p) \tag{1}$$

for every t in the domain of f. If p is the smallest positive number for which (1) is true, then p is called the **period** of the function f. Thus the foregoing properties imply that the cosine and the sine functions are periodic. To see that the period of $\sin t$ is actually 2π, we note that there is only one point P_t on the unit circle with y-coordinate 1, namely, $(0, 1)$. That is,

$$\sin t = 1 \text{ only for } t = \frac{\pi}{2}, \frac{\pi}{2} \pm 2\pi, \frac{\pi}{2} \pm 4\pi, \text{ and so on.}$$

Thus the smallest possible positive value of p is 2π, and the sine function has period 2π. The verification that the cosine function has period 2π is left as an exercise (see Problem 31).

ADDITIONAL PROPERTIES

For any real number t satisfying $0 < t < \pi/2$, the corresponding point P_t on the unit circle lies in the first quadrant. As shown in Figure 6, the points P_t and P_{-t} are symmetric with respect to the x-axis. Since their x-coordinates are equal, we have that

$$\cos (-t) = \cos t.$$

Again referring to Figure 6, we see that the y-coordinate of P_{-t} is the negative of the y-coordinate of P_t. It follows that

$$\sin (-t) = -\sin t.$$

The fact that $\cos (-t) = \cos t$ and $\sin (-t) = -\sin t$ hold for any real number t is left as an exercise (see Problem 32). These properties imply that the cosine function is even and the sine function is odd.

The following additional properties can be verified by considering appropriate points on the unit circle.

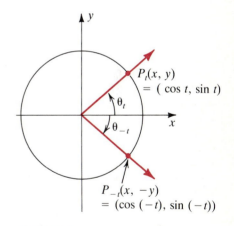

FIGURE 6

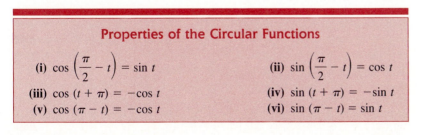

Properties of the Circular Functions

(i) $\cos \left(\dfrac{\pi}{2} - t \right) = \sin t$ **(ii)** $\sin \left(\dfrac{\pi}{2} - t \right) = \cos t$

(iii) $\cos (t + \pi) = -\cos t$ **(iv)** $\sin (t + \pi) = -\sin t$

(v) $\cos (\pi - t) = -\cos t$ **(vi)** $\sin (\pi - t) = \sin t$

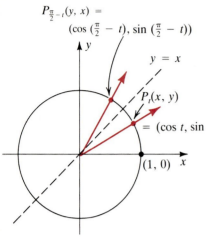

$P_{\frac{\pi}{2}-t}(y, x) =$
$(\cos(\frac{\pi}{2} - t), \sin(\frac{\pi}{2} - t))$

$y = x$

$P_t(x, y)$

$= (\cos t, \sin t)$

$(1, 0)$ x

FIGURE 7

For example, to justify properties (i) and (ii) for $0 < t < \pi/2$, consider Figure 7. Since P_t and $P_{(\pi/2)-t}$ are symmetric with respect to the line $y = x$, we can obtain the coordinates of $P_{(\pi/2)-t}$ by interchanging the coordinates of P_t. Thus,

$$\cos t = x = \sin\left(\frac{\pi}{2} - t\right) \quad \text{and} \quad \sin t = y = \cos\left(\frac{\pi}{2} - t\right).$$

In Section 4.1 we will present an entirely different approach for verifying properties (i)–(vi) for any real number t.

■ EXAMPLE 2

Use periodicity to evaluate $\sin(13\pi/3)$.

Solution Since

$$\frac{13\pi}{3} = 4\pi + \frac{\pi}{3},$$

it follows from $\sin t = \sin(t + 2n\pi)$, with $n = 2$, that

$$\sin\frac{13\pi}{3} = \sin\left(\frac{\pi}{3} + 4\pi\right) = \sin\frac{\pi}{3} = \frac{\sqrt{3}}{2}.$$ ■

■ EXAMPLE 3

Evaluate $\sin(-\pi/6)$.

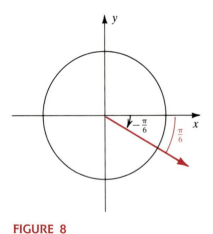

y

$-\frac{\pi}{6}$ $\frac{\pi}{6}$ x

FIGURE 8

Solution Applying $\sin(-t) = -\sin t$, we obtain

$$\sin\left(-\frac{\pi}{6}\right) = -\sin\frac{\pi}{6} = -\frac{1}{2}.$$

Alternatively, this problem can be solved by means of a reference angle. Using this approach (from Section 2.4) we consider the angle of $-\pi/6$ radian. As shown in Figure 8, this angle has a reference angle of $\pi/6$. Thus we find $\sin(\pi/6) = \frac{1}{2}$. Since the sine function is negative for an angle whose terminal side lies in quadrant IV, we have

$$\sin\left(-\frac{\pi}{6}\right) = -\frac{1}{2}.$$ ■

■ EXAMPLE 4

You may have observed that $\cos(\pi/3) = \sin(\pi/6)$. This is a special case of $\sin(\pi/2 - t) = \cos t$ since $\sin(\pi/6) = \sin(\pi/2 - \pi/3)$.   ■

The reciprocal, quotient, and Pythagorean identities discussed in Sections 2.2 and 2.4 for trigonometric functions of angles hold as well for trigonometric functions of real numbers. This is the case because the trigonometric function of the real number t is simply the trigonometric function of the angle of t radians. Therefore, $\sec t = 1/\cos t$, $\csc t = 1/\sin t$, $\tan t = \sin t/\cos t$, and so on.

The next four properties of the tangent function follow directly from the corresponding properties of the sine and the cosine functions.

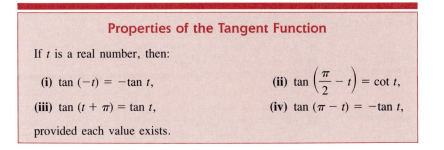

Properties of the Tangent Function

If t is a real number, then:

(i) $\tan (-t) = -\tan t$,

(ii) $\tan \left(\dfrac{\pi}{2} - t\right) = \cot t$,

(iii) $\tan (t + \pi) = \tan t$,

(iv) $\tan (\pi - t) = -\tan t$,

provided each value exists.

■■■ **EXAMPLE 5**

Show that $\tan (-t) = \tan t$.

Solution We proceed as follows:

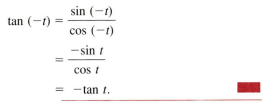

$$\tan (-t) = \frac{\sin (-t)}{\cos (-t)}$$

$$= \frac{-\sin t}{\cos t}$$

$$= -\tan t.$$

The property $\tan (t + \pi) = \tan t$ implies that the tangent function is periodic with period $p \leq \pi$. Since $\tan t = \sin t/\cos t$, $\tan t = 0$ only when $\sin t = 0$. Thus, $\tan t = 0$ only if $t = 0, \pm\pi, \pm 2\pi$, and so on. Therefore, the smallest positive number p for which $\tan (t + p) = \tan t$ is $p = \pi$.

It is important to recognize that the properties just discussed for the sine, cosine, and tangent of a real number t also hold for an angle θ measured in degrees, provided that π is replaced by $180°$ whenever π appears in a formula. For example, from $\sin (-t) = -\sin t$, it follows that $\sin (-45°) = -\sin 45°$; and from $\tan (\pi/2 - t) = \cot t$, we have that $\tan (90° - \theta) = \cot \theta$.

EXERCISE 3.1

In Problems 1–4, for the given real number t, (a) locate the point $(\cos t, \sin t)$ on the unit circle and (b) find the exact value of the coordinates $\cos t$ and $\sin t$.

1. $7\pi/6$ **2.** $2\pi/3$ **3.** $-\pi/2$ **4.** 2π

In Problems 5–8, for the given real number t, (a) locate the point $(\cos t, \sin t)$ on the unit circle and (b) use a calculator to approximate the coordinates $\cos t$ and $\sin t$.

5. 2.5 **6.** 15.3 **7.** -7.2 **8.** 0.1

In Problems 9–12, use periodicity to find the given trigonometric function value. Do *not* use a calculator.

9. $\sin (13\pi/6)$ **10.** $\cos (61\pi/3)$
11. $\tan (3\pi/4)$ **12.** $\sin (-5\pi/3)$

In Problems 13–20, find the given trigonometric function value. Do *not* use a calculator.

13. $\sin (-11\pi/3)$ **14.** $\cot (17\pi/6)$
15. $\tan 5\pi$ **16.** $\sin (-19\pi/2)$
17. $\cos (-7\pi/4)$ **18.** $\tan (41\pi/3)$
19. $\sec (23\pi/3)$ **20.** $\csc (19\pi/6)$

In Problems 21–30, justify the given statement with one of the properties of the trigonometric functions.

21. $\sin \pi = \sin 3\pi$ **22.** $\cos (\pi/4) = \sin (\pi/4)$
23. $\tan (-1.402) = -\tan 1.402$
24. $\cos 16.8\pi = \cos 14.8\pi$
25. $\cos (2.5 + \pi) = -\cos 2.5$
26. $\tan 0.3\pi = \cot 0.2\pi$
27. $\sin (-3 - \pi) = -\sin (3 + \pi)$
28. $\tan 9\pi = \tan 8\pi$
29. $\cos 0.43 = \cos (-0.43)$ **30.** $\sin (2\pi/3) = \sin (\pi/3)$
31. Show that the period of the cosine function is 2π.
32. Verify that $\cos (-t) = \cos t$ and $\sin (-t) = -\sin t$.
33. For $0 < t < \pi/2$, verify that $\cos (t + \pi) = -\cos t$ and $\sin (t + \pi) = -\sin t$. [*Hint*: P_t and $P_{t+\pi}$ are symmetric with respect to the origin.]
34. For $0 < t < \pi/2$, verify that $\cos (\pi - t) = -\cos t$ and $\sin (\pi - t) = \sin t$. [*Hint*: P_t and $P_{\pi-t}$ are symmetric with respect to the y-axis.]

In Problems 35–37, verify the given property of the tangent function using the corresponding properties of the sine and the cosine functions.

35. $\tan \left(\dfrac{\pi}{2} - t \right) = \cot t$

36. $\tan (t + \pi) = \tan t$
37. $\tan (\pi - t) = -\tan t$
38. For which real numbers t does $\sin t = \sqrt{2}/2$?
39. For which real numbers t does $\cos t = -1/2$?

3.2

Graphs of the Trigonometric Functions

One way to further your understanding of the trigonometric functions is to examine their graphs.

In Section 3.1 we saw that the domain of the sine function $f(t) = \sin t$ is all real numbers and that the interval $[-1, 1]$ is its range. Since the sine function has period 2π, we begin by sketching its graph on the interval $[0, 2\pi]$. We obtain a rough sketch of the graph (see Figure 9(b)) by consider-

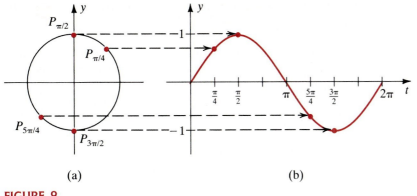

(a)   (b)

FIGURE 9

ing various positions of the point P_t on the unit circle, as shown in Figure 9(a). As t varies from 0 to $\pi/2$, the value $y = \sin t$ increases from 0 to its maximum value 1. But as t varies from $\pi/2$ to $3\pi/2$, the value $\sin t$ decreases from 1 to its minimum value -1. We note that $\sin t$ changes from positive to negative at $t = \pi$. For t between $3\pi/2$ and 2π, we see that the corresponding values of $\sin t$ increase from -1 to 0.

Using the values of the sine function for $0 \leq t \leq 2\pi$ obtained in Problem 35 of Exercise 2.4, we plot points and join them with a smooth curve, as shown in color in Figure 10(a). Since $\sin (t + 2\pi) = \sin t$, the

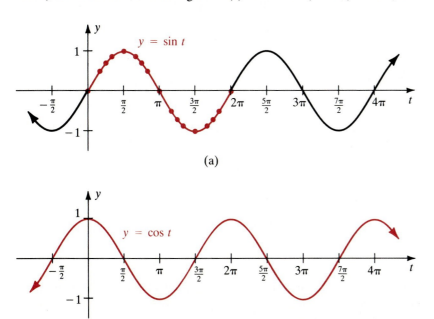

FIGURE 10

graph of $y = \sin t$ for $2\pi \le t \le 4\pi$ is the same as the graph for $0 \le t \le 2\pi$. Using the periodicity of the sine function, we can extend the graph in either direction, as shown in Figure 10(a).

Recall from Section 3.1 that the sine function is an *odd* function since $f(-t) = \sin(-t) = -\sin t = -f(t)$. Thus, as we can see in Figure 10(a), its graph is symmetric with respect to the origin.

Working again with the unit circle, we obtain the graph of the cosine function, $g(t) = \cos t$, shown in Figure 10(b). As is standard practice, this graph is labeled $y = \cos t$. We note that, in this context, the symbol y does *not* represent the y-coordinate of a point on the unit circle; rather, it is the y-coordinate of a point on the graph of the function $g(t) = \cos t$.

We see from Figure 10(b) that the graph of the cosine function is symmetric with respect to the y-axis. This is a result of

$$g(-t) = \cos(-t) = \cos t = g(t);$$

that is, the cosine function is an *even* function.

You may have observed that the graph of the sine function is identical to the graph of the cosine function but shifted $\pi/2$ units to the right. This is a consequence of the property

$$\sin t = \cos\left(t - \frac{\pi}{2}\right)$$

(see Problem 34). For a detailed discussion of shifted sine and cosine graphs, see Section 3.3.

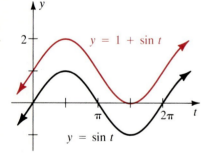

FIGURE 11

■ EXAMPLE 1

Graph $y = 1 + \sin t$.

Solution Recall from Section 1.6 that the graph of $y = 1 + \sin t$ can be obtained by shifting the graph of $y = \sin t$ up one unit. The result is shown in color in Figure 11. ■

■ EXAMPLE 2

Graph $y = \sin t + \cos t$.

Solution Recall from Section 1.6 that the graph of the sum of two functions can be obtained by addition of the y-coordinates. To obtain the graph of $y = \sin t + \cos t$, we first graph $y = \sin t$ and $y = \cos t$ on the same axes. Then, as shown in Figure 12, to obtain a point on the graph of $y = \sin t +$

cos t for any t, we add the y-coordinates sin t and cos t. After plotting a sufficient number of points, we obtain the graph shown in color in Figure 12.

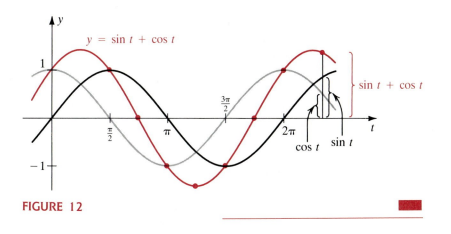

FIGURE 12

We now consider the graph of the tangent function, $h(t) = \tan t$. Since $\tan t = \sin t/\cos t$, the domain of the tangent function consists of all real numbers t for which $\cos t \neq 0$. After we sketch the graph of the tangent function, we will see that its range is R. Because $h(t) = \tan t$ has period π, we only need graph it on an interval of length π. A convenient interval to choose is $(-\pi/2, \pi/2)$.

To sketch the graph of the tangent function, we begin by considering its values for the special angles 0, $\pi/6$, $\pi/4$, and $\pi/3$ (shown in the accompanying table) and plotting the corresponding points (see Figure 13).

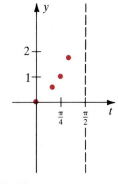

FIGURE 13

t	0	$\dfrac{\pi}{6}$	$\dfrac{\pi}{4}$	$\dfrac{\pi}{3}$
$\tan t$	0	$\dfrac{\sqrt{3}}{3} \approx 0.6$	1	$\sqrt{3} \approx 1.7$

We know that tan t is undefined for $t = \pi/2$ (since cos $(\pi/2) = 0$), but we can consider values of tan t for t close to but less than $\pi/2$. Since

$$\tan t = \frac{\sin t}{\cos t},$$

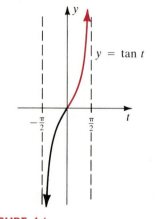

FIGURE 14

we examine sin t and cos t for $0 < t < \pi/2$. As t approaches $\pi/2$, the numerator sin t approaches 1 and the denominator cos t approaches 0. Thus, as t increases toward $\pi/2$, the values of tan t grow larger and larger. This behavior is illustrated in the following table, where the values of sin t, cos t, and tan t were obtained from a calculator. The values chosen for t are increasing toward $\pi/2 \approx \frac{1}{2}(3.1416) = 1.5708$. Thus the line $t = \pi/2$ is a vertical asymptote for the graph. From this observation we obtain the portion of the curve shown in color in Figure 14. From

$$h(-t) = \tan(-t) = -\tan t = -h(t),$$

we conclude that the tangent function is odd and its graph is symmetric with respect to the origin. Hence we obtain the portion of the graph shown in black in Figure 14.

t	1	1.5	1.57	1.5707	1.57075
sin t	0.84	0.997	0.99999968	0.999999995	0.999999998
cos t	0.54	0.071	0.0008	0.000096	0.000046
tan t	1.6	14.1	1255.8	10,381.3	21,585.8

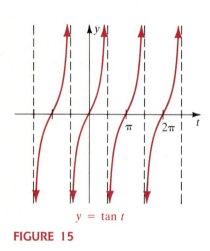

$y = \tan t$

FIGURE 15

It follows that the line $t = -\pi/2$ is another vertical asymptote of the graph. Using the fact that the period of the tangent function is π, we complete the graph by repeating the same pattern (see Figure 15). We observe that the domain of the tangent function is all real numbers *except* odd multiples of $\pi/2$ and the range is R.

▮▮▮ EXAMPLE 3

Graph **(a)** $y = -\tan t$ and **(b)** $y = \tan(t + \pi/2)$.

Solution   We obtain these graphs using the techniques of shifting and reflecting discussed in Section 1.6.

(a) The graph of $y = -\tan t$ is the reflection of the graph of $y = \tan t$ in the t-axis. See Figure 16(a).

(b) The graph of $y = \tan(t + \pi/2)$ can be obtained by shifting the graph of $y = \tan t$ to the left $\pi/2$ units. See Figure 16(b).

■ EXAMPLE 4

Sketch the graph of $f(t) = \csc t$ and determine its domain and its range.

Solution Since $\csc t = 1/\sin t$, the graph of the cosecant function will have vertical asymptotes wherever $\sin t = 0$, namely, $t = 0, \pm\pi, \pm2\pi. \ldots$ Furthermore, we can obtain the y-coordinate of a point on the graph of the cosecant function by taking the reciprocal of a nonzero y-coordinate of a point on the graph of the sine function.

As shown in Figure 17, it is convenient to sketch the graph of the sine function (shown in black), then locate the vertical asymptotes, and finally take the reciprocals of the y-coordinates to obtain points on the cosecant graph. The domain of the cosecant function consists of all real numbers t such that $\sin t \neq 0$; that is, $\{t \mid t \neq n\pi, n = 0, \pm1, \pm2, \ldots\}$. From Figure 17, we see that the range of the cosecant function consists of the union of the intervals $(-\infty, -1]$ and $[1, \infty)$.

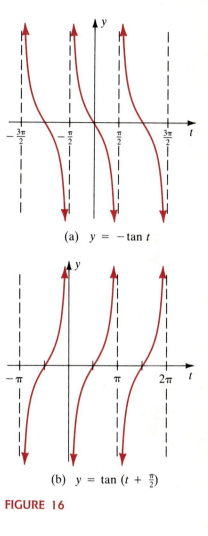

(a) $y = -\tan t$

(b) $y = \tan(t + \frac{\pi}{2})$

FIGURE 16

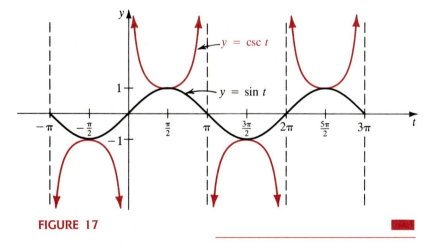

FIGURE 17

From the range of the cosecant function found in Example 4, it follows that

$$|\csc t| \geq 1$$

for all real numbers t in the domain of the cosecant function. Note too that the cosecant is an odd function; its graph is symmetric with respect to the origin. (See Problems 31–33.)

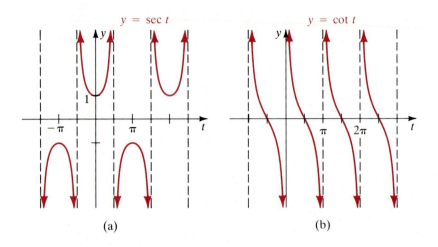

FIGURE 18   (a)   (b)

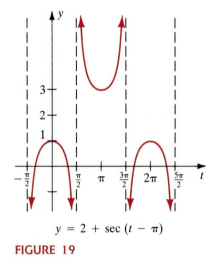

$y = 2 + \sec (t - \pi)$

FIGURE 19

Using the fact that sec $t = 1/\cos t$ and cot $t = 1/\tan t$, we can obtain the graphs of the secant function, as in Figure 18(a), and the cotangent function, as in Figure 18(b), by an approach similar to that used in Example 4. The verification of Figure 18 is left as an exercise (see Problems 35 and 37).

An important property of the secant function,

$$|\sec t| \geq 1,$$

is evident from the graph of $y = \sec t$. See Figure 18(b).

■ **EXAMPLE 5**

Graph $y = 2 + \sec (t - \pi)$.

Solution   The graph of $y = 2 + \sec (t - \pi)$ can be obtained by shifting the graph of $y = \sec t$, as in Figure 18(a), π units to the right and 2 units up (see Figure 19).

EXERCISE 3.2

In Problems 1–16, use the techniques of shifting and reflecting to graph the given function.

1. $y = \dfrac{1}{2} + \cos t$

2. $y = -1 + \sin t$

3. $y = -\sin t$

4. $y = -\cos t$

5. $y = 2 - \sin t$

6. $y = -(3 + \cos t)$

7. $y = \sin (t - \pi)$

8. $y = \cos \left(t - \dfrac{\pi}{4} \right)$

9. $y = 2 + \tan t$

10. $y = 1 - \tan \left(t + \dfrac{\pi}{2} \right)$

11. $y = -\cot t$

12. $y = 2.5 + \sec t$

13. $y = \dfrac{\pi}{6} - \sec t$

14. $y = -\csc (t + \pi)$

15. $y = \csc\left(t - \dfrac{\pi}{3}\right)$

16. $y = -2 + \cot\left(t - \dfrac{5\pi}{6}\right)$

In Problems 17–22, use addition of *y*-coordinates to graph the given function.

17. $y = \cos t - \sin t$ **18.** $y = \sin t + \cos t + 1$

19. $y = t + \sin t$ **20.** $y = t - \tan t$

21. $y = -\sin t - \cos t$ **22.** $y = \sin t - \cos t + 1$

23. Compare the values of tan 1.57 and tan 1.58 obtained from a calculator set in radian mode. Explain the great difference in these two values.

24. Compare the values of cot 3.14 and cot 3.15 obtained from a calculator set in radian mode. [*Hint*: On most calculators you use the tangent and reciprocal keys to obtain values of the cotangent function.] Explain the great difference in these two values.

25. Can $9 \csc t = 1$ for any real number t?

26. Can $10 \sec t + 7 = 0$ for any real number t?

27. For which real numbers t is **(a)** $\sin t \leq \csc t$?
(b) $\sin t < \csc t$?

28. For which real numbers t is **(a)** $\sec t \leq \cos t$?
(b) $\sec t < \cos t$?

29. The graphs of which trigonometric functions do not have vertical asymptotes?

30. The graphs of which trigonometric functions do not intersect
(a) the *y*-axis? **(b)** the *x*-axis?

31. For every t in the domain of the function, prove that
(a) $\csc(-t) = -\csc t$,   **(b)** $\sec(-t) = \sec t$,   and
(c) $\cot(-t) = -\cot t$.

32. Using the results of Problem 31, discuss the symmetry of the graphs of **(a)** $y = \csc t$, **(b)** $y = \sec t$, and **(c)** $y = \cot t$.

33. Use the results of Problem 31 to answer each of the following.
(a) Is the cotangent function even or odd?
(b) Is the secant function even or odd?
(c) Is the cosecant function even or odd?

34. Show that $\sin t = \cos(t - \pi/2)$ by writing $\cos(t - \pi/2)$ as $\cos(-(\pi/2 - t))$ and using the properties $\cos(-t) = \cos t$ and $\cos(\pi/2 - t) = \sin t$ from Section 3.1.

35. Use the graph of $y = \tan t$ and the fact that $\cot t = 1/\tan t$ to graph $y = \cot t$.

36. The graph of the cotangent function can be obtained by shifting and reflecting the graph of the tangent function. Use the properties $\tan(-t) = -\tan t$ and $\tan(\pi/2 - t) = \cot t$ from Section 3.1 to obtain the graph of the cotangent function.

37. Use the graph of $y = \cos t$ and the fact that $\sec t = 1/\cos t$ to graph $y = \sec t$.

38. Using Figure 17, determine the domain, the range, the asymptotes, and the period of the cosecant function.

39. Using Figure 18(a), determine the domain, the range, the asymptotes, and the period of the secant function.

40. Using Figure 18(b), determine the domain, the range, the asymptotes, the period, and the *t*-intercepts of the cotangent function.

3.3

Harmonic Motion; Shifted Sine and Cosine Graphs

HARMONIC MOTION

Many physical objects vibrate or oscillate in a regular manner, repeatedly moving back and forth in a definite time interval. Some examples are clock pendulums, sound waves, the strings on a guitar when plucked, the human heart, tides, and alternating electric current. Since all sounds and, in particu-

lar, musical tones are produced by vibrations, any oscillatory motion is called **harmonic motion.** Oscillatory motion described by either of the functions

$$f(t) = a \sin (bt + c) \quad \text{and} \quad g(t) = a \cos (bt + c), \tag{2}$$

where a, b, and c are real numbers, is called **simple harmonic motion.** In this section we will study the properties and the graphs of these functions. A number of applications are described in Problems 55–60.

We begin by considering

$$y = a \sin t \quad \text{and} \quad y = a \cos t,$$

which are a special case of (2) with $b = 1$ and $c = 0$. For example, as Figure 20(a) shows, we obtain the graph of $y = 2 \sin t$ by doubling each y-coordinate on the graph of $y = \sin t$. Note that the maximum and the minimum values of $y = 2 \sin t$ occur at the same t-values as the maximum and the minimum values of $y = \sin t$, respectively. However, as we see in Figure 20(b), this situation is reversed for $y = -2 \sin t$; that is, a minimum value occurs when $y = \sin t$ has a maximum value, and conversely. We also observe that the graph of $y = -2 \sin t$ is the reflection in the horizontal axis of the graph of $y = 2 \sin t$. In general, the graph of $y = a \sin t$ can be obtained by multiplying the y-coordinates of the graph of $y = \sin t$ by the number a. Similar remarks can be made for $y = a \cos t$.

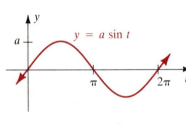

$a > 0$

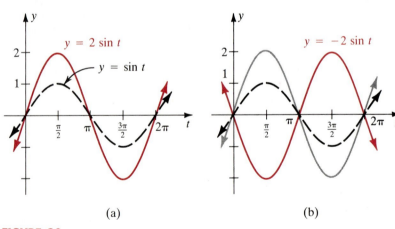

(a)   (b)

FIGURE 20

AMPLITUDE

It follows from the preceding discussion that the maximum distance from any point on the graph of $y = a \sin t$ or $y = a \cos t$ to the t-axis is $|a|$ (see Figure 21). The number $|a|$ is called the **amplitude** of the functions

$$f(t) = a \sin t \quad \text{and} \quad g(t) = a \cos t$$

or of their graphs.

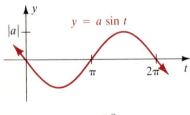

$a < 0$

FIGURE 21

▰ EXAMPLE 1

Graph $y = 2 \cos t$ and $y = \frac{1}{2} \cos t$ on the same axes. Determine the maximum and the minimum values for $0 \le t < 2\pi$.

Solution The given functions have amplitudes 2 and $\frac{1}{2}$, respectively. Limiting our attention to the interval $0 \le t < 2\pi$, we know that the cosine attains its maximum at $t = 0$ and its minimum at $t = \pi$. Thus,

$$y = 2 \cos 0 = 2 \quad \text{and} \quad y = 2 \cos \pi = -2$$

are the maximum and the minimum values of $y = 2 \cos t$. For $y = \frac{1}{2} \cos t$, we find the maximum and the minimum values to be

$$y = \frac{1}{2} \cos 0 = \frac{1}{2} \quad \text{and} \quad y = \frac{1}{2} \cos \pi = -\frac{1}{2}.$$

The graphs are shown on the same axes in Figure 22.

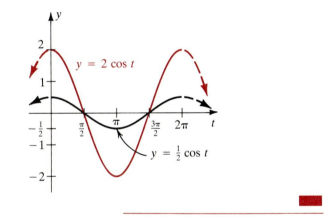

FIGURE 22

Note in Example 1 that for both $y = 2 \cos t$ and $y = \frac{1}{2} \cos t$,

$$\frac{1}{2}(\text{maximum value} - \text{minimum value}) = \text{amplitude}.$$

This result holds for all shifted sine and cosine graphs. For example, the amplitude of the graph of $y = 1 + \sin t$ (Figure 11) is $\frac{1}{2}(2 - 0) = 1$, since the maximum value is 2 and the minimum value is 0.

PERIOD AND CYCLE

We now consider the graph of $y = \sin bt$, for $b > 0$. The graph will have amplitude 1 since $a = 1$. Recall that $y = \sin t$ has period 2π. Thus, starting at $t = 0$, $y = \sin bt$ will repeat its values beginning at $bt = 2\pi$, or $t = 2\pi/b$. It follows that $y = \sin bt$ has the **period** $2\pi/b$; this means that the graph will repeat itself every $2\pi/b$ units. For this reason, we say that the graph of

$y = \sin bt$ over an interval of length $2\pi/b$ is a **cycle** of the sine curve. For example, the period of $y = \sin 2t$ is $2\pi/2 = \pi$, and therefore one cycle of the graph is completed in the interval $0 \le t \le \pi$. We obtain the graph of $y = \sin 2t$ on the interval (shown in color) using the data in the table accompanying Figure 23. The extension of this graph (shown in black) is obtained by periodicity. For comparison the figure also shows the graph of $y = \sin t$.

t	0	$\dfrac{\pi}{4}$	$\dfrac{\pi}{2}$	$\dfrac{3\pi}{4}$	π
$2t$	0	$\dfrac{\pi}{2}$	π	$\dfrac{3\pi}{2}$	2π
$\sin 2t$	0	1	0	-1	0

FIGURE 23

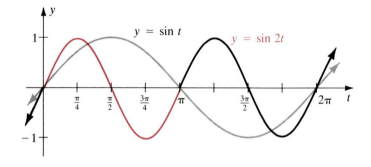

Similar remarks hold for the cosine function; that is, for $b > 0$, the function $g(t) = \cos bt$ has period $2\pi/b$.

■ EXAMPLE 2

Find the period of $y = \cos 4t$ and graph the function.

Solution   Since $b = 4$, we see that the period is $2\pi/4 = \pi/2$. Thus one cycle of the graph is completed in any interval of length $\pi/2$. To graph the function, we draw one cycle of the cosine curve on the interval $0 \le t \le \pi/2$. Then, as shown in Figure 24, we use periodicity to extend the graph.

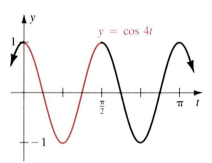

FIGURE 24

Combining the results of the preceding discussions, we find that the graphs of $y = a \sin bt$ and $y = a \cos bt$, $b > 0$, each have

$$\text{amplitude} = |a| \quad \text{and} \quad \text{period} = \frac{2\pi}{b}.$$

If $b < 0$ in either $y = a \sin bt$ or $y = a \cos bt$, we use the properties

$$\sin(-t) = -\sin t \quad \text{or} \quad \cos(-t) = \cos t$$

to rewrite the expression in the required form with positive b. This is illustrated in the following example.

▦ EXAMPLE 3

Find the amplitude and the period of $y = \sin\left(-\frac{1}{2}t\right)$. Graph.

Solution Since we require $b > 0$, we use $\sin(-t) = -\sin t$ to write

$$y = \sin\left(-\tfrac{1}{2}t\right) = -\sin\tfrac{1}{2}t.$$

It follows that the amplitude is $|a| = |-1| = 1$. Since $b = 1/2$, the period is $2\pi/\frac{1}{2} = 4\pi$. Hence the graph of the given function completes one cycle on the interval $0 \le t \le 4\pi$. In Figure 25, the color curve is the graph of $y = -\sin\frac{1}{2}t$ and is a reflection in the horizontal axis of the graph of $y = \sin\frac{1}{2}t$, shown in black. ▦

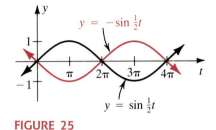

FIGURE 25

▦ EXAMPLE 4

Graph $y = \frac{5}{2}\sin 2\pi t$.

Solution The amplitude is $\frac{5}{2}$ and the period is $2\pi/2\pi = 1$. Thus the function completes one cycle on the interval $0 \le t \le 1$ (see Figure 26). The maximum value $y = \frac{5}{2}$ occurs at $t = \frac{1}{4}$, and the minimum value $y = -\frac{5}{2}$ corresponds to $t = \frac{3}{4}$. ▦

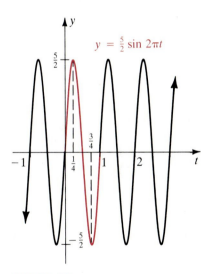

FIGURE 26

We note that there is no amplitude associated with $y = \tan t$, $y = \cot t$, $y = \sec t$, or $y = \csc t$, since their graphs are unbounded. However, the y-coordinates of points on the graph of $y = a \tan t$ can still be obtained by multiplying the y-coordinates of points on the graph of $y = \tan t$ by a. This is also true for $y = a \cot t$, $y = a \sec t$, and $y = a \csc t$.

The functions $y = \sec bt$ and $y = \csc bt$, $b > 0$, each have period $2\pi/b$, whereas the functions $y = \tan bt$ and $y = \cot bt$, $b > 0$, each have the period π/b (see Problems 51–54).

PHASE SHIFT

We now consider the graph of $y = a \sin(bt + c)$, for $b > 0$. Since the values of $\sin(bt + c)$ range from -1 to 1, it follows that the **amplitude** of $y = a \sin(bt + c)$ is $|a|$. As $bt + c$ varies from 0 to 2π, the graph will complete one cycle. By solving $bt + c = 0$ and $bt + c = 2\pi$, we find that one cycle is completed as t varies from $-c/b$ to $(2\pi - c)/b$. Therefore, $y = a \sin(bt + c)$ has **period**

$$\frac{2\pi - c}{b} - \left(-\frac{c}{b}\right) = \frac{2\pi}{b}.$$

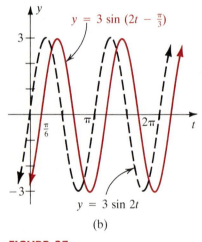

$y = 3 \sin 2t$

(a)

$y = 3 \sin \left(2t - \frac{\pi}{3}\right)$

$y = 3 \sin 2t$

(b)

FIGURE 27

If we rewrite $y = a \sin (bt + c)$ as $y = a \sin b(t + c/b)$, we see that the graph of $y = a \sin (bt + c)$ can be obtained by *shifting* the graph of $y = a \sin bt$ horizontally a distance of $|c|/b$. If $c < 0$, the shift is to the right, whereas if $c > 0$, the shift is to the left. The number $-c/b$ is called the **phase shift** of the graph of $y = a \sin (bt + c)$.

■ EXAMPLE 5

Graph $y = 3 \sin (2t - \pi/3)$ by shifting the graph of $y = 3 \sin 2t$.

Solution The amplitude of $y = 3 \sin 2t$ is $|a| = 3$, and the period is $2\pi/2 = \pi$. Thus one cycle is completed on the interval $0 \le t \le \pi$. Then we extend the graph beyond this interval by periodicity, as shown in Figure 27(a).

Rewriting $y = 3 \sin (2t - \pi/3)$ as

$$y = 3 \sin 2(t - \pi/6),$$

we see that we can obtain the graph of this function by shifting the graph of $y = 3 \sin 2t$ to the right $\pi/6$ units (see Figure 27(b)).   ■

An analysis similar to the one above can be made for the graph of $y = a \cos (bt + c)$. We summarize the results in the following manner.

Shifted SINE and COSINE Graphs

The graphs of $y = a \sin (bt + c)$ and $y = a \cos (bt + c)$, for $b > 0$, have

$$\text{amplitude} = |a|,$$

$$\text{period} = \frac{2\pi}{b}, \text{ and}$$

$$\text{phase shift} = -\frac{c}{b}.$$

The graphs have the same form as those of $y = a \sin bt$ and $y = a \cos bt$, *except* they are shifted to the right or to the left $|c|/b$ units depending on whether $c < 0$ or $c > 0$, respectively.

■ EXAMPLE 6

Determine the amplitude, the period, the phase shift, and the direction of shift for each of the following.

(a) $y = 15 \cos \left(5t - \frac{3\pi}{2}\right)$ **(b)** $y = 10 \sin \left(-2t - \frac{\pi}{6}\right)$

Solution

(a) We first make the identifications $a = 15$, $b = 5$, and $c = -3\pi/2$. Thus the amplitude is $|a| = 15$, the period is $2\pi/b = 2\pi/5$, and the phase shift is $-c/b = -(-3\pi/2)/5 = 3\pi/10$. Since $c = -3\pi/2 < 0$, we know that the graph of $y = 15 \cos (5t - 3\pi/2)$ is the graph of $y = 15 \cos 5t$ shifted $3\pi/10$ units to the right.

(b) Since we require that $b > 0$, we first use $\sin(-t) = -\sin t$ to write

$$y = 10 \sin \left(-2t - \frac{\pi}{6}\right) = -10 \sin \left(2t + \frac{\pi}{6}\right).$$

Now with $a = -10$, $b = 2$, and $c = \pi/6$, we find that the amplitude is $|a| = 10$, the period is $2\pi/2 = \pi$, and the phase shift is $-(\pi/6)/2 = -\pi/12$. Since $c = \pi/6 > 0$, the graph of $y = -10 \sin 2t$ is shifted $\pi/12$ units to the left. ■

■ EXAMPLE 7

Graph $y = -2 \cos \left(t + \dfrac{\pi}{3}\right)$.

Solution Since $b = 1$ and $c = \pi/3$, the phase shift is $-\pi/3$. Thus the graph shown in Figure 28 is obtained by shifting the graph of $y = -2 \cos t$ to the left $\pi/3$ units, since c is positive. (By now you should recognize that the graph of $y = -2 \cos t$ is easy to obtain by reflecting the graph of $y = 2 \cos t$ in the horizontal axis.) ■

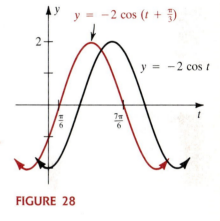

FIGURE 28

■ EXAMPLE 8

The current i (in amperes) in a wire of an alternating-current circuit is given by

$$i(t) = 30 \sin 120\pi t,$$

where time t is measured in seconds. Sketch one cycle of the graph. What is the maximum value of the current?

Solution The graph has amplitude 30 and period $2\pi/120\pi = 1/60$. Therefore, we sketch one cycle of the sine curve on the interval $[0, 1/60]$, as shown in Figure 29. From the figure it is evident that the maximum value of the current is $i = 30$ amperes and it occurs at $t = 1/240$ second. ■

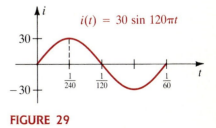

FIGURE 29

EXERCISE 3.3

In Problems 1–6, determine the amplitude and the period, and sketch the graphs of the given pair of functions on the same set of axes. (Be sure to observe the similarities and differences in each pair of graphs.)

1. (a) $y = 4 \sin t$; (b) $y = \dfrac{1}{4} \sin t$

2. (a) $y = 3 \cos t$; (b) $y = \dfrac{1}{3} \cos t$

3. (a) $y = \sin 4t$; (b) $y = \sin \dfrac{1}{4}t$

4. (a) $y = \cos 3t$; (b) $y = \cos \dfrac{1}{3}t$

5. (a) $y = \cos 2\pi t$; (b) $y = \cos 2t$

6. (a) $y = \sin \dfrac{\pi}{4}t$; (b) $y = \sin \dfrac{1}{4}t$

In Problems 7–12, one cycle of the graph of either $y = a \sin bt$ or $y = a \cos bt$ is shown. Identify the function.

7.

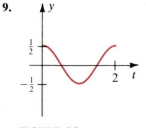

FIGURE 30

8.

FIGURE 31

9.

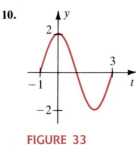

FIGURE 32

10.

FIGURE 33

11.

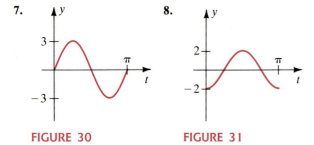

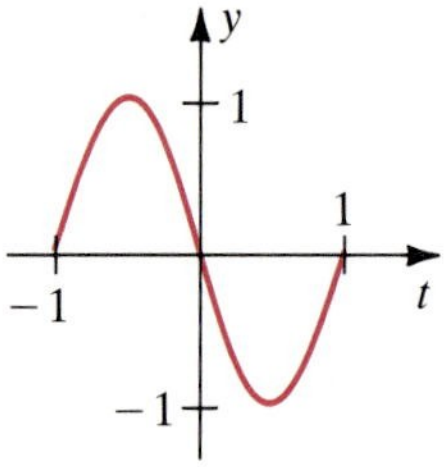

FIGURE 34

12.

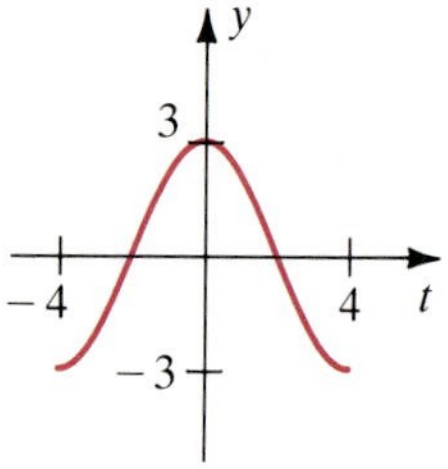

FIGURE 35

In Problems 13–26, determine the amplitude and the period, and sketch the graph of each function.

13. $y = 4 \cos t$

14. $y = -2 \cos t$

15. $y = -\dfrac{1}{2} \sin t$

16. $y = \dfrac{3}{2} \sin t$

17. $y = 2 \cos 2t$

18. $y = \cos \dfrac{1}{4}t$

19. $y = 4 \sin (-t)$

20. $y = 2 \sin 4t$

21. $y = 5 \cos 2\pi t$

22. $y = \dfrac{1}{2} \cos \pi t$

23. $y = \sin \dfrac{2}{3}t$

24. $y = \cos \left(-\dfrac{1}{2}t\right)$

25. $y = -3 \sin (-2t)$

26. $y = -4 \cos \left(-\dfrac{\pi}{2}t\right)$

In Problems 27–30, find shifted sine and shifted cosine functions that satisfy the given conditions. Graph the functions.

27. Amplitude 3, period $2\pi/3$, phase shift $\pi/3$
28. Amplitude 2/3, period π, phase shift $-\pi/4$
29. Amplitude 0.7, period 0.5, phase shift 4
30. Amplitude 5/4, period 4, phase shift $-1/2\pi$

In Problems 31–46, sketch the graph of the given function. If appropriate, determine the amplitude, the period, and the phase shift.

31. $y = \sin \left(t - \dfrac{\pi}{6}\right)$

32. $y = \sin \left(3t - \dfrac{\pi}{4}\right)$

33. $y = \cos \left(t + \dfrac{\pi}{4}\right)$

34. $y = -2 \cos \left(2t - \dfrac{\pi}{6}\right)$

35. $y = 4 \cos \left(2t - \dfrac{3\pi}{2} \right)$ **36.** $y = 3 \sin \left(2t + \dfrac{\pi}{4} \right)$

37. $y = 3 \sin \left(\dfrac{t}{2} - \dfrac{\pi}{3} \right)$ **38.** $y = -\cos \left(\dfrac{t}{2} - \pi \right)$

39. $y = 5 \cos \left(\dfrac{2t}{3} - \dfrac{\pi}{12} \right)$ **40.** $y = 2 \sin \left(-t + \dfrac{\pi}{8} \right)$

41. $y = 2 \cos \left(-4t - \dfrac{4\pi}{3} \right)$ **42.** $y = 5 \sin (\pi t - 1)$

43. $y = \tan \pi t$ **44.** $y = \cot \dfrac{\pi}{2} t$

45. $y = 3 \sec \dfrac{1}{2} t$ **46.** $y = -\csc 2t$

47. Write the function $y = -4 \sin 3t$ in the form $y = a \cos (bt + c)$, $a > 0$.

48. Is the function $y = t \sin t$ periodic?

49. Find the period of $y = \sin \dfrac{1}{2} t + \sin 2t$.

50. For what value of d is the graph of $y = a \cos (bt + d)$ the same as the graph of $y = a \sin (bt + c)$?

51. Verify that the period of $y = \tan bt$ for $b > 0$ is π/b.

52. Verify that the period of $y = \cot bt$ for $b > 0$ is π/b.

53. Verify that the period of $y = \sec bt$ for $b > 0$ is $2\pi/b$.

54. Verify that the period of $y = \csc bt$ for $b > 0$ is $2\pi/b$.

55. The angular displacement θ of a pendulum bob from the vertical at time t seconds is given by $\theta = \theta_0 \cos \omega t$, where θ_0 is the initial displacement at $t = 0$ seconds (see Figure 36). For $\omega = 2$ radians/second and $\theta_0 = \pi/10$, sketch two cycles of the graph of the resulting function.

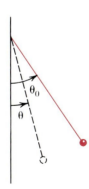

FIGURE 36

56. In a certain electrical circuit, the current I measured in amperes at time t seconds is given by

$$I(t) = 10 \cos \left(120\pi t + \dfrac{\pi}{3} \right).$$

Sketch two cycles of the graph of I as a function of time t.

57. The depth y of water at the entrance to a small harbor at time t is given by

$$y = a \sin b\left(t - \dfrac{\pi}{2} \right) + k,$$

where a is one half the difference between the high- and low-tide depths, $2\pi/b$ is the tidal period, and k is the average depth. Assume that the tidal period is 12 hr, the depth at high tide is 18 ft, and the depth at low tide is 6 ft. Sketch two cycles of the graph.

58. After a weight has been attached to a spring, it will stretch the spring and attain a position of equilibrium, or rest. If the weight is pulled A cm below the equilibrium position and released from rest at $t = 0$ sec, then, under ideal conditions, the weight would bounce back and forth A cm on either side of the equilibrium position, taking $\pi/2$ sec to complete one cycle. From Figure 37, determine the distance y from the equilibrium position as a function of time t.

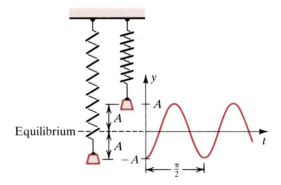

FIGURE 37

59. Trigonometric functions having the form $y = a + b \sin \omega(t - t_0)$, where a, b, ω, and t_0 are real constants, are often used to simulate variation in temperature. Suppose

$$F(t) = 60 + 10 \sin \dfrac{\pi}{12}(t - 8), \quad 0 \le t \le 24,$$

gives the Fahrenheit temperature F at t hours after midnight on a certain day.

(a) What is the temperature at 8 A.M.? at 12:00 noon?

(b) At what time(s) does $F(t) = 60$?

(c) Sketch the graph of F.

(d) Find the maximum and the minimum temperatures and the times at which they occur.

60. The sensation of sound is produced when the human ear detects periodic variations in air pressure produced by a sound wave at the eardrum. If this variation is simple harmonic variation, then the sound is perceived as a **pure tone.** The sound produced by a tuning fork vibrating at 256 cycles per second is identified as middle C. If the pressure amplitude is 0.2 dynes per square centimeter, the sound wave can be described by

$$y = 0.2 \sin 512\pi t,$$

where y is the difference (in dynes per square centimeter) between atmospheric pressure and the air pressure at the eardrum at t seconds. Sketch two cycles of the graph.

61. Most lightning flashes move from cloud to cloud; only a few go from cloud to ground. The ratio of the number of occurrences of these two types of lightning seems to depend on latitude. Empirical studies have found that the ratio of the number of cloud-to-cloud flashes in a storm, N_c, to the number of cloud-to-ground flashes, N_g, is approximated by

$$\frac{N_c}{N_g} = 4.16 + 2.16 \cos 3\phi,$$

where ϕ is the latitude (restricted to nonpolar regions $0° \leq \phi \leq 60°$). Graph this ratio for the latitudes $0° \leq \phi \leq 60°$.

3.4 Inverse Trigonometric Functions

From Section 1.6 we know that a function has an inverse if and only if it is one-to-one. Inspection of the graphs of the various trigonometric functions clearly shows that these functions are not one-to-one. However, by suitably restricting each of their domains, we can ensure that the resulting functions are one-to-one.

ARCSINE

From Figure 38 we see that the function $y = \sin x$ is one-to-one on the closed interval $[-\pi/2, \pi/2]$, since on this interval any horizontal line intersects the graph at most once. Thus when restricted to this particular interval, the sine function has an inverse. We denote this inverse by

$$\sin^{-1} x, \quad \text{or} \quad \arcsin x.$$

These symbols are read "inverse sine of x" and "arcsine of x," respectively.

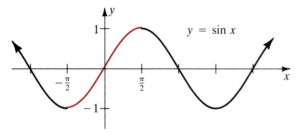

FIGURE 38

We make the following definition.

DEFINITION 2

> The **arcsine,** or **inverse sine,** function is defined by
>
> $$y = \arcsin x \quad \text{if and only if} \quad x = \sin y,$$
>
> where $-\pi/2 \leq y \leq \pi/2$ and $-1 \leq x \leq 1$.

In other words,

the arcsine of the number x is that number (or radian-measured angle) y between $-\pi/2$ and $\pi/2$ whose sine is x.

Note that the "-1" in $\sin^{-1} x$ is *not* an exponent; rather, it denotes an inverse function, that is,

$$(\sin x)^{-1} = \frac{1}{\sin x} \neq \sin^{-1} x.$$

To avoid this possible confusion, some prefer the terminology "arcsin x," which refers to the "arc" or angle whose sine is x. However, $\sin^{-1} x$ and arcsin x are generally used interchangeably in mathematics and its applications. Therefore, we will continue to alternate their use so as to familiarize you with both notations.

Recall from Section 1.6 that the graph of an inverse function is the reflection of the graph of the given function in the line $y = x$. This technique is used in Figure 39(a) to obtain the graph of $y = \arcsin x$. From Figure 39(b) (showing $y = \arcsin x$ only), we see that the domain of arcsin x is $[-1, 1]$ and the range is $[-\pi/2, \pi/2]$.

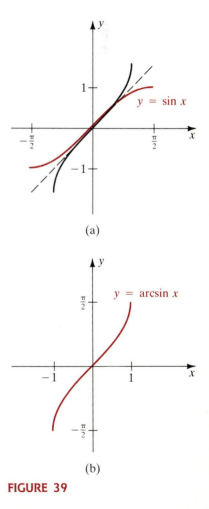

FIGURE 39

EXAMPLE 1

Find **(a)** $\arcsin \frac{1}{2}$, **(b)** $\sin^{-1} (-\frac{1}{2})$, and **(c)** $\sin^{-1} (-1)$.

Solution

(a) If we let $y = \arcsin \frac{1}{2}$, then $-\pi/2 \le y \le \pi/2$ and $\sin y = \frac{1}{2}$. It follows that $y = \pi/6$.

(b) If we let $y = \sin^{-1} (-\frac{1}{2})$, then $\sin y = -\frac{1}{2}$. Since we must choose y such that $-\pi/2 \le y \le \pi/2$, we find that $y = -\pi/6$.

(c) Letting $y = \sin^{-1} (-1)$, we have that $\sin y = -1$ and $-\pi/2 \le y \le \pi/2$. Hence $y = -\pi/2$.

Note of Caution: In part (b) of Example 1 we were careful to choose y so that $-\pi/2 \le y \le \pi/2$. It is a *common error* to think that since $\sin (11\pi/6) = -\frac{1}{2}$, $\sin^{-1} (-\frac{1}{2})$ is $11\pi/6$. Remember: If $y = \sin^{-1} x$, then y is subject to the restriction $-\pi/2 \le y \le \pi/2$.

EXAMPLE 2

Without using a calculator, find $\tan (\sin^{-1} \frac{1}{4})$.

Solution We must find the tangent of the angle of t radians with sine equal to $\frac{1}{4}$, that is, $\tan t$ where $t = \sin^{-1} \frac{1}{4}$. The angle t is shown in Figure 40. Since

$$\tan t = \frac{\sin t}{\cos t} = \frac{\frac{1}{4}}{\cos t}, \tag{2}$$

we want to determine $\cos t$. From Figure 40 and the identity $\cos^2 t + \sin^2 t = 1$, we see that

$$\cos^2 t + \left(\frac{1}{4}\right)^2 = 1,$$

or

$$\cos t = \frac{\sqrt{15}}{4}.$$

Substituting this value into (11), we have

$$\tan t = \frac{1/4}{\sqrt{15}/4} = \frac{\sqrt{15}}{15},$$

and thus

$$\tan \left(\sin^{-1} \frac{1}{4} \right) = \tan t = \frac{\sqrt{15}}{15}.$$

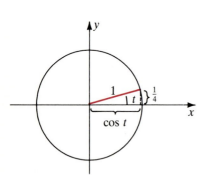

FIGURE 40

Another approach to finding $\tan\left(\sin^{-1}\frac{1}{4}\right)$ is to sketch a right triangle containing an angle t for which $t = \sin^{-1}\frac{1}{4}$, or $\sin t = \frac{1}{4}$. This is shown in Figure 41, where we have let hyp = 4 and opp = 1. We can then find adj by the Pythagorean theorem:

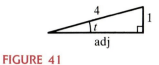

FIGURE 41

$$4^2 = 1^2 + (\text{adj})^2$$
$$\text{adj} = \sqrt{4^2 - 1^2} = \sqrt{15}.$$

Thus
$$\tan\left(\sin^{-1}\frac{1}{4}\right) = \tan t = \frac{\text{opp}}{\text{adj}} = \frac{1}{\sqrt{15}} = \frac{\sqrt{15}}{15}.$$

ARCCOSINE

If we restrict the domain of the cosine function to the closed interval $[0, \pi]$, the resulting function is one-to-one and thus has an inverse. We denote this inverse by

$$\cos^{-1} x, \quad \text{or} \quad \arccos x,$$

which gives us the following definition.

DEFINITION 3

The **arccosine,** or **inverse cosine,** function is defined by

$$y = \arccos x \quad \text{if and only if} \quad x = \cos y,$$

where $0 \leq y \leq \pi$ and $-1 \leq x \leq 1$.

The graphs of $y = \cos x$ on $[0, \pi]$ and $y = \arccos x$ are shown in Figure 42. The domain of arccos x is $[-1, 1]$, and the range is $[0, \pi]$.

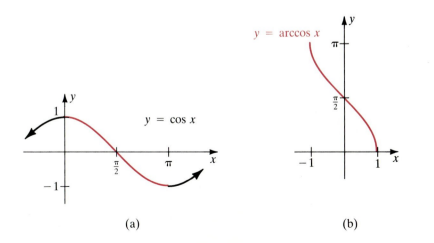

(a)   (b)

FIGURE 42

■ EXAMPLE 3

Find **(a)** arccos $(\sqrt{2}/2)$ and **(b)** $\cos^{-1}(-\sqrt{3}/2)$.

Solution

(a) If we let $y = $ arccos $(\sqrt{2}/2)$, then $\cos y = \sqrt{2}/2$, and $0 \le y \le \pi$. Thus, $y = \pi/4$.

(b) Letting $y = \cos^{-1}(-\sqrt{3}/2)$, we have that $\cos y = -\sqrt{3}/2$, and we must find y such that $0 \le y \le \pi$. Therefore, $y = 5\pi/6$, since $\cos(5\pi/6) = -\sqrt{3}/2$. ■

■ EXAMPLE 4

Write $\sin(\cos^{-1} x)$ as an algebraic expression in x.

Solution In Figure 43 we have constructed an angle of t radians with cosine equal to x. Then $t = \cos^{-1} x$, or $x = \cos t$, where $0 \le t \le \pi$. Now to find $\sin(\cos^{-1} x) = \sin t$, we use the identity $\cos^2 t + \sin^2 t = 1$. Thus

$$x^2 + \sin^2 t = 1$$
$$\sin^2 t = 1 - x^2$$
$$\sin t = \sqrt{1 - x^2}.$$

We use the *positive* square root of $1 - x^2$, since the range of $\cos^{-1} x$ is $[0, \pi]$, and the sine of an angle in the first or second quadrant is positive. ■

FIGURE 43

ARCTANGENT

If we restrict the domain of $\tan x$ to the open interval $(-\pi/2, \pi/2)$, then the resulting function is one-to-one and thus has an inverse. This inverse is denoted by

$$\tan^{-1} x, \quad \text{or} \quad \arctan x.$$

DEFINITION 4

> The **arctangent,** or **inverse tangent,** function is defined by
>
> $$y = \arctan x \quad \text{if and only if} \quad x = \tan y,$$
>
> where $-\pi/2 < y < \pi/2$ and $-\infty < x < \infty$.

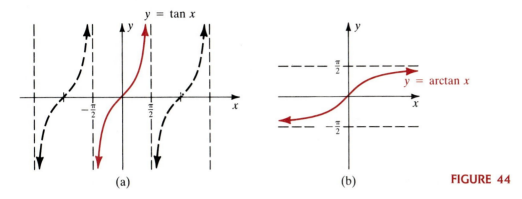

$y = \tan x$

(a)

$y = \arctan x$

(b)

FIGURE 44

The graphs of $y = \tan x$ and $y = \arctan x$ are shown in Figure 44. The domain of arctan x is $(-\infty, \infty)$, and the range is $(-\pi/2, \pi/2)$.

■ EXAMPLE 5

Find $\tan^{-1}(-1)$.

Solution If $\tan^{-1}(-1) = y$, then $\tan y = -1$, where $-\pi/2 < y < \pi/2$. It follows that $\tan^{-1}(-1) = y = -\pi/4$. ■

■ EXAMPLE 6

Without using a calculator, find $\sin(\tan^{-1}(-\tfrac{5}{3}))$.

Solution If we let $t = \tan^{-1}(-\tfrac{5}{3})$, then $\tan t = -\tfrac{5}{3}$. We use the identity $1 + \tan^2 t = \sec^2 t$ to find $\sec t$:

$$1 + \left(-\frac{5}{3}\right)^2 = \sec^2 t$$

$$\sec t = \sqrt{\frac{25}{9} + 1} = \sqrt{\frac{34}{9}} = \frac{\sqrt{34}}{3}.$$

We take the positive square root, since the range of $\tan^{-1} t$ is $(-\pi/2, \pi/2)$ and the secant of an angle in the first or fourth quadrant is positive. Now we can find $\cos t$ from the reciprocal identity:

$$\cos t = \frac{1}{\sec t} = \frac{1}{\dfrac{\sqrt{34}}{3}} = \frac{3}{\sqrt{34}}.$$

Finally we can use the identity $\tan t = \sin t / \cos t$ to compute $\sin (\tan^{-1} (-\frac{5}{3}))$. We find that

$$\sin t = \tan t \cos t = -\frac{5}{3} \left(\frac{3}{\sqrt{34}} \right) = -\frac{5\sqrt{34}}{34}.$$

■

INVERSE PROPERTIES

Recall from Definition 6 in Chapter 1 that $f^{-1}(f(x)) = x$ and $f(f^{-1}(x)) = x$ hold for any function f and its inverse under suitable restrictions on x. Thus for the inverse trigonometric functions, we have the following properties.

Inverse Properties

(i) $\sin^{-1} (\sin x) = \arcsin (\sin x) = x$ if $-\dfrac{\pi}{2} \le x \le \dfrac{\pi}{2}$

(ii) $\sin (\sin^{-1} x) = \sin (\arcsin x) = x$ if $-1 \le x \le 1$

(iii) $\cos^{-1} (\cos x) = \arccos (\cos x) = x$ if $0 \le x \le \pi$

(iv) $\cos (\cos^{-1} x) = \cos (\arccos x) = x$ if $-1 \le x \le 1$

(v) $\tan^{-1} (\tan x) = \arctan (\tan x) = x$ if $-\dfrac{\pi}{2} < x < \dfrac{\pi}{2}$

(vi) $\tan (\tan^{-1} x) = \tan (\arctan x) = x$ if $-\infty < x < \infty$

■ **EXAMPLE 7**

Without using a calculator, evaluate each of the following.

(a) $\sin^{-1} \left(\sin \dfrac{\pi}{12} \right)$ **(b)** $\cos \left(\cos^{-1} \dfrac{1}{3} \right)$ **(c)** $\tan^{-1} \left(\tan \dfrac{3\pi}{4} \right)$

Solution

(a) By (i), $\sin^{-1} \left(\sin \dfrac{\pi}{12} \right) = \dfrac{\pi}{12}.$

(b) By (iv), $\cos (\cos^{-1} \frac{1}{3}) = \frac{1}{3}$.

(c) We *cannot* use the inverse property (v), since $3\pi/4$ is *not* in $(-\pi/2, \pi/2)$. If we first evaluate $\tan (3\pi/4) = -1$, then we have

$$\tan^{-1} \left(\tan \dfrac{3\pi}{4} \right) = \tan^{-1} (-1) = -\dfrac{\pi}{4}.$$

■

USE OF A CALCULATOR

A calculator can be used to obtain approximations to the values of the inverse trigonometric functions. In this context the calculator should be set in radian mode, since the values of the inverse trigonometric functions were defined as real numbers (or radian-measured angles), not as degree-measured angles. (If you set your calculator in degree mode to compute, say, $\tan^{-1} x$, you will obtain the angle in degrees, between $-90°$ and $90°$, whose tangent is x). If a calculator is not available, tables can be "read backward" to provide approximations. Details are given in Appendix C.

■ EXAMPLE 8

Evaluate $\sin (\tan^{-1} 3.75)$.

Solution With a calculator set in radian mode, we enter 3.75, press $\boxed{\text{I N V}}$ and then $\boxed{\text{t a n}}$ (or press $\boxed{\text{a r c t a n}}$ or $\boxed{\text{t a n}^{-1}}$ if your calculator has one of these keys). At this point the display should read 1.3101939. We complete the solution by pressing $\boxed{\text{s i n}}$ to obtain

$$\sin (\tan^{-1} 3.75) \approx 0.9662. \qquad ■$$

■ EXAMPLE 9

Graph each of the following.

(a) $y = \sin (\sin^{-1} x)$ **(b)** $y = \sin^{-1} (\sin x)$

Solution

(a) From the inverse property (ii),

$$\sin (\sin^{-1} x) = x, \quad \text{if } -1 \le x \le 1.$$

However, for $|x| \ge 1$, $\sin^{-1} x$ is not defined, hence, $\sin (\sin^{-1} x)$ is not defined. Therefore, the graph of $y = \sin (\sin^{-1} x)$ is the line segment $y = x$ for $-1 \le x \le 1$. See Figure 45(a).

(b) From the inverse property (i),

$$\sin^{-1} (\sin x) = x, \quad \text{if } -\frac{\pi}{2} \le x \le \frac{\pi}{2}.$$

Thus we obtain the line segment shown in color in Figure 45(b). Next we observe that $\sin^{-1} (\sin x)$ is defined for any real number since $-1 \le \sin x \le 1$. The accompanying table gives values of $y = \sin^{-1} (\sin x)$ for certain values of x in the interval $[\pi/2, 3\pi/2]$. Using these data, we obtain the line segment shown in black in Figure 45(b). Since the sine function is 2π-periodic, it follows that

$$\sin^{-1}(\sin x) = \sin^{-1}(\sin(x + 2\pi)).$$

Hence we complete the graph (the dashed lines shown in Figure 45(b) by using periodicity.

Observe that $-\pi/2 \le \sin^{-1}(\sin x) \le \pi/2$ for every real number x.

x	$\dfrac{\pi}{2}$	$\dfrac{2\pi}{3}$	$\dfrac{3\pi}{4}$	$\dfrac{5\pi}{6}$	π	$\dfrac{7\pi}{6}$	$\dfrac{5\pi}{4}$	$\dfrac{4\pi}{3}$	$\dfrac{3\pi}{2}$
$\sin x$	1	$\dfrac{\sqrt{3}}{2}$	$\dfrac{\sqrt{2}}{2}$	$\dfrac{1}{2}$	0	$-\dfrac{1}{2}$	$-\dfrac{\sqrt{2}}{2}$	$-\dfrac{\sqrt{3}}{2}$	-1
$\sin^{-1}(\sin x)$	$\dfrac{\pi}{2}$	$\dfrac{\pi}{3}$	$\dfrac{\pi}{4}$	$\dfrac{\pi}{6}$	0	$-\dfrac{\pi}{6}$	$-\dfrac{\pi}{4}$	$-\dfrac{\pi}{3}$	$-\dfrac{\pi}{2}$

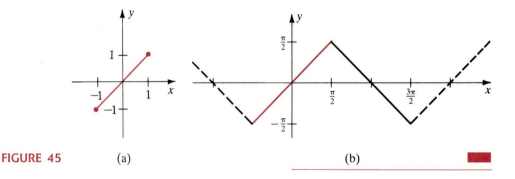

FIGURE 45   (a)   (b)

ARCCSC, ARCSEC, AND ARCCOT

The functions $\cot x$, $\sec x$, and $\csc x$ also have inverses when restricted to the domains $(0, \pi)$, $[0, \pi/2) \cup (\pi/2, \pi]$, and $[-\pi/2, 0) \cup (0, \pi/2]$, respectively. (See Problems 49–52.) Domain restrictions other than those discussed here are sometimes used in defining the inverse cotangent, secant, and cosecant functions. (See Problem 52.)

These functions are not used as often as arctan, arccos, and arcsin, so most scientific calculators do not have keys for them. However, any calculator that computes arcsin, arccos, and arctan can be used to obtain values for **arccsc, arcsec,** and **arccot.** Unlike the fact that $\sec x = 1/\cos x$, $\sec^{-1} x \ne 1/\cos^{-1} x$; rather, $\sec^{-1} x = \cos^{-1}(1/x)$ for $|x| \ge 1$. Similar relationships hold for $\csc^{-1} x$ and $\cot^{-1} x$ (see Problems 81–83).

EXERCISE 3.4

In Problems 1–14, find the indicated value without using a calculator.

1. $\sin^{-1} 0$

2. $\tan^{-1} \sqrt{3}$

3. $\arccos(-1)$

4. $\arcsin \dfrac{\sqrt{3}}{2}$

5. $\arccos \dfrac{1}{2}$

6. $\arctan(-\sqrt{3})$

7. $\sin^{-1}\left(-\dfrac{\sqrt{3}}{2}\right)$

8. $\cos^{-1} \dfrac{\sqrt{3}}{2}$

9. $\tan^{-1} 1$

10. $\sin^{-1} \dfrac{\sqrt{2}}{2}$

11. $\arctan\left(-\dfrac{\sqrt{3}}{3}\right)$

12. $\arccos\left(-\dfrac{1}{2}\right)$

13. $\sin^{-1}\left(-\dfrac{\sqrt{2}}{2}\right)$

14. $\arctan 0$

In Problems 15–32, find the indicated value without using a calculator.

15. $\sin\left(\cos^{-1} \dfrac{3}{5}\right)$

16. $\cos\left(\sin^{-1} \dfrac{1}{3}\right)$

17. $\tan\left(\arccos\left(-\dfrac{2}{3}\right)\right)$

18. $\sin\left(\arctan \dfrac{1}{4}\right)$

19. $\cos(\arctan(-2))$

20. $\tan\left(\sin^{-1}\left(-\dfrac{1}{6}\right)\right)$

21. $\csc\left(\sin^{-1} \dfrac{3}{5}\right)$

22. $\sec(\tan^{-1} 4)$

23. $\sin\left(\sin^{-1} \dfrac{1}{5}\right)$

24. $\cos\left(\cos^{-1}\left(-\dfrac{4}{5}\right)\right)$

25. $\tan(\tan^{-1} 1.2)$

26. $\sin(\arcsin 0.75)$

27. $\arcsin\left(\sin \dfrac{\pi}{16}\right)$

28. $\arccos\left(\cos \dfrac{2\pi}{3}\right)$

29. $\tan^{-1}(\tan \pi)$

30. $\sin^{-1}\left(\sin \dfrac{5}{6}\pi\right)$

31. $\cos^{-1}\left(\cos\left(-\dfrac{\pi}{4}\right)\right)$

32. $\arctan\left(\tan \dfrac{\pi}{7}\right)$

In Problems 33–40, write the given expression as an algebraic expression in x.

33. $\sin(\tan^{-1} x)$

34. $\cos(\tan^{-1} x)$

35. $\tan(\arcsin x)$

36. $\sec(\arccos x)$

37. $\cot(\sin^{-1} x)$

38. $\cos(\sin^{-1} x)$

39. $\csc(\arctan x)$

40. $\tan(\arccos x)$

In Problems 41–48, use a calculator set in radian mode to obtain the value of the expression.

41. $\sin^{-1} 0.7033$

42. $\cos^{-1} 0.2675$

43. $\tan^{-1} 5.798$

44. $\sin(\cos^{-1} 0.7317)$

45. $\tan(\arcsin 0.1296)$

46. $\cos(\arctan 1.369)$

47. $\sin(\tan^{-1} 2.066)$

48. $\cos(\sin^{-1} 0.5227)$

49. The **inverse cotangent** function can be defined by $y = \text{arccot } x$ (or $y = \cot^{-1} x$) if and only if $x = \cot y$, where $0 < y < \pi$. Graph $y = \text{arccot } x$, and give the domain and the range of this function.

50. The **inverse cosecant** function can be defined by $y = \text{arccsc } x$ (or $y = \csc^{-1} x$) if and only if $x = \csc y$, where $-\pi/2 \leq y \leq \pi/2$ and $y \neq 0$. Graph $y = \text{arccsc } x$, and give the domain and the range of this function.

51. One definition of the **inverse secant** function is $y = \text{arcsec } x$ (or $y = \sec^{-1} x$) if and only if $x = \sec y$, where $0 \leq y \leq \pi$ and $y \neq \pi/2$. (See Problem 52 for an alternative definition.) Graph $y = \text{arcsec } x$, and give the domain and the range of arcsec x for this definition.

52. An alternative definition of the inverse secant function can be made by restricting the domain of the secant function to $[0, \pi/2) \cup [\pi, 3\pi/2)$. Under this restriction, define the arcsecant function, state its domain and its range, and graph $y = \text{arcsec } x$.

In Problems 53–64, find the indicated values without using a calculator.

53. $\sec^{-1} 2$

54. $\cot^{-1}(-\sqrt{3})$

55. $\csc^{-1} \sqrt{2}$

56. $\text{arccsc}^{-1}(-1)$

57. $\text{arccot}\left(-\dfrac{\sqrt{3}}{3}\right)$

58. $\text{arcsec}(-\sqrt{2})$

59. $\sin(\sec^{-1} 2)$

60. $\cos(\cot^{-1} \tfrac{1}{2})$

61. $\cot(\cot^{-1}(-3))$

62. $\sec^{-1}(\sec 1.6)$

63. $\text{arccsc}\left(\csc \dfrac{3\pi}{2}\right)$

64. $\cot(\text{arccot}(-5))$

In Problems 65 and 66, write the given expression as an algebraic expression in x.

65. $\cos(\sec^{-1} x)$

66. $\tan(\cot^{-1} x)$

67. Using a calculator set in radian mode, evaluate $\arctan(\tan 1.8)$, $\arccos(\cos 1.8)$, and $\arcsin(\sin 1.8)$. Explain the results.

68. Using a calculator set in radian mode, evaluate $\tan^{-1}(\tan(-1))$, $\cos^{-1}(\cos(-1))$, and $\sin^{-1}(\sin(-1))$. Explain the results.

In Problems 69–78, sketch the graph of the given function.

69. $y = \arctan|x|$ **70.** $y = \dfrac{\pi}{2} - \arctan x$

71. $y = |\arcsin x|$ **72.** $y = \sin^{-1}(x + 1)$
73. $y = 2\cos^{-1}x$ **74.** $y = \cos^{-1}2x$
75. $y = \cos(\cos^{-1}x)$ **76.** $y = \cos^{-1}(\cos x)$
77. $y = \arccos(x - 1)$ **78.** $y = \cos(\arcsin x)$
79. For what values of x is it true that **(a)** $\cot(\text{arccot}\, x) = x$ and **(b)** $\text{arccot}(\cot x) = x$?
80. For what values of x is it true that **(a)** $\csc(\text{arccsc}\, x) = x$ and **(b)** $\text{arccsc}(\csc x) = x$?
81. Verify that

$$\text{arccot}\, x = \frac{\pi}{2} - \arctan x.$$

82. Verify that

$$\text{arcsec}\, x = \arccos\left(\frac{1}{x}\right) \quad \text{for } |x| \ge 1.$$

83. Verify that

$$\text{arccsc}\, x = \arcsin\left(\frac{1}{x}\right) \quad \text{for } |x| \ge 1.$$

In Problems 84–89, use the results of Problems 81–83 and a calculator to find the indicated value.

84. $\sec^{-1}2.5$ **85.** $\cot^{-1}0.75$
86. $\csc^{-1}(-1.3)$ **87.** $\text{arccsc}(-1.5)$
88. $\text{arccot}(-0.3)$ **89.** $\text{arcsec}(-1.2)$
90. Under certain conditions the current i in an electrical circuit at time t is given by

$$i = I\sin(\omega t + \theta + \phi).$$

Solve for t.
91. The departure angle θ for a bullet to hit a target at a distance R (assuming that the target and the gun are at the same height) satisfies

$$R = \frac{v_0^2\sin 2\theta}{g},$$

where v_0 is the muzzle velocity and g is acceleration due to gravity. If the target is 800 ft from the gun and the muzzle velocity is 200 ft/sec, find the departure angle. Use $g = 32$ ft/sec^2. [*Hint:* There are two solutions.]

92. For the olympic event the hammer throw, it can be shown that the maximum distance is achieved for the release angle θ (measured from the horizontal) that satisfies

$$\cos 2\theta = \frac{gh}{v^2 + gh},$$

where h is the height of the hammer above the ground at release, v is the initial velocity, and g is acceleration due to gravity. For $v = 13.7$ m/sec and $h = 2.25$ m, find the optimal release angle. Use $g = 9.81$ m/sec^2.
93. In the design of highways and railroads, curves are banked to provide centripetal force for safety. The optimal banking angle θ (see Figure 46) is given by

$$\tan \theta = \frac{v^2}{Rg},$$

where v is the speed of the vehicle, R is the radius of the curve, and g is acceleration due to gravity. As the formula indicates, for a given radius there is no one correct angle for all speeds. Consequently, curves are banked for the average speed of the traffic over them. Find the correct banking angle for a curve of radius 600 ft on a country road where speeds average 30 mph. Use $g = 32$ ft/sec^2. [*Hint:* Use consistent units.]

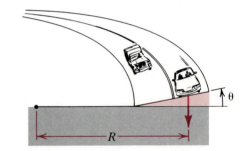

FIGURE 46

94. If μ is the coefficient of friction between the car and the road, then the maximum velocity v_m that a car can travel around a curve without slipping is given by

$$v_m^2 = gR\tan(\theta + \tan^{-1}\mu),$$

where θ is the banking angle of the curve. Find v_m for the country road in Problem 93 if $\mu = 0.26$.
95. Viewed from the side, a volcanic cinder cone usually looks like a symmetric trapezoid (see Figure 47). Studies of cinder cones less than 50,000 years old indicate that cone height

H_{co} and crater width W_{cr} are related to the cone width W_{co} by the equations

$$H_{co} = 0.18W_{co}$$

and

$$W_{cr} = 0.40W_{co}.$$

If $W_{co} = 1.00$, use these equations to determine the base angle ϕ of the trapezoid in Figure 47.

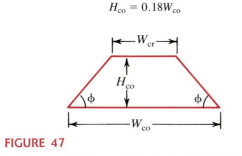

FIGURE 47

IMPORTANT CONCEPTS

Circular functions	cycle	Inverse trigonometric functions
Unit circle	amplitude	arcsine
Periodic functions	phase shift	arccosine
Graphs of trigonometric functions	Harmonic motion	arctangent
period		

CHAPTER 3 REVIEW EXERCISE

In Problems 1–10, answer true or false.

1. The line $x = \pi/2$ is an asymptote for the graph of $y = \tan x$.____

2. The graph of $y = \csc t$ does not intersect the y-axis.____

3. The line $y = 1$ is an asymptote for the graph of $y = \sin t$.____

4. $\arcsin x = 1/\sin x$____

5. The graphs of $f(t) = 3 \sin (-2t)$ and $g(t) = -3 \cos (2t - \pi/2)$ are identical.____

6. The unit circle is the circle with equation $x^2 + y^2 = 1$.____

7. For any real number x, $\tan (\tan^{-1} x) = x$.____

8. There exists a real number t such that $\sin t = 3$.____

9. The tangent function is an odd function.____

10. An object is said to exhibit simple harmonic motion if its displacement from an equilibrium point at time t can be described by $y = a \sin (bt + c)$, where a, b, and c are real numbers.____

In Problems 11–26, graph the given function. If appropriate, determine the amplitude, the period, and the phase shift.

11. $y = 5(1 + \sin t)$

12. $y = -\dfrac{4}{3} \cos t$

13. $y = \tan \left(t + \dfrac{\pi}{2}\right)$

14. $y = \pi - \cot t$

15. $y = -\sec t$

16. $y = \csc (t - \pi)$

17. $y = t + \cos t$

18. $y = \sin t - \cos t$

19. $y = -2 \cos \dfrac{1}{4} t$

20. $y = \dfrac{1}{4} \cos \pi t$

21. $y = -\sin (3t + \pi)$

22. $y = 4 \cos \left(-2t - \dfrac{\pi}{3}\right)$

23. $y = 2 \sec 4t$

24. $y = 1 - \tan \pi t$

25. $y = 10 \cos \left(-3t + \dfrac{\pi}{2}\right)$

26. $y = -4 \sin \left(\dfrac{1}{4} t - \pi\right)$

4.1

Trigonometric Identities

Recall from Section 1.1 that an equation such as

$$2(x - 1) = 2x - 2,$$

which is valid for all real numbers x, is called an **identity.** An equation such as

$$\frac{x^2 - 4x}{x} = x - 4$$

is also called an identity, since it is valid for all real numbers for which both sides of the equation are defined, in this case, all $x \neq 0$. The trigonometric equation

$$\frac{\sin t}{\tan t} = \cos t$$

is an identity, because

$$\frac{\sin t}{\tan t} = \frac{\sin t}{\dfrac{\sin t}{\cos t}} = \sin t \left(\frac{\cos t}{\sin t} \right) = \cos t$$

for all real numbers t for which $\tan t$ is defined and $\tan t \neq 0$.

There are numerous identities involving the trigonometric functions. The most important of these are the fundamental identities, which were introduced in Sections 2.2 and 2.4 and which are restated here. In this summary we also include the even–odd identities, which were discussed in Sections 3.1 and 3.2. The variable t in each identity can represent either a real number or the degree or radian measure of an angle.

FUNDAMENTAL IDENTITIES		
PYTHAGOREAN	QUOTIENT	RECIPROCAL
$\cos^2 t + \sin^2 t = 1$	$\tan t = \dfrac{\sin t}{\cos t}$	$\sec t = \dfrac{1}{\cos t}$
$1 + \tan^2 t = \sec^2 t$	$\cot t = \dfrac{\cos t}{\sin t}$	$\csc t = \dfrac{1}{\sin t}$
$\cot^2 t + 1 = \csc^2 t$		$\cot t = \dfrac{1}{\tan t}$

EVEN–ODD IDENTITIES

$$\sin(-t) = -\sin t, \quad \cos(-t) = \cos t, \quad \tan(-t) = -\tan t,$$
$$\csc(-t) = -\csc t, \quad \sec(-t) = \sec t, \quad \cot(-t) = -\cot t$$

These identities can be used to simplify complicated trigonometric expressions.

EXAMPLE 1

Write as a single trigonometric function:

$$\sin t \sec t.$$

Solution Using the reciprocal identity $\sec t = 1/\cos t$, we find

$$\sin t \sec t = \sin t \frac{1}{\cos t} = \frac{\sin t}{\cos t} = \tan t.$$

We will often encounter variations of the Pythagorean identities. For example, from $\cos^2 t + \sin^2 t = 1$, we obtain

$$\sin^2 t = 1 - \cos^2 t \quad \text{and} \quad \cos^2 t = 1 - \sin^2 t.$$

Alternative ways of writing the other Pythagorean identities are

$$\tan^2 t = \sec^2 t - 1 \quad \text{and} \quad \cot^2 t = \csc^2 t - 1.$$

EXAMPLE 2

Simplify $(1 + \sin t)(1 + \sin(-t))$.

Solution We have

$$(1 + \sin t)(1 + \sin(-t)) = (1 + \sin t)(1 - \sin t) \quad \Leftarrow \boxed{\textbf{By } \sin(-t) = -\sin t}$$

$$= 1 - \sin^2 t \quad \Leftarrow \boxed{\textbf{By algebra}}$$

$$= \cos^2 t. \quad \Leftarrow \boxed{\textbf{From } \cos^2 t + \sin^2 t = 1}$$

■ EXAMPLE 3

Simplify $\csc^2 t - \cot^2 t$.

Solution From $\cot^2 t + 1 = \csc^2 t$, we see that

$$\csc^2 t - \cot^2 t = 1.$$

We can use these fundamental identities and algebraic manipulation to verify an identity by showing that the given expressions are equivalent. The preferred method of verifying an identity is to show that one side of the equation is equivalent to the other, as in the following two examples.

■ EXAMPLE 4

Verify the identity

$$\sec^2 t + \csc^2 t = \sec^2 t \csc^2 t.$$

Solution We show that the left-hand side of the equation is equivalent to the right-hand side:

$$\sec^2 t + \csc^2 t = \frac{1}{\cos^2 t} + \frac{1}{\sin^2 t} \qquad \Longleftarrow \textbf{Fundamental identities}$$

$$= \frac{1}{\cos^2 t}\left(\frac{\sin^2 t}{\sin^2 t}\right) + \frac{1}{\sin^2 t}\left(\frac{\cos^2 t}{\cos^2 t}\right) \qquad \Longleftarrow \textbf{Least common denominator}$$

$$= \frac{\sin^2 t + \cos^2 t}{\cos^2 t \sin^2 t} \qquad \Longleftarrow \textbf{Adding}$$

$$= \frac{1}{\cos^2 t \sin^2 t} \qquad \Longleftarrow \textbf{Fundamental identity}$$

$$= \left(\frac{1}{\cos t}\right)^2\left(\frac{1}{\sin t}\right)^2 \qquad \Longleftarrow \textbf{By algebra}$$

$$= \sec^2 t \csc^2 t. \qquad \Longleftarrow \textbf{Fundamental identities}$$

Implicit in Example 4 is the assumption that the identity is valid only for those values of t for which both sides of the identity are defined. In Example 4, for t a real number, we must require $t \neq n\pi$ and $t \neq \pi/2 + n\pi$, where n is an integer. In the following examples we will not mention the restrictions on the variable.

EXAMPLE 5

Verify the identity

$$\sin \theta \cos \theta = \frac{1}{\tan \theta + \cot \theta}.$$

Solution We show that the right-hand side of the equation is equivalent to the left-hand side. (You should supply the reason for each step.)

$$\frac{1}{\tan \theta + \cot \theta} = \frac{1}{\dfrac{\sin \theta}{\cos \theta} + \dfrac{\cos \theta}{\sin \theta}}$$

$$= \frac{1}{\dfrac{\sin^2 \theta + \cos^2 \theta}{\sin \theta \cos \theta}}$$

$$= \frac{\sin \theta \cos \theta}{\sin^2 \theta + \cos^2 \theta}$$

$$= \sin \theta \cos \theta$$

There is no general method for showing that a trigonometric equation is an identity. We list below a few techniques that may be useful.

Suggestions for Verifying Identities

(i) Simplify the more complicated side of the equation first.

(ii) Find least common denominators for sums or differences of fractions.

(iii) If the two preceding techniques fail, express all trigonometric functions in terms of sines and cosines and then try to simplify.

EXAMPLE 6

Verify the identity

$$\sin x + \sin x \cot^2 x = \cos x \csc x \sec x.$$

Solution We begin by simplifying the left-hand side:

$$\sin x + \sin x \cot^2 x = \sin x \, (1 + \cot^2 x)$$

$$= \sin x (\csc^2 x)$$

$$= \frac{1}{\csc x} \csc^2 x$$

$$= \csc x.$$

Since we have arrived at such a simple expression, we try to reduce the right-hand side to the same quantity:

$$\cos x \csc x \sec x = \cos x \csc x \frac{1}{\cos x}$$

$$= \csc x.$$

Since both sides of the given equation are equivalent to csc x, they are equivalent to each other. Therefore, the equation is an identity.

As Example 6 illustrates, another technique for verifying an identity is to reduce each side of the equation separately to the same expression.

Note of Caution: In order to verify a trigonometric identity, we are required to show that the given expressions are equivalent. Note that in the preceding three examples we worked *independently* with the expressions on each side of the equation to show that they are equivalent. This is standard practice in verifying trigonometric identities. The same algebraic operation should *not* be performed on both sides of the equation *simultaneously*. In other words, do not treat a trigonometric equation as an identity until after you have proven that it is really true.

To show that an equation is not an identity, we need only find one value in the domain of the variable for which the equation is not true. As the following example indicates, this is often a trial-and-error process.

EXAMPLE 7
Show that

$$(\sin t + \cos t)^2 = 1$$

is *not* an identity.

Solution For $t = 0$, both sides of the equation are defined and we obtain $(0 + 1)^2 = 1$, which is true. This neither proves nor disproves that the equation is an identity. Since we are expected to show that it is *not* an identity, we try another value for t. If we let $t = \pi/4$, we find that the left-hand side is

$$\left(\sin \frac{\pi}{4} + \cos \frac{\pi}{4}\right)^2 = \left(\frac{\sqrt{2}}{2} + \frac{\sqrt{2}}{2}\right)^2 = (\sqrt{2})^2 = 2,$$

which does not equal the right-hand side, 1. Consequently, the equation is not an identity, since we have shown that it is not true for at least one value of t, namely, $t = \pi/4$.

Although most of the identities considered in this section are not themselves particularly important, what is important is the facility you gain in simplifying and manipulating trigonometric expressions. This ability is essential for more advanced work in mathematics, science, and engineering.

EXERCISE 4.1

In Problems 1–10, use the fundamental identities and the even-odd identities to simplify each expression.

1. $\sec t \cos t$

2. $\tan \alpha \cos \alpha$

3. $\dfrac{\sin \theta}{\csc \theta} + \dfrac{\cos \theta}{\sec \theta}$

4. $\dfrac{\csc^2 x - 1}{\cot x}$

5. $\tan^2 t - \sec^2 t$

6. $1 + \tan^2 (-\theta)$

7. $\sin (-t) + \sin t$

8. $\cos^2 t + \dfrac{1}{\csc^2 t}$

9. $\sec (-x) \cos x$

10. $1 + \dfrac{\cot \beta}{\tan \beta}$

In Problems 11–20, reduce the given expression to a single trigonometric function.

11. $\dfrac{\sin t + \sin t \cos t}{1 + \cos t}$

12. $\cos x + \cos x \tan^2 x$

13. $\dfrac{\sec^2 \alpha - 1}{\tan \alpha}$

14. $\dfrac{\tan t + \cot t}{\csc t}$

15. $\sin x + \cos x \cot x$

16. $\sin \theta \tan \theta \csc^2 \theta - \sin \theta \tan \theta$

17. $\dfrac{\sec^2 \alpha}{\cos \alpha + \cos \alpha \tan^2 \alpha}$

18. $\dfrac{\sin^2 \theta \cos \theta + \cos^3 \theta - \cos \theta + \sin \theta}{\cos \theta}$

19. $\sin t \cos t \tan t \sec t \cot t$ **20.** $\dfrac{\sin \alpha \tan \alpha}{\csc \alpha} + \dfrac{\sin \alpha}{\sec \alpha}$

In Problems 21–60, verify the identity.

21. $\dfrac{\sin t}{\csc t} = 1 - \dfrac{\cos t}{\sec t}$

22. $\dfrac{1 + \sin x}{\cos x} = \sec x + \tan x$

23. $1 - \cos^4 \theta = (2 - \sin^2 \theta) \sin^2 \theta$

24. $\dfrac{1 + \tan t}{\tan t} = \cot t + \sec^2 t - \tan^2 t$

25. $1 - 2 \sin^2 t = 2 \cos^2 t - 1$

26. $\tan^2 \beta - \sin^2 \beta = \tan^2 \beta \sin^2 \beta$

27. $\dfrac{\sec z - \csc z}{\sec z + \csc z} = \dfrac{\tan z - 1}{\tan z + 1}$

28. $\dfrac{\sin t + \tan t}{1 + \cos t} = \tan t$

29. $\dfrac{\sec^4 t - \tan^4 t}{1 + 2 \tan^2 t} = 1$

30. $\dfrac{1 + \sin t}{\cos t} + \dfrac{\cos t}{1 + \sin t} = 2 \sec t$

31. $\sin^2 x \cot^2 x + \cos^2 x \tan^2 x = 1$

32. $\dfrac{\sin \alpha + \tan \alpha}{\cot \alpha + \csc \alpha} = \sin^2 \alpha \sec \alpha$

33. $\sec t - \dfrac{\cos t}{1 + \sin t} = \tan t$

34. $\dfrac{1}{\sec t - \tan t} = \sec t + \tan t$

35. $\dfrac{\tan^2 \beta}{1 + \cos \beta} = \dfrac{\sec \beta - 1}{\cos \beta}$

36. $\dfrac{\tan^2 t - 1}{\sin t + \cos t} = \dfrac{\sin t - \cos t}{\cos^2 t}$

37. $(\csc t - \cot t)^2 = \dfrac{1 - \cos t}{1 + \cos t}$

38. $\cos \theta - \sin \theta + \csc \theta = \dfrac{\sin \theta + \cos \theta}{\tan \theta}$

39. $1 + \dfrac{1}{\cos x} = \dfrac{\tan^2 x}{\sec x - 1}$

40. $\dfrac{\tan t + \cot t}{\cos^2 t} - \sin t \sec^3 t = \sec t \csc t$

41. $\dfrac{\cot t - \tan t}{\cot t + \tan t} = 1 - 2\sin^2 t$

42. $\dfrac{1 + \sec t}{\sin t + \tan t} = \csc t$

43. $\cos(-t)\csc(-t) = -\cot t$

44. $\dfrac{\tan(-t)}{\sin(-t)} = \sec t$

45. $\sqrt{\dfrac{1 + \sin\theta}{1 - \sin\theta}} = \dfrac{1 + \sin\theta}{|\cos\theta|}$

46. $\sqrt{\dfrac{1 + \cos\alpha}{1 - \cos\alpha}} = \dfrac{|\sin\alpha|}{1 - \cos\alpha}$

47. $\sqrt{\dfrac{\sec t + \tan t}{\sec t - \tan t}} = \left|\dfrac{1 + \sin t}{\cot t}\right|$

48. $\sqrt{\dfrac{\csc t - \cot t}{\csc t + \cot t}} = \left|\dfrac{1}{\csc t + \cot t}\right|$

49. $\sin^4 x + \cos^4 x = 1 - 2\sin^2 x \cos^2 x$

50. $\sin^4 x - \cos^4 x = 1 - 2\cos^2 x$

51. $\left(\dfrac{\sin^2\theta}{\cot^4\theta}\right)^4 \cdot \left(\dfrac{\csc\theta}{\tan^2\theta}\right)^8 = 1$

52. $\dfrac{\cos^3 x + \sin^3 x}{\cos x + \sin x} = 1 - \cos x \sin x$

53. $(\tan^2 t + 1)(\cos^2 t - 1) = 1 - \sec^2 t$

54. $\dfrac{1}{1 - \cos\alpha} + \dfrac{1}{1 + \cos\alpha} = 2\csc^2\alpha$

55. $(1 - \tan\beta)^2(1 + \tan\beta)^2 + 4\tan^2\beta = \sec^4\beta$

56. $\dfrac{\cos(-t)}{1 + \tan(-t)} - \dfrac{\sin(-t)}{1 + \cot(-t)} = \sin t + \cos t$

57. $\dfrac{\sin\theta}{1 - \cot\theta} + \dfrac{\cos\theta}{1 - \tan\theta} = \cos\theta + \sin\theta$

58. $\sin^6 t + \cos^6 t = 1 - 3\sin^2 t \cos^2 t$

59. $\csc^4 t - \csc^2 t = \cot^4 t + \cot^2 t$

60. $\dfrac{\tan x - \cot x}{\sin x \cos x} = \sec^2 x - \csc^2 x$

In Problems 61–70, show that the given trigonometric equation is *not* an identity.

61. $\sin t = \sqrt{1 - \cos^2 t}$ **62.** $\sqrt{\cos^2 t} = \cos t$

63. $1 + \sec^2\theta = \tan^2\theta$ **64.** $\sin x = 1 - \cos x$

65. $\cot^2 t + \csc^2 t = 1$ **66.** $1 + \sin^2\alpha = \cos^2\alpha$

67. $\sin(\csc\theta) = 1$ **68.** $\tan(\cot\alpha) = \alpha$

69. $\cos(-x) = -\cos x$ **70.** $\sin^2\beta = \sin(\beta^2)$

4.2

Addition and Subtraction Formulas

The formulas that we will derive in this section enable us to express certain types of trigonometric expressions in simpler or more useful forms. These formulas are very important in science and engineering. Although the derivations make use of angles, the applications to other fields generally involve trigonometric functions of real numbers.

The **addition** and **subtraction formulas** for the cosine and sine functions reduce $\cos(u + v)$, $\cos(u - v)$, $\sin(u + v)$, and $\sin(u - v)$ to expressions involving $\cos u$, $\cos v$, $\sin u$, and $\sin v$. We first derive the formula for $\cos(u - v)$ and then use it to obtain the others.

To derive the formula for $\cos(u - v)$, we let u and v be angles as shown in Figure 1(a). If we place the angle $u - v$ in standard position, as shown in Figure 1(b), then the distance d from R to S equals the distance from P to Q

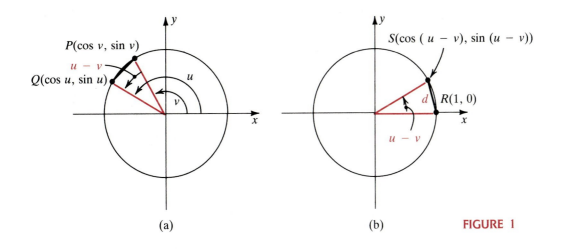

(a) (b) FIGURE 1

shown in Figure 1(a). The squares of these distances are equal, that is,

$$[d(P, Q)]^2 = [d(R, S)]^2.$$

Using the distance formula gives

$$(\cos u - \cos v)^2 + (\sin u - \sin v)^2 = (\cos (u - v) - 1)^2 + \sin^2 (u - v),$$

or

$$\cos^2 u - 2 \cos u \cos v + \cos^2 v + \sin^2 u - 2 \sin u \sin v + \sin^2 v =$$
$$\cos^2 (u - v) - 2 \cos (u - v) + 1 + \sin^2 (u - v).$$

Since $\cos^2 u + \sin^2 u = 1$, $\cos^2 v + \sin^2 v = 1$, and $\cos^2 (u - v) + \sin^2 (u - v) = 1$, the preceding equation simplifies to

$$2 - 2 \cos u \cos v - 2 \sin u \sin v = 2 - 2 \cos (u - v),$$

from which we obtain the following result.

Subtraction Formula for COSINE

$$\cos (u - v) = \cos u \cos v + \sin u \sin v$$

To obtain the addition formula for $\cos (u + v)$, we write

$$\cos (u + v) = \cos (u - (-v))$$
$$= \cos u \cos (-v) + \sin u \sin (-v).$$

Using the even–odd identities

$$\cos(-v) = \cos v \quad \text{and} \quad \sin(-v) = -\sin v,$$

we now have the following.

Addition Formula for COSINE

$$\cos(u + v) = \cos u \cos v - \sin u \sin v$$

In Section 3.1 we discussed the following cofunction formulas for $0 < t < \pi/2$ by considering the coordinates of a point on a unit circle.

Cofunction Formulas

(i) $\cos\left(\dfrac{\pi}{2} - t\right) = \sin t$ **(ii)** $\sin\left(\dfrac{\pi}{2} - t\right) = \cos t$

(iii) $\tan\left(\dfrac{\pi}{2} - t\right) = \cot t$

We now prove these formulas for any real number t (or angle t measured in radians). Recall that if the angle t is measured in degrees, then $\pi/2$ must be replaced by $90°$ in the cofunction formulas.

To verify (i) we apply the subtraction formula for $\cos(u - v)$ to $\cos(\pi/2 - t)$:

$$\cos\left(\frac{\pi}{2} - t\right) = \cos\frac{\pi}{2}\cos t + \sin\frac{\pi}{2}\sin t = 0 \cdot \cos t + 1 \cdot \sin t$$

$$= \sin t.$$

To prove (ii) we replace t by $\pi/2 - t$ in (i) and obtain

$$\cos\left[\frac{\pi}{2} - \left(\frac{\pi}{2} - t\right)\right] = \sin\left(\frac{\pi}{2} - t\right),$$

or

$$\cos t = \sin\left(\frac{\pi}{2} - t\right),$$

as desired.

And finally,

$$\tan\left(\frac{\pi}{2} - t\right) = \frac{\sin\left(\frac{\pi}{2} - t\right)}{\cos\left(\frac{\pi}{2} - t\right)} = \frac{\cos t}{\sin t} = \cot t.$$

Using the cofunction formula (i), we can now derive the addition and subtraction formulas for the sine:

$$\sin(u + v) = \cos\left[\frac{\pi}{2} - (u + v)\right] \longleftarrow \boxed{\text{Cofunction formula}}$$

$$= \cos\left[\left(\frac{\pi}{2} - u\right) - v\right]$$

$$= \cos\left(\frac{\pi}{2} - u\right)\cos v + \sin\left(\frac{\pi}{2} - u\right)\sin v \longleftarrow \boxed{\begin{array}{l}\text{Subtraction formula} \\ \text{for the cosine}\end{array}}$$

$$= \sin u \cos v + \cos u \sin v.$$

This result is summarized as follows.

Addition Formula for SINE

$$\sin(u + v) = \sin u \cos v + \cos u \sin v$$

Writing $u - v$ as $u + (-v)$ and using the addition formula for the sine, we obtain

$$\sin(u - v) = \sin(u + (-v)) = \sin u \cos(-v) + \cos u \sin(-v)$$

$$= \sin u \cos v - \cos u \sin v.$$

This verifies the following.

Subtraction Formula for SINE

$$\sin(u - v) = \sin u \cos v - \cos u \sin v$$

These addition and subtraction formulas can be used to find the *exact* values of the sine and cosine functions of angles or numbers that can be represented as sums or differences of $\pi/3$, $\pi/4$, $\pi/6$, $2\pi/3$, and so on. For example, we can compute the precise values of the sine and cosine for angles such as $\pi/12 = 15°$, $5\pi/12 = 75°$, and $7\pi/12 = 105°$. (Recall that a calculator or trigonometric tables provide only decimal approximations to these values.)

■ EXAMPLE 1

Find the exact value of $\cos(7\pi/12)$.

Solution We have no way of evaluating $\cos(7\pi/12)$ directly. However, we can write $\cos(7\pi/12) = \cos(\pi/3 + \pi/4)$. Then by the addition formula for the cosine, it follows that

$$\cos\left(\frac{\pi}{3} + \frac{\pi}{4}\right) = \cos\frac{\pi}{3}\cos\frac{\pi}{4} - \sin\frac{\pi}{3}\sin\frac{\pi}{4}$$

$$= \frac{1}{2}\frac{\sqrt{2}}{2} - \frac{\sqrt{3}}{2}\frac{\sqrt{2}}{2} = \frac{\sqrt{2}}{4}(1 - \sqrt{3}).$$

We can also write the answer as $(\sqrt{2} - \sqrt{6})/4$. Note that $\cos(7\pi/12) < 0$ as expected. ■

■ EXAMPLE 2

Evaluate $\sin(7\pi/12)$.

Solution We use the addition formula for the sine as follows:

$$\sin\frac{7\pi}{12} = \sin\left(\frac{\pi}{3} + \frac{\pi}{4}\right) = \sin\frac{\pi}{3}\cos\frac{\pi}{4} + \cos\frac{\pi}{3}\sin\frac{\pi}{4}$$

$$= \frac{\sqrt{3}}{2}\frac{\sqrt{2}}{2} + \frac{1}{2}\frac{\sqrt{2}}{2} = \frac{\sqrt{2}}{4}(1 + \sqrt{3}). \tag{1}$$

This result can also be written as $(\sqrt{2} + \sqrt{6})/4$. Alternatively, we can obtain the value of $\sin(7\pi/12)$ from

$$\cos^2\frac{7\pi}{12} + \sin^2\frac{7\pi}{12} = 1.$$

Using the value of $\cos(7\pi/12)$ from the preceding example, we find that

$$\sin\frac{7\pi}{12} = \sqrt{1 - \cos^2\frac{7\pi}{12}} = \sqrt{1 - \left[\frac{\sqrt{2}}{4}(1 - \sqrt{3})\right]^2}$$

$$= \sqrt{1 - \frac{1}{8}(1 - 2\sqrt{3} + 3)} = \sqrt{\frac{4 + 2\sqrt{3}}{8}}.$$

Although this number does not look like the result obtained in (1), the values are the same, since

$$\sqrt{\frac{4 + 2\sqrt{3}}{8}} = \sqrt{\frac{2}{16}(1 + 2\sqrt{3} + 3)} = \sqrt{\frac{2}{16}(1 + \sqrt{3})^2}$$
$$= \frac{\sqrt{2}}{4}(1 + \sqrt{3}).$$

For any function f, the expression

$$\frac{f(x + h) - f(x)}{h}$$

is called a **difference quotient.** The identity in the following example involves a difference quotient where $f(x) = \sin x$.

EXAMPLE 3

Verify that

$$\frac{\sin (x + h) - \sin x}{h} = \cos x \left(\frac{\sin h}{h} \right) + \sin x \left(\frac{\cos h - 1}{h} \right).$$

Solution Applying the addition formula for the sine to $\sin (x + h)$, we have

$$\frac{\sin (x + h) - \sin x}{h} = \frac{\sin x \cos h + \cos x \sin h - \sin x}{h}$$
$$= \frac{\cos x \sin h + \sin x (\cos h - 1)}{h}$$
$$= \cos x \left(\frac{\sin h}{h} \right) + \sin x \left(\frac{\cos h - 1}{h} \right).$$

There are addition and subtraction formulas for the tangent function as well.

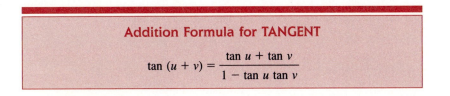

Addition Formula for TANGENT

$$\tan (u + v) = \frac{\tan u + \tan v}{1 - \tan u \tan v}$$

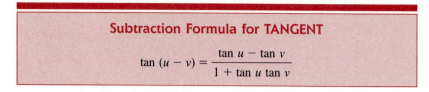

Subtraction Formula for TANGENT

$$\tan (u - v) = \frac{\tan u - \tan v}{1 + \tan u \tan v}$$

We derive the addition formula by using the addition formulas for the sine and cosine, as follows:

$$\tan (u + v) = \frac{\sin (u + v)}{\cos (u + v)} = \frac{\sin u \cos v + \cos u \sin v}{\cos u \cos v - \sin u \sin v}.$$

We can divide the numerator and the denominator by $\cos u \cos v$, provided that $\cos u \cos v \neq 0$. (If $\cos u \cos v = 0$, then either $\cos u = 0$ or $\cos v = 0$. In this case, either $\tan u$ or $\tan v$ is undefined. It follows that the expression on the right-hand side of the addition formula for the tangent is undefined.) Performing the division, we obtain

$$\tan (u + v) = \frac{\left(\dfrac{\sin u}{\cos u}\right)\left(\dfrac{\cos v}{\cos v}\right) + \left(\dfrac{\cos u}{\cos u}\right)\left(\dfrac{\sin v}{\cos v}\right)}{\left(\dfrac{\cos u}{\cos u}\right)\left(\dfrac{\cos v}{\cos v}\right) - \left(\dfrac{\sin u}{\cos u}\right)\left(\dfrac{\sin v}{\cos v}\right)}$$

$$= \frac{\tan u + \tan v}{1 - \tan u \tan v}.$$

Consequently, we have shown that the addition formula for the tangent is valid for all values of u and v for which both sides of the equation are defined.

The derivation of the subtraction formula for the tangent is left as an exercise. (See Problem 56.)

■ EXAMPLE 4

Evaluate $\tan (\pi/12)$.

Solution   In order to use the addition or subtraction formulas to obtain an exact value, we must express $\pi/12$ as a sum or difference of the special angles 0, $\pi/6$, $\pi/4$, $\pi/3$, or $\pi/2$. By trial and error, we find that $\pi/12 = \pi/4 - \pi/6$ and that $\pi/12 = \pi/3 - \pi/4$. Using the first of these expressions and the subtraction formula for the tangent, we obtain

$$\tan \frac{\pi}{12} = \tan \left(\frac{\pi}{4} - \frac{\pi}{6}\right) = \frac{\tan \dfrac{\pi}{4} - \tan \dfrac{\pi}{6}}{1 + \tan \dfrac{\pi}{4} \tan \dfrac{\pi}{6}}$$

$$= \frac{1 - \dfrac{1}{\sqrt{3}}}{1 + 1 \cdot \dfrac{1}{\sqrt{3}}} = \frac{\sqrt{3} - 1}{\sqrt{3} + 1}$$

$$= \frac{\sqrt{3} - 1}{\sqrt{3} + 1} \cdot \frac{\sqrt{3} - 1}{\sqrt{3} - 1} \quad \longleftarrow \boxed{\textbf{Rationalizing the denominator}}$$

$$= \frac{(\sqrt{3} - 1)^2}{2} = \frac{4 - 2\sqrt{3}}{2} = 2 - \sqrt{3}.$$

You should rework this example using $\pi/12 = \pi/3 - \pi/4$ to see that the result is the same.

■ EXAMPLE 5

Find the exact value of $\tan 105°$.

Solution Since $105° = 45° + 60°$, we use the addition formula for the tangent to obtain

$$\tan 105° = \tan (45° + 60°)$$

$$= \frac{\tan 45° + \tan 60°}{1 - \tan 45° \tan 60°}$$

$$= \frac{1 + \sqrt{3}}{1 - (1)(\sqrt{3})} = \frac{(1 + \sqrt{3})^2}{(1 - \sqrt{3})(1 + \sqrt{3})} = \frac{(1 + \sqrt{3})^2}{-2}$$

$$= \frac{1 + 2\sqrt{3} + 3}{-2} = \frac{4 + 2\sqrt{3}}{-2} = -2 - \sqrt{3}. \quad ■$$

Many of the special properties of the trigonometric functions discussed in Section 3.1 can be verified by the addition and subtraction formulas.

■ EXAMPLE 6

Verify each of the following.

(a) $\sin (\pi - t) = \sin t$ **(b)** $\tan (u + \pi) = \tan u$

Solution

(a) The subtraction formula for the sine gives us

$$\sin (\pi - t) = \sin \pi \cos t - \cos \pi \sin t$$

$$= (0)(\cos t) - (-1)(\sin t) = \sin t.$$

(b) From the addition formula for the tangent, we have

$$\tan (u + \pi) = \frac{\tan u + \tan \pi}{1 - \tan u \tan \pi} = \tan u,$$

since $\tan \pi = 0$.

A linear combination of a sine and a cosine of the same value can be converted to an expression involving only the sine, as follows.

Reduction of $a_1 \sin bt + a_2 \cos bt$ to $A \sin (bt + \phi)$

For any real numbers a_1, a_2, b, and t,

$$a_1 \sin bt + a_2 \cos bt = A \sin (bt + \phi), \qquad \textbf{(2)}$$

where $A = \sqrt{a_1^2 + a_2^2}$, $\sin \phi = a_2/A$, and $\cos \phi = a_1/A$.

To verify this, we use the addition formula for $\sin (u + v)$ with $u = bt$ and $v = \phi$, obtaining

$$A \sin (bt + \phi) = A \sin bt \cos \phi + A \cos bt \sin \phi$$

$$= (A \sin \phi) \cos bt + (A \cos \phi) \sin bt. \qquad \textbf{(3)}$$

Now, as shown in Figure 2, the angle ϕ is defined to be the angle in standard position with the point (a_1, a_2) on its terminal side. It follows that

$$\sin \phi = \frac{a_2}{\sqrt{a_1^2 + a_2^2}} = \frac{a_2}{A} \quad \text{and} \quad \cos \phi = \frac{a_1}{\sqrt{a_1^2 + a_2^2}} = \frac{a_1}{A}.$$

Thus, (3) becomes

$$A \sin (bt + \phi) = A\left(\frac{a_2}{A}\right) \cos bt + A\left(\frac{a_1}{A}\right) \sin bt$$

$$= a_1 \sin bt + a_2 \cos bt.$$

It is also possible to write the sum of a sine and a cosine of the same value as a single cosine. See Problem 49.

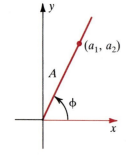

FIGURE 2

▰ EXAMPLE 7

Sketch the graph of $y = -\sqrt{3} \sin 2t + \cos 2t$.

Solution   We could add y-coordinates to obtain the graph. However, since $2t$ appears in both terms, we will use (2) to express y as a single sine function.

With $a_1 = -\sqrt{3}$, $a_2 = 1$, and $b = 2$, we have

$$A = \sqrt{a_1^2 + a_2^2} = \sqrt{4} = 2,$$

$$\sin \phi = \frac{a_2}{A} = \frac{1}{2}, \quad \text{and} \quad \cos \phi = \frac{a_1}{A} = \frac{-\sqrt{3}}{2}.$$

Thus ϕ is the quadrant II angle $5\pi/6$. Therefore $y = -\sqrt{3} \sin 2t + \cos 2t$ can be rewritten as

$$y = 2 \sin \left(2t + \frac{5\pi}{6} \right).$$

Hence the graph is the shifted sine curve with amplitude 2, period $2\pi/2 = \pi$, and phase shift $-5\pi/12$, as shown in Figure 3.

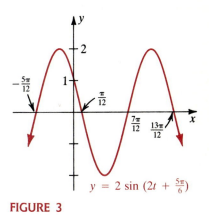

$y = 2 \sin (2t + \frac{5\pi}{6})$

FIGURE 3

As stated earlier, the trigonometric formulas presented in this section have a variety of applications in engineering and the biological and physical sciences. A number of these applications are discussed in Problems 51–54.

EXERCISE 4.2

In Problems 1–24, use an appropriate addition or subtraction formula to find the exact value of the expression.

1. $\cos \dfrac{\pi}{12}$ **2.** $\sin \dfrac{\pi}{12}$

3. $\sin 75°$ **4.** $\cos 75°$

5. $\sin \dfrac{5\pi}{12}$ **6.** $\cos \dfrac{11\pi}{12}$

7. $\tan \dfrac{5\pi}{12}$ **8.** $\cos \dfrac{5\pi}{12}$

9. $\sin \left(-\dfrac{\pi}{12} \right)$ **10.** $\tan \dfrac{11\pi}{12}$

11. $\sin \dfrac{11\pi}{12}$ **12.** $\tan \dfrac{7\pi}{12}$

13. $\cos 165°$ **14.** $\sin 165°$

15. $\tan 165°$ **16.** $\cos 195°$

17. $\sin 195°$ **18.** $\tan 195°$

19. $\cos 345°$ **20.** $\sin 345°$

21. $\cos \dfrac{13\pi}{12}$ **22.** $\tan \dfrac{17\pi}{12}$

23. $\sin \left(-\dfrac{7\pi}{12} \right)$ **24.** $\cos \left(-\dfrac{5\pi}{12} \right)$

In Problems 25–38, use an appropriate addition or subtraction formula to verify the identity.

25. $\cos (t + 2\pi) = \cos t$ **26.** $\sin (t + 2\pi) = \sin t$

27. $\sin \left(t + \dfrac{\pi}{2} \right) = \cos t$ **28.** $\cos \left(t + \dfrac{\pi}{2} \right) = -\sin t$

29. $\tan \left(t + \dfrac{\pi}{2} \right) = -\cot t$ **30.** $\cos (t + \pi) = -\cos t$

31. $\sin (t + \pi) = -\sin t$ **32.** $\cos (\pi - t) = -\cos t$

33. $\tan (\pi - t) = -\tan t$ **34.** $\cot (\pi - t) = -\cot t$

35. $\sin \left(t + \dfrac{\pi}{4} \right) = \dfrac{\sqrt{2}}{2} (\sin t + \cos t)$

36. $\cos \left(t + \dfrac{\pi}{4} \right) = \dfrac{\sqrt{2}}{2} (\cos t - \sin t)$

37. $\cot (t + \pi) = \cot t$

38. $\sin \left(t - \dfrac{3\pi}{2} \right) = \cos t$

39. Verify that

$$\frac{\cos (x + h) - \cos x}{h} =$$
$$\cos x \left(\frac{\cos h - 1}{h} \right) - \sin x \left(\frac{\sin h}{h} \right).$$

40. Using the results of Problem 1 and the identity $\cos^2 t + \sin^2 t = 1$, find the exact value of $\sin (\pi/12)$. Show that your result is the same as that obtained in Problem 2.

41. Use the results of Problem 5 and the identity $\cos^2 t + \sin^2 t = 1$ to find the exact value of $\cos (5\pi/12)$. Compare your answer to the result obtained from applying the addition formula to $\cos (\pi/6 + \pi/4)$.

42. If $\sin u = \frac{3}{5}$ and $\cos v = \frac{12}{13}$, where $0 \le u \le \pi/2$ and $0 \le v \le \pi/2$, find **(a)** $\sin (u + v)$, **(b)** $\cos (u + v)$, and **(c)** $\tan (u - v)$.

43. If P_u and P_v are points in quadrant II on the terminal side of u and v, respectively, with $\cos u = -\frac{1}{3}$ and $\sin v = \frac{2}{3}$, find **(a)** $\sin (u + v)$, **(b)** $\cos (u + v)$, **(c)** $\sin (u - v)$, and **(d)** $\cos (u - v)$. In what quadrants do P_{u+v} and P_{u-v} lie?

44. If u is a quadrant II angle, v is a quadrant III angle, $\sin u = \frac{8}{17}$, and $\tan v = \frac{3}{4}$, find **(a)** $\sin (u + v)$, **(b)** $\sin (u - v)$, **(c)** $\cos (u + v)$, and **(d)** $\cos (u - v)$. In which quadrant does the terminal side of $u + v$ lie? the terminal side of $u - v$?

In Problems 45–48, rewrite the given equation in the form $y = A \sin (bt + \phi)$. Sketch the graph and state the amplitude, the period, and the phase shift.

45. $y = \cos \pi t - \sin \pi t$

46. $y = \sin \dfrac{\pi}{2}t - \sqrt{3} \cos \dfrac{\pi}{2}t$

47. $y = \sqrt{3} \sin 2t + \cos 2t$

48. $y = \sqrt{3} \cos 4t - \sin 4t$

49. Show that

$$y = a_1 \sin bt + a_2 \cos bt = A \cos (bt + \phi),$$

where $A = \sqrt{a_1^2 + a_2^2}$, $\sin \phi = -a_1/A$, and $\cos \phi = a_2/A$.

50. Using Problem 49, rewrite each equation given in Problems 45–48 in the form $y = A \cos (bt + \phi)$.

51. By means of differential equations, it can be shown that under certain conditions the motion of a weight on a spring is given by

$$y(t) = \tfrac{2}{3} \cos 8t - \tfrac{1}{6} \sin 8t,$$

where y is the distance in feet below the equilibrium (at rest) position at time t seconds. Determine a and ϕ such that $y(t) = a \sin (8t + \phi)$.

52. Under certain conditions, the equation of motion of a vibrating string stretched between two points on the x-axis is

$$y = A \sin (\omega t - kx) - A \sin (\omega t + kx),$$

where t is time and A, ω, and k are constants. Show that y can be represented in the equivalent form

$$y = -2A \cos \omega t \sin kx.$$

53. Consider a shock wave traveling in a plane. One method for determining the direction in which a shock wave moves is to measure its arrival time at a system of tripartite stations, assuming that the shock wave travels at a constant velocity. Suppose that the shock wave arrives at station A at time t_1, at station B at t_2, and at station C at t_3. Let α, β, and γ be the angles in the triangle formed by the three stations, as shown in Figure 4. The angle ϕ that the wave front makes with the line from A to B can be determined as follows.

(a) Show that

$$R = \frac{b \sin (\phi + \alpha)}{c \sin \phi}, \quad \text{where } R = \frac{t_3 - t_1}{t_2 - t_1}.$$

(b) Conclude that

$$\cot \phi = \frac{R \sin \gamma}{\sin \alpha \sin \beta} - \cot \alpha.$$

(Tripartite stations are sometimes used to locate the source of **microseisms,** which are small ground movements not caused by earthquakes. Hurricanes are one source of microseisms.)

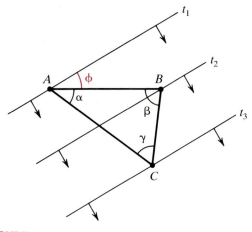

FIGURE 4

54. A mathematical model for blood flow predicts that the optimal values of the angles θ_1 and θ_2, which represent the (positive) angles of the daughter branches with respect to the axis of the parent branch, are given by

$$\cos \theta_1 = \frac{A_0^2 + A_1^2 - A_2^2}{2A_0A_1} \quad \text{and} \quad \cos \theta_2 = \frac{A_0^2 - A_1^2 + A_2^2}{2A_0A_2},$$

where A_0 is the cross-sectional area of the parent branch and A_1 and A_2 are the cross-sectional areas of the daughter branches (see Figure 5). Let $\psi = \theta_1 + \theta_2$ be the junction angle, as shown in the figure.

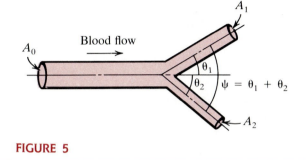

Blood flow

FIGURE 5

(a) Show that

$$\cos \psi = \frac{A_0^2 - A_1^2 - A_2^2}{2A_1A_2}.$$

(b) Show that for the optimal values of θ_1 and θ_2, the cross-sectional area of the daughter branches, $A_1 + A_2$, is greater than or equal to that of the parent branch. (Therefore, the blood must slow down in the daughter branches.)

55. Explain why you get an error message from your calculator when you try to evaluate

$$\frac{\tan 35° + \tan 55°}{1 - \tan 35° \tan 55°}.$$

56. Derive the subtraction formula for the tangent function.

4.3

Double- and Half-Angle Formulas

Many useful formulas can be derived from the addition formulas discussed in the previous section. In this section we use them to derive the **double-** and **half-angle formulas,** so called because they express trigonometric functions of $2t$ and $t/2$ in terms of trigonometric functions of t. The formulas apply to any real number t as well as to any angle t measured in degrees or radians.

THE DOUBLE-ANGLE FORMULAS

The following two formulas are special cases of the addition formulas for the sine and the cosine. If we set $v = u$ in

$$\cos (u + v) = \cos u \cos v - \sin u \sin v,$$

then, since $\cos (u + u) = \cos 2u$, we obtain the next result.

Double-Angle Formula for COSINE

$$\cos 2u = \cos^2 u - \sin^2 u$$

Similarly, replacing $v = u$ in the formula for $\sin (u + v)$, we have

$$\sin (u + u) = \sin u \cos u + \cos u \sin u.$$

Simplifying, we get the following.

Double-Angle Formula for SINE

$$\sin 2u = 2 \sin u \cos u$$

■ EXAMPLE 1

If $\sin t = -\frac{1}{4}$ and $\pi < t < 3\pi/2$, find the exact values of $\cos 2t$ and $\sin 2t$.

Solution   First, we compute $\cos t$ from

$$\cos^2 t + \sin^2 t = 1$$
$$\cos^2 t = 1 - \sin^2 t.$$

Since $\pi < t < 3\pi/2$, we choose the negative square root:

$$\cos t = -\sqrt{1 - \sin^2 t}$$
$$= -\sqrt{1 - \left(-\frac{1}{4}\right)^2} = -\frac{\sqrt{15}}{4}.$$

From the double-angle formulas, we obtain

$$\cos 2t = \cos^2 t - \sin^2 t$$
$$= \left(-\frac{\sqrt{15}}{4}\right)^2 - \left(-\frac{1}{4}\right)^2$$
$$= \frac{15}{16} - \frac{1}{16}$$
$$= \frac{14}{16} = \frac{7}{8},$$

and $$\sin 2t = 2 \sin t \cos t$$
$$= 2\left(-\frac{1}{4}\right)\left(-\frac{\sqrt{15}}{4}\right) = \frac{\sqrt{15}}{8}.$$ ■

■ EXAMPLE 2

Express $\sin 3\theta$ in terms of $\sin \theta$.

Solution   Since $3\theta = 2\theta + \theta$, we first use the addition formula for the sine and then the double-angle formulas:

$$\sin 3\theta = \sin (2\theta + \theta) = \sin 2\theta \cos \theta + \cos 2\theta \sin \theta$$

$$= 2 \sin \theta \cos \theta \cos \theta + (\cos^2 \theta - \sin^2 \theta) \sin \theta$$
$$= 2 \sin \theta \cos^2 \theta + \cos^2 \theta \sin \theta - \sin^3 \theta$$
$$= 3 \cos^2 \theta \sin \theta - \sin^3 \theta$$
$$= 3(1 - \sin^2 \theta) \sin \theta - \sin^3 \theta = 3 \sin \theta - 4 \sin^3 \theta.$$

Note that in the last line $\cos^2 \theta$ was replaced by $1 - \sin^2 \theta$ by using $\cos^2 \theta + \sin^2 \theta = 1$.

Two alternative forms for the double-angle formula for the cosine can be obtained from $\cos 2u = \cos^2 u - \sin^2 u$ by substituting, in turn, $1 - \sin^2 u$ for $\cos^2 u$ and $1 - \cos^2 u$ for $\sin^2 u$. The resulting forms are given below.

Alternative Forms:
Double-Angle Formulas for COSINE

(i) $\cos 2u = 1 - 2 \sin^2 u$
(ii) $\cos 2u = 2 \cos^2 u - 1$

EXAMPLE 3

Express $\cos 4x$ in terms of $\cos x$.

Solution By form (ii) with $u = 2x$, we have
$$\cos 4x = \cos 2(2x) = 2 \cos^2 2x - 1 = 2(\cos 2x)^2 - 1.$$

Now using form (ii) with $u = x$, we obtain
$$\cos 4x = 2(2 \cos^2 x - 1)^2 - 1 = 2(4 \cos^4 x - 4 \cos^2 x + 1) - 1$$
$$= 8 \cos^4 x - 8 \cos^2 x + 1.$$

The alternative forms of the double-angle formula for the cosine are the source of two **half-angle formulas.** Solving (i) for $\sin^2 u$ gives
$$2 \sin^2 u = 1 - \cos 2u$$
$$\sin^2 u = \tfrac{1}{2}(1 - \cos 2u).$$

Then letting $t = 2u$, we obtain the following.

Half-Angle Formula for SINE

$$\sin^2 \frac{t}{2} = \frac{1}{2}(1 - \cos t)$$

Similarly, the half-angle formula for the cosine can be derived from (ii) (see Problem 43).

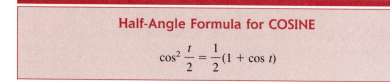

Half-Angle Formula for COSINE

$$\cos^2 \frac{t}{2} = \frac{1}{2}(1 + \cos t)$$

EXAMPLE 4

Find the exact value of $\sin (5\pi/8)$ and $\cos (5\pi/8)$.

Solution　If $t = 5\pi/4$, then $t/2 = 5\pi/8$ and the half-angle formulas yield

$$\sin^2 \frac{5\pi}{8} = \frac{1}{2}\left(1 - \cos \frac{5\pi}{4}\right) = \frac{1}{2}\left[1 - \left(-\frac{\sqrt{2}}{2}\right)\right] = \frac{1}{2}\left(1 + \frac{\sqrt{2}}{2}\right),$$

and

$$\cos^2 \frac{5\pi}{8} = \frac{1}{2}\left(1 + \cos \frac{5\pi}{4}\right) = \frac{1}{2}\left[1 + \left(-\frac{\sqrt{2}}{2}\right)\right] = \frac{1}{2}\left(1 - \frac{\sqrt{2}}{2}\right).$$

Since $5\pi/8$ radians is a quadrant II angle, $\sin (5\pi/8) > 0$ and $\cos (5\pi/8) < 0$. Therefore,

$$\sin \frac{5\pi}{8} = \sqrt{\frac{1}{2}\left(1 + \frac{\sqrt{2}}{2}\right)} = \frac{1}{2}\sqrt{2 + \sqrt{2}}$$

and

$$\cos \frac{5\pi}{8} = -\sqrt{\frac{1}{2}\left(1 - \frac{\sqrt{2}}{2}\right)} = -\frac{1}{2}\sqrt{2 - \sqrt{2}}.$$

Alternative forms for the half-angle formulas can be obtained by taking the square root of each side in the half-angle formulas for the sine and the cosine.

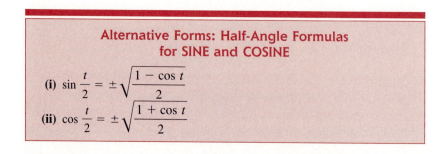

Alternative Forms: Half-Angle Formulas for SINE and COSINE

(i) $\sin \dfrac{t}{2} = \pm\sqrt{\dfrac{1 - \cos t}{2}}$

(ii) $\cos \dfrac{t}{2} = \pm\sqrt{\dfrac{1 + \cos t}{2}}$

In these forms (as in Example 4) the choice of the algebraic sign depends on the quadrant in which the terminal side of the angle $t/2$ lies.

If we let $u = v$ in the addition formula for $\tan (u + v)$, we have

$$\tan (u + u) = \frac{\tan u + \tan u}{1 - \tan u \tan u} = \frac{2 \tan u}{1 - \tan^2 u}.$$

This result is the double-angle formula for the tangent function.

Double-Angle Formula for TANGENT

$$\tan 2u = \frac{2 \tan u}{1 - \tan^2 u}$$

▮ EXAMPLE 5

Verify the identity

$$\tan 2t = \frac{2 \tan t}{2 - \sec^2 t}.$$

Solution   We use the double-angle formula for the tangent to rewrite the left-hand side of the identity:

$$\tan 2t = \frac{2 \tan t}{1 - \tan^2 t} = \frac{2 \tan t}{1 - (\sec^2 t - 1)} = \frac{2 \tan t}{2 - \sec^2 t}. \qquad ▮$$

The first of the following half-angle formulas for the tangent can be obtained by dividing the corresponding formulas for the sine and the cosine and simplifying.

Half-Angle Formulas for TANGENT

(i) $\tan^2 \dfrac{t}{2} = \dfrac{1 - \cos t}{1 + \cos t}$

Alternative Forms

(ii) $\tan \dfrac{t}{2} = \dfrac{1 - \cos t}{\sin t}$

(iii) $\tan \dfrac{t}{2} = \dfrac{\sin t}{1 + \cos t}$

To obtain (ii), we first write

$$\tan \frac{t}{2} = \frac{\sin \dfrac{t}{2}}{\cos \dfrac{t}{2}}. \tag{4}$$

We want to show that this equals $(1 - \cos t)/\sin t$. We will obtain $\sin t$ in the denominator of (4) if we multiply $\cos (t/2)$ by $2 \sin (t/2)$ and use the double-angle formula for the sine:

$$\tan \frac{t}{2} = \frac{\sin \dfrac{t}{2}}{\cos \dfrac{t}{2}}$$

$$= \frac{2 \sin \dfrac{t}{2} \sin \dfrac{t}{2}}{2 \sin \dfrac{t}{2} \cos \dfrac{t}{2}}$$

$$= \frac{2 \sin^2 \dfrac{t}{2}}{\sin t}.$$

Finally, from the half-angle formula $\sin^2 (t/2) = \frac{1}{2}(1 - \cos t)$, we have

$$\tan \frac{t}{2} = \frac{2 \sin^2 \dfrac{t}{2}}{\sin t} = \frac{2 \left[\dfrac{1}{2}(1 - \cos t) \right]}{\sin t}$$

$$= \frac{1 - \cos t}{\sin t}.$$

■ EXAMPLE 6

Find the exact value of $\tan 22.5°$.

Solution Using formula (ii) for $\tan (t/2)$ and the fact that $22.5° = \frac{1}{2}(45°)$, we find

$$\tan 22.5° = \frac{1 - \cos 45°}{\sin 45°} = \frac{1 - \sqrt{2}/2}{\sqrt{2}/2}$$

$$= \sqrt{2} - 1.$$

EXAMPLE 7

Given $\cos \theta = -\frac{3}{5}$, $90° < \theta < 180°$, find the exact values of $\tan 2\theta$ and $\tan (\theta/2)$.

Solution First we use $\cos^2 \theta + \sin^2 \theta = 1$ to find $\sin \theta$ from

$$\sin^2 \theta = 1 - \cos^2 \theta = 1 - (-\tfrac{3}{5})^2$$
$$= \tfrac{16}{25}.$$

Since $90° < \theta < 180°$, we take the positive square root:

$$\sin \theta = \tfrac{4}{5}.$$

Therefore, $\tan \theta = \sin \theta / \cos \theta = -\frac{4}{3}$. Now using the double-angle formula for the tangent, we obtain

$$\tan 2\theta = \frac{2 \tan \theta}{1 - \tan^2 \theta} = \frac{2\left(-\dfrac{4}{3}\right)}{1 - \left(-\dfrac{4}{3}\right)^2} = \frac{24}{7}.$$

From the half-angle formula (iii) for the tangent function, we find

$$\tan \frac{\theta}{2} = \frac{\sin \theta}{1 + \cos \theta} = \frac{\dfrac{4}{5}}{1 + \left(-\dfrac{3}{5}\right)} = 2.$$

EXERCISE 4.3

In Problems 1–6, use the double-angle formulas to write the given expression as a single trigonometric function of twice the angle.

1. $2 \cos \beta \sin \beta$

2. $\cos^2 2t - \sin^2 2t$

3. $1 - 2 \sin^2 \dfrac{\pi}{5}$

4. $2 \cos^2 \left(\dfrac{19}{2}x\right) - 1$

5. $\dfrac{\tan 3t}{1 - \tan^2 3t}$

6. $2 \sin \dfrac{y}{2} \cos \dfrac{y}{2}$

In Problems 7–12, use the given information to find
(a) $\cos 2t$, (b) $\sin 2t$, and (c) $\tan 2t$.

7. $\sin t = \sqrt{2}/3$, $\pi/2 < t < \pi$

8. $\cos t = \sqrt{3}/5$, $3\pi/2 < t < 2\pi$

9. $\tan t = 1/2$, $\pi < t < 3\pi/2$

10. $\csc t = -3$, $\pi < t < 3\pi/2$

11. $\sec t = -13/5$, $\pi/2 < t < \pi$

12. $\cot t = 4/3$, $0 < t < \pi/2$

In Problems 13–22, find the exact value of the given expression.

13. $\cos (\pi/12)$

14. $\sin (\pi/8)$

15. $\sin (3\pi/8)$

16. $\tan (\pi/12)$

17. $\cos 67.5°$

18. $\sin 15°$

19. $\tan 105°$

20. $\cot 157.5°$

21. $\csc (13\pi/12)$

22. $\sec (-3\pi/8)$

In Problems 23–28, use the given information to find (a) cos $(t/2)$, (b) sin $(t/2)$, and (c) tan $(t/2)$.

23. $\sin t = 12/13,\ \pi/2 < t < \pi$
24. $\cos t = 4/5,\ 3\pi/2 < t < 2\pi$
25. $\tan t = 2,\ \pi < t < 3\pi/2$
26. $\csc t = 9,\ 0 < t < \pi/2$
27. $\sec t = 3/2,\ 0 < t < 90°$
28. $\cot t = -1/4,\ 90° < t < 180°$

In Problems 29–42, verify the given identity.

29. $\sin 4u = 4 \cos u\ (\sin u - 2 \sin^3 u)$
30. $\cos 3v = 4 \cos^3 v - 3 \cos v$
31. $(\sin t + \cos t)^2 = 1 + \sin 2t$
32. $\cos 2x = \cos^4 x - \sin^4 x$
33. $\cot 2\theta = \dfrac{1}{2}(\cot \theta - \tan \theta)$
34. $\sec 2x = \dfrac{1}{2 \cos^2 x - 1}$
35. $\tan (\theta/2) = \csc \theta - \cot \theta$ **36.** $\dfrac{1 - \cos 2\theta}{\sin 2\theta} = \tan \theta$
37. $\dfrac{2 \tan x}{1 + \tan^2 x} = \sin 2x$ **38.** $\dfrac{1 - \tan^2 t}{\cos 2t} = \dfrac{2 \tan t}{\sin 2t}$
39. $\cot 2z = \dfrac{\csc^2 z - 2}{2 \cot z}$ **40.** $\dfrac{\sin^2 2t}{(1 + \cos 2t)^2} = \sec^2 t - 1$
41. $\tan 4t = \dfrac{4 \tan t - 4 \tan^3 t}{1 - 6 \tan^2 t + \tan^4 t}$
42. $\dfrac{\cot t - \tan t}{\cot t + \tan t} = \cos 2t$
43. Derive the half-angle formula for the cosine.
44. Show that
$$\tan \frac{u}{2} = \frac{\sin u}{1 + \cos u}.$$

In Problems 45–48, state the amplitude and the period of the given function. Graph. [*Hint*: Rewrite the function in the form $y = a \sin bt$ or $y = a \cos bt$.]

45. $y = 4 \cos^2 t - 2$
46. $y = \sin (t/2) \cos (t/2)$
47. $y = 2 \sin 2t \cos 2t$
48. $y = 5 \cos^2 4t - 5 \sin^2 4t$

49. A particle is moving back and forth along the x-axis with its distance d from the origin at time t seconds given by $d = 4 - 8 \sin^2 4t$.
(a) Show that the motion is simple harmonic motion by expressing d in the form $d = a \sin (bt + c)$.
(b) Determine the amplitude and the period of the motion.

50. For the points $P(x, y)$ and $Q(x_1, 0)$ with $x_1 < x$, as illustrated in Figure 6, show that
$$d(O, Q) + d(P, Q) = y \tan (\alpha/2) + x.$$

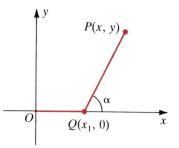

FIGURE 6

51. The ratio of the speed of an airplane to the speed of sound is called the Mach number M of the plane. If $M > 1$, the plane makes sound waves that form a (moving) cone, as shown in Figure 7. A sonic boom is heard at the intersection of the cone with the ground. If the vertex angle of the cone is θ, then
$$\sin (\theta/2) = 1/M.$$
If $\theta = \pi/6$, find the exact value of the Mach number.

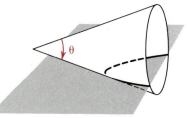

FIGURE 7

52. Show that the area of an isosceles triangle with equal sides of length x is
$$A = (x^2/2) \sin \theta,$$
where θ is the angle formed by the two equal sides. [*Hint*: Consider $\theta/2$, as shown in Figure 8.]

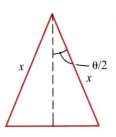

FIGURE 8

53. If a projectile, such as a shotput, is released from a height h, upward at an angle ϕ with velocity v_0, the range R at which it strikes the ground is given by the formula

$$R = \frac{v_0^2 \cos \phi}{g}(\sin \phi + \sqrt{\sin^2 \phi + 2gh/v_0^2}),$$

where g is the acceleration due to gravity (see Figure 9). It can be shown that the maximum range $R_{\max}$ is achieved if the angle ϕ satisfies the equation

$$\cos 2\phi = \frac{gh}{v_0^2 + gh}.$$

Using these expressions for R and $\cos 2\phi$ and the half-angle formulas for the sine and the cosine with $t = 2\phi$, show that

$$R_{\max} = \frac{v_0\sqrt{v_0^2 + 2gh}}{g}.$$

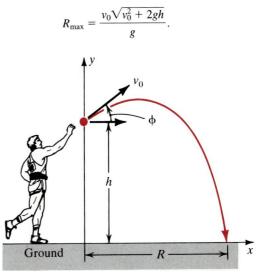

FIGURE 9

4.4

Product and Sum Formulas

We now discuss **product** and **sum formulas,** which enable us to write certain products of sines and cosines as sums of sines and cosines, and vice versa.

Product Formulas

(i) $\sin u \sin v = \frac{1}{2}[\cos (u - v) - \cos (u + v)]$

(ii) $\cos u \cos v = \frac{1}{2}[\cos (u - v) + \cos (u + v)]$

(iii) $\sin u \cos v = \frac{1}{2}[\sin (u + v) + \sin (u - v)]$

(iv) $\cos u \sin v = \frac{1}{2}[\sin (u + v) - \sin (u - v)]$

We will prove (i) here and leave the verification of (ii)–(iv) as an exercise (see Problem 37). We begin by applying the addition and subtraction formulas to the right-hand side of (i) and then we simplify:

$\frac{1}{2}[\cos(u - v) - \cos(u + v)]$

$$= \tfrac{1}{2}[\cos u \cos v + \sin u \sin v - (\cos u \cos v - \sin u \sin v)]$$

$$= \tfrac{1}{2}[2 \sin u \sin v] = \sin u \sin v.$$

■ EXAMPLE 1

Use a product formula to rewrite each of the following.

(a) $\sin 45° \cos 15°$ (b) $\cos 2x \cos 3x$

Solution

(a) Using the product formula (iii) with $u = 45°$ and $v = 15°$, we have

$$\sin 45° \cos 15° = \frac{1}{2}[\sin(45° + 15°) + \sin(45° - 15°)]$$

$$= \frac{1}{2}[\sin 60° + \sin 30°] = \frac{1}{2}\left(\frac{\sqrt{3}}{2} + \frac{1}{2}\right) = \frac{\sqrt{3} + 1}{4}.$$

(b) From the product formula (ii) with $u = 2x$ and $v = 3x$, we obtain

$$\cos 2x \cos 3x = \tfrac{1}{2}[\cos(2x - 3x) + \cos(2x + 3x)]$$

$$= \tfrac{1}{2}[\cos(-x) + \cos 5x] = \tfrac{1}{2}[\cos x + \cos 5x].$$

■ EXAMPLE 2

Sketch the graph of

$$y = 2 \cos \frac{x}{2} \cos \frac{3x}{2}.$$

Solution   Using the product formula (ii) with $u = x/2$ and $v = 3x/2$, we find that

$$2 \cos \frac{x}{2} \cos \frac{3x}{2} = (2)\left(\frac{1}{2}\right)\left[\cos\left(\frac{x}{2} - \frac{3x}{2}\right) + \cos\left(\frac{x}{2} + \frac{3x}{2}\right)\right]$$

$$= \cos(-x) + \cos 2x = \cos x + \cos 2x.$$

Therefore, the original equation becomes

$$y = \cos x + \cos 2x.$$

Once the expression is written as a sum, the graph can be sketched by addition of y-coordinates, as shown in Figure 10.

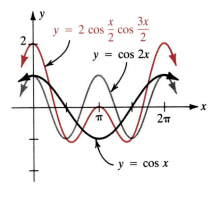

FIGURE 10

The following formulas enable us to write sums as products.

Sum Formulas

(i) $\sin x + \sin y = 2 \sin \left(\dfrac{x + y}{2} \right) \cos \left(\dfrac{x - y}{2} \right)$

(ii) $\sin x - \sin y = 2 \cos \left(\dfrac{x + y}{2} \right) \sin \left(\dfrac{x - y}{2} \right)$

(iii) $\cos x + \cos y = 2 \cos \left(\dfrac{x + y}{2} \right) \cos \left(\dfrac{x - y}{2} \right)$

(iv) $\cos x - \cos y = -2 \sin \left(\dfrac{x + y}{2} \right) \sin \left(\dfrac{x - y}{2} \right)$

We can obtain these formulas from the product formulas by using the substitutions

$$x = u + v \quad \text{and} \quad y = u - v,$$

from which we find that $x + y = (u + v) + (u - v) = 2u$, or

$$u = \frac{x + y}{2}.$$

Similarly, $x - y = (u + v) - (u - v) = 2v$, so that

$$v = \frac{x - y}{2}.$$

With these substitutions in the product formula (i), we obtain

$$\sin \left(\frac{x + y}{2}\right) \sin \left(\frac{x - y}{2}\right) = \frac{1}{2}[\cos y - \cos x],$$

or

$$-2 \sin \left(\frac{x + y}{2}\right) \sin \left(\frac{x - y}{2}\right) = \cos x - \cos y,$$

which is the sum formula (iv). In a similar manner, each of the remaining product formulas together with these substitutions yields one of the sum formulas. (See Problem 38.)

▬ EXAMPLE 3

Use a sum formula to rewrite each of the following.

(a) $\sin 75° + \sin 15°$ **(b)** $\cos t - \cos 5t$

Solution

(a) Using the sum formula (i) with $x = 75°$ and $y = 15°$ gives us

$$\sin 75° + \sin 15° = 2 \sin \left(\frac{75° + 15°}{2}\right) \cos \left(\frac{75° - 15°}{2}\right)$$

$$= 2 \sin 45° \cos 30°$$

$$= 2\left(\frac{\sqrt{2}}{2}\right)\left(\frac{\sqrt{3}}{2}\right) = \frac{\sqrt{6}}{2}.$$

(b) From the sum formula (iv) with $x = t$ and $y = 5t$, we find

$$\cos t - \cos 5t = -2 \sin \left(\frac{t + 5t}{2}\right) \sin \left(\frac{t - 5t}{2}\right)$$

$$= -2 \sin 3t \sin (-2t) = 2 \sin 3t \sin 2t. \quad ▬$$

▬ EXAMPLE 4

Verify the identity

$$\frac{\cos 10t - \cos 12t}{\sin 10t + \sin 12t} = \tan t.$$

Solution   We use the sum formulas (iv) and (i) to obtain

$$\frac{\cos 10t - \cos 12t}{\sin 10t + \sin 12t} = \frac{-2 \sin \dfrac{10t + 12t}{2} \sin \dfrac{10t - 12t}{2}}{2 \sin \dfrac{10t + 12t}{2} \cos \dfrac{10t - 12t}{2}}$$

$$= \frac{-2 \sin 11t \sin(-t)}{2 \sin 11t \cos(-t)} = \frac{-\sin(-t)}{\cos(-t)}$$

$$= \frac{\sin t}{\cos t} = \tan t.$$

For easy reference a complete list of the trigonometric formulas that we have considered can be found inside the cover of this text.

EXERCISE 4.4

In Problems 1–10, use a product formula to rewrite the given expression.

1. $\cos \dfrac{5\pi}{12} \sin \dfrac{\pi}{12}$ **2.** $\sin \dfrac{5\pi}{8} \cos \dfrac{\pi}{8}$

3. $\sin 75° \sin 15°$ **4.** $\cos 15° \cos 45°$

5. $\sin 2x \cos 4x$ **6.** $\sin x \sin 3x$

7. $\cos 5\theta \sin 3\theta$ **8.** $\cos 5x \cos x$

9. $\sin \dfrac{4x}{3} \cos \dfrac{x}{3}$ **10.** $\sin \dfrac{x}{2} \sin \dfrac{5x}{2}$

In Problems 11–16, write the product as a sum and then sketch the graph of the given function by addition of y-coordinates.

11. $y = \sin t \cos 2t$ **12.** $y = 2 \cos 5x \sin 4x$

13. $y = 4 \sin \dfrac{x}{2} \sin \dfrac{5x}{2}$ **14.** $y = -\cos \pi\theta \cos 2\pi\theta$

15. $y = -5 \cos 6\theta \sin 2\theta$ **16.** $y = 6 \sin \dfrac{7t}{2} \cos \dfrac{3t}{2}$

In Problems 17–26, use a sum formula to rewrite the given expression.

17. $\sin \dfrac{7\pi}{12} + \sin \dfrac{\pi}{12}$ **18.** $\sin \dfrac{\pi}{12} - \sin \dfrac{5\pi}{12}$

19. $\cos 105° - \cos 15°$ **20.** $\cos 15° + \cos 75°$

21. $\sin y - \sin 5y$ **22.** $\cos 3\theta - \cos \theta$

23. $\cos 2x + \cos 6x$ **24.** $\sin 5t + \sin 3t$

25. $\sin \omega_1 t + \sin \omega_2 t$ **26.** $\cos(\theta + \phi) - \cos \theta$

In Problems 27–36, make use of the product and sum formulas to verify the given statement.

27. $2 \sin\left(x + \dfrac{\pi}{4}\right) \sin\left(x - \dfrac{\pi}{4}\right) = -\cos 2x$

28. $\dfrac{\sin \theta + \sin 7\theta}{\cos \theta + \cos 7\theta} = \tan 4\theta$ **29.** $\dfrac{\sin 6\beta + \sin 2\beta}{\cos 2\beta - \cos 6\beta} = \cot 2\beta$

30. $\cos(\alpha + \beta) \cos(\alpha - \beta) = \frac{1}{2}(\cos 2\alpha + \cos 2\beta)$

31. $2 \sin\left(t + \dfrac{\pi}{2}\right) \cos\left(t - \dfrac{\pi}{2}\right) = \sin 2t$

32. $\dfrac{\cos x - \cos y}{\sin y - \sin x} = \tan\left(\dfrac{x + y}{2}\right)$

33. $\dfrac{\sin(x + h) - \sin x}{h} = \dfrac{2}{h} \sin \dfrac{h}{2} \cos\left(x + \dfrac{h}{2}\right)$

34. $\dfrac{\sin \theta + \sin 2\theta + \sin 3\theta}{\cos \theta + \cos 2\theta + \cos 3\theta} = \tan 2\theta$

35. $\dfrac{\sin 2\alpha + \sin 2\beta}{\sin 2\alpha - \sin 2\beta} = \dfrac{\tan(\alpha + \beta)}{\tan(\alpha - \beta)}$

36. $2 \cos 2t \cos t - \cos 3t = \cos t$

37. Verify the product formulas (ii)–(iv).

38. Verify the sum formulas (i)–(iii).

39. A note produced by a certain musical instrument results in a sound wave described by

$$f(t) = 0.03 \sin 500\pi t + 0.03 \sin 1000\pi t,$$

where $f(t)$ is the difference between atmospheric pressure and air pressure in dynes per square centimeter at the eardrum after t seconds. Express f as the product of a sine and a cosine function.

40. If two piano wires struck by the same key are slightly out of tune, the difference between the atmospheric pressure and air pressure at the eardrum can be represented by

$$f(t) = a \cos 2\pi b_1 t + a \cos 2\pi b_2 t,$$

where the value of the constant b_1 is close to the constant b_2. The variations in loudness that occur are called **beats.** (See Figure 11.) The two strings can be tuned to the same frequency by tightening one of them while sounding both until the beats disappear.

(a) Use a sum formula to write $f(t)$ as a product.

(b) Show that $f(t)$ can be considered a cosine function with period $2/(b_1 + b_2)$ and variable amplitude $2a \cos \pi(b_1 - b_2)t$.

41. The term $\sin \omega t \sin (\omega t + \phi)$ is encountered in the derivation of an expression for the power in an alternating-current circuit. Show that this term can be written as

$$\tfrac{1}{2}[\cos \phi - \cos (2\omega t + \phi)].$$

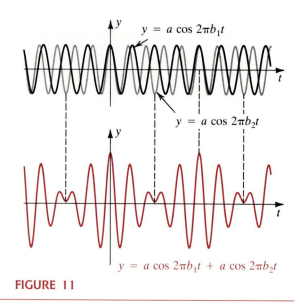

FIGURE 11

4.5

Trigonometric Equations

In this section we discuss techniques for solving equations involving trigonometric functions. Equations, such as

$$\sin t = \frac{\sqrt{2}}{2} \tag{5}$$

and

$$4 \sin^2 t - 8 \sin t + 3 = 0,$$

are called **trigonometric equations.**

Usually trigonometric equations are conditional equations. Recall from Section 1.1 that, unlike identities, conditional equations are true only for certain values in the domain of the variable.

We consider the problem of finding *all* real numbers t satisfying $\sin t = \sqrt{2}/2$. As Figure 12 indicates, there exists an infinite number of solutions

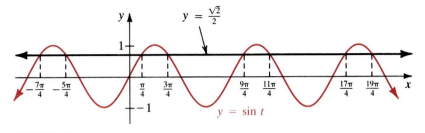

FIGURE 12

$$\ldots, -\frac{7\pi}{4}, \frac{\pi}{4}, \frac{9\pi}{4}, \frac{17\pi}{4}, \ldots$$

and

$$\ldots, -\frac{5\pi}{4}, \frac{3\pi}{4}, \frac{11\pi}{4}, \frac{19\pi}{4}, \ldots$$

(6)

Note that in each list in (6), every solution can be obtained by adding 2π to the preceding solution. Of course, this is a consequence of the periodicity of the sine function. It is generally true that trigonometric equations have infinitely many solutions because of the periodicity of the trigonometric functions.

To obtain the solutions of an equation such as $\sin t = \sqrt{2}/2$, it is more convenient to use a unit circle and reference angles rather than the graph of $y = \sin t$. Since $\sin t = \sqrt{2}/2$, the reference angle for t is $\pi/4$ radian. The fact that the value $\sin t = \sqrt{2}/2$ is positive implies that the angle of t radians can be in either the first or the second quadrant. Thus, as shown in Figure 13, the only solutions between 0 and 2π are

$$t = \frac{\pi}{4} \quad \text{and} \quad t = \frac{3\pi}{4}.$$

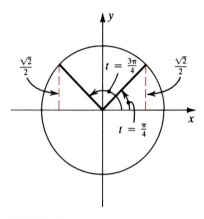

FIGURE 13

Since the sine function is periodic with period 2π, all other solutions can be obtained by adding integer multiples of 2π to these solutions:

$$t = \frac{\pi}{4} + 2n\pi \quad \text{or} \quad t = \frac{3\pi}{4} + 2n\pi, \quad n = 0, \pm 1, \pm 2, \ldots.$$

When we are faced with a more complicated equation, such as

$$4 \sin^2 t - 8 \sin t + 3 = 0,$$

the basic approach is to solve for a single trigonometric function (in this case, it would be $\sin t$) by using methods similar to those for solving algebraic equations. Then the values of the angle are determined using the unit circle and reference angles. The following example illustrates this technique.

EXAMPLE 1

Find all solutions of $4 \sin^2 t - 8 \sin t + 3 = 0$.

Solution   We first observe that this is a quadratic equation in $\sin t$, and that it factors as

$$(2 \sin t - 3)(2 \sin t - 1) = 0.$$

This implies that either

$$\sin t = \tfrac{3}{2} \quad \text{or} \quad \sin t = \tfrac{1}{2}.$$

The first equation has no solution since $|\sin t| \leq 1$. As we see in Figure 14, the two angles between 0 and 2π with sine equal to $\dfrac{1}{2}$ are

$$t = \frac{\pi}{6} \quad \text{and} \quad t = \frac{5\pi}{6}.$$

Therefore, by the periodicity of the sine function, the solutions are

$$t = \frac{\pi}{6} + 2n\pi \quad \text{and} \quad t = \frac{5\pi}{6} + 2n\pi, \quad n = 1, \pm 1, \pm 2, \ldots \ .$$

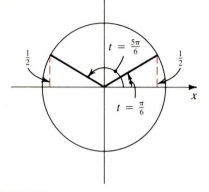

FIGURE 14

EXAMPLE 2

Solve $\sin t = \cos t.$ (7)

Solution   Dividing both sides of the equation by $\cos t$ gives

$$\tan t = 1.$$ (8)

This equation is equivalent to (7) provided that $\cos t \neq 0$.

We observe that if $\cos t = 0$, then

$$t = \frac{\pi}{2} + 2n\pi \quad \text{or} \quad t = \frac{3\pi}{2} + 2n\pi$$

for any integer n. Since

$$\sin\left(\frac{\pi}{2} + 2n\pi\right) \neq 0 \quad \text{and} \quad \sin\left(\frac{3\pi}{2} + 2n\pi\right) \neq 0,$$

these values of t do not satisfy the original equation. Thus we will find *all* the solutions to (7) by solving equation (8).

Now $\tan t = 1$ implies that the reference angle for t is $\pi/4$ radian. Since $\tan t = 1 > 0$, the angle of t radians can lie in either the first or the third

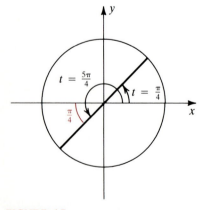

FIGURE 15

quadrant, as shown in Figure 15. Thus the solutions are

$$t = \frac{\pi}{4} + 2n\pi \quad \text{and} \quad t = \frac{5\pi}{4} + 2n\pi, \quad n = 0, \pm 1, \pm 2, \ldots ,$$

or, equivalently,

$$t = \frac{\pi}{4} + n\pi, \quad n = 0, \pm 1, \pm 2, \ldots .$$

Note of Caution: If you divide by an expression containing a variable, it is essential to determine whether the values that make the expression zero are solutions of the original equation. If they are and you have not checked, these solutions may be lost. Note that in Example 2, when we divided by cos t, we took care to check that no solutions were lost.

Whenever possible, it is preferable to avoid dividing by a variable expression. This can frequently be accomplished by collecting all nonzero terms on one side of the equation and then factoring. Example 3 illustrates this technique.

EXAMPLE 3

Solve

$$2 \sin t \cos^2 t = -\frac{\sqrt{3}}{2} \cos t.$$

Solution Instead of dividing by cos t, we write the equation as

$$2 \sin t \cos^2 t + \frac{\sqrt{3}}{2} \cos t = 0$$

and factor:

$$\cos t \left(2 \sin t \cos t + \frac{\sqrt{3}}{2} \right) = 0.$$

Thus either

$$\cos t = 0 \quad \text{or} \quad 2 \sin t \cos t + \frac{\sqrt{3}}{2} = 0.$$

Now the cosine is zero for all odd multiples of $\pi/2$, that is,

$$t = (2n + 1)\frac{\pi}{2} = \frac{\pi}{2} + n\pi, \quad n = 0, \pm 1, \pm 2, \ldots .$$

In the second equation we use the double-angle formula $\sin 2t = 2 \sin t \cos t$ to obtain

$$\sin 2t + \frac{\sqrt{3}}{2} = 0 \quad \text{or} \quad \sin 2t = -\frac{\sqrt{3}}{2}.$$

Thus the reference angle for $2t$ is $\pi/3$. Since the sine is negative, the angle $2t$ must be in either the third or the fourth quadrant. As Figure 16 illustrates, either

$$2t = \frac{4\pi}{3} + 2n\pi \quad \text{or} \quad 2t = \frac{5\pi}{3} + 2n\pi,$$

and hence

$$t = \frac{2\pi}{3} + n\pi \quad \text{or} \quad t = \frac{5\pi}{6} + n\pi.$$

Therefore, the solutions are

$$t = \frac{\pi}{2} + n\pi, \quad t = \frac{2\pi}{3} + n\pi, \quad \text{and} \quad t = \frac{5\pi}{6} + n\pi, \quad n = 0, \pm 1, \pm 2, \ldots .$$

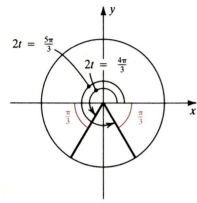

FIGURE 16

In Example 3 had we simplified the equation by dividing by $\cos t$ and not checked the values of t for which $\cos t = 0$, we would have lost the solutions $t = \pi/2 + n\pi$, $n = 0, \pm 1, \pm 2, \ldots .$

■ **EXAMPLE 4**

Solve $3 \cos^2 x - \cos 2x = 1$.

Solution We observe that the given equation involves the cosines of x and of $2x$. Consequently, we use the double-angle formula

$$\cos 2x = 2 \cos^2 x - 1$$

to replace the equation by an equivalent equation that involves $\cos x$ only. We find that

$$3 \cos^2 x - (2 \cos^2 x - 1) = 1,$$

or $$\cos^2 x = 0.$$

Therefore, $$\cos x = 0,$$

and the solutions are

$$x = (2n + 1)\frac{\pi}{2}, \quad n = 0, \pm 1, \pm 2, \ldots .$$

So far in this section we have viewed the variable in the trigonometric equation as representing either a real number or an angle measured in radians. If the variable represents an angle measured in degrees, the technique for solving is the same.

■ EXAMPLE 5

Solve $\cos 2\theta = -\frac{1}{2}$, where θ is an angle measured in degrees.

Solution Since $\cos 2\theta = -\frac{1}{2}$, the reference angle for 2θ is $60°$ and the angle 2θ must be in either the second or the third quadrant. As Figure 17 illustrates, either

$$2\theta = 120° \quad \text{or} \quad 2\theta = 240°.$$

Any angle that is coterminal with one of these angles will also satisfy $\cos 2\theta = -\frac{1}{2}$. These angles are obtained by adding any multiple of $360°$. Thus we have

$$2\theta = 120° + 360°n \quad \text{or} \quad 2\theta = 240° + 360°n, \quad n = 0, \pm1, \pm2, \ldots,$$

which simplifies to

$$\theta = 60° + 180°n \quad \text{or} \quad \theta = 120° + 180°n, \quad n = 0, \pm1, \pm2, \ldots.$$

Thus the solutions are

$$\theta = 60° + 180°n \quad \text{and} \quad \theta = 120° + 180°n, \quad n = 0, \pm1, \pm2, \ldots.$$

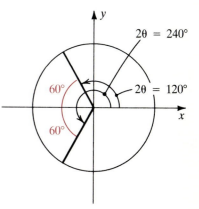

FIGURE 17

■ EXAMPLE 6

Find all solutions of

$$1 + \tan \alpha = \sec \alpha,$$

where α is an angle measured in degrees.

Solution The equation does not factor, but we see that if we square both sides, we can use a fundamental identity:

$$(1 + \tan \alpha)^2 = (\sec \alpha)^2$$
$$1 + 2\tan \alpha + \tan^2 \alpha = \sec^2 \alpha$$
$$1 + 2\tan \alpha + \tan^2 \alpha = 1 + \tan^2 \alpha$$
$$2\tan \alpha = 0$$
$$\tan \alpha = 0.$$

The values of α in $[0°, 360°)$ for which $\tan \alpha = 0$ are

$$\alpha = 0° \quad \text{and} \quad \alpha = 180°.$$

Since we squared each side of the original equation, we may have introduced extraneous solutions. Therefore, it is important that we check all solutions in the original equation. Substituting $\alpha = 0°$ into $1 + \tan \alpha = \sec \alpha$, we obtain the true statement $0 + 1 = 1$. But after substituting $\alpha = 180°$, we obtain the *false* statement $0 + 1 = -1$. Therefore, $180°$ is an extraneous solution and $\alpha = 0°$ is the only solution in the interval $[0°, 360°)$. Thus the solutions are

$$\alpha = 0° + 360°n = 360°n, \quad n = 0, \pm 1, \pm 2, \ldots,$$

that is, all angles that are coterminal with $0°$.   ■

Recall from Section 1.5 that finding the x-intercepts of the graph of a function $y = f(x)$ is equivalent to solving the equation $f(x) = 0$. The following example makes use of this fact.

■ EXAMPLE 7

Find, without graphing, the first three positive t-intercepts of the graph of

$$f(t) = \sin 2t \cos t.$$

Solution   We must solve $f(t) = 0$, that is,

$$\sin 2t \cos t = 0.$$

It follows that either $\sin 2t = 0$ or $\cos t = 0$. From $\sin 2t = 0$, we obtain

$$2t = n\pi, \quad n = 0, \pm 1, \pm 2, \ldots,$$

or

$$t = \frac{n\pi}{2}, \quad n = 0, \pm 1, \pm 2, \ldots.$$

From $\cos t = 0$, we find

$$t = \frac{\pi}{2} + 2n\pi, \quad n = 0, \pm 1, \pm 2, \ldots.$$

Thus the first three positive t-intercepts are

$$t = \frac{\pi}{2}, \ \pi, \text{ and } \frac{3\pi}{2}.$$ ■

Up to this point all the trigonometric equations we dealt with had solutions that were related by reference angles to the special angles 0, $\pi/6$, $\pi/4$, $\pi/3$, or $\pi/2$. If this is not the case, we can use inverse trigonometric functions and calculators or tables to find solutions, as illustrated in the next example.

▰▰ EXAMPLE 8

Find the solutions of

$$4 \cos^2 x - 3 \cos x - 2 = 0$$

in the interval $[0, \pi]$.

Solution We recognize that this is a quadratic equation in $\cos x$. Since it does not readily factor, we apply the quadratic formula to obtain

$$\cos x = \frac{3 \pm \sqrt{37}}{8}.$$

At this point we discard $(3 + \sqrt{37})/8 \approx 1.14$, since it is greater than 1. Thus we have only

$$\cos x = \frac{3 - \sqrt{37}}{8} \approx -0.3853$$

or

$$x = \cos^{-1}\left(\frac{3 - \sqrt{37}}{8}\right) \approx 1.97.$$

Interestingly, if we had attempted to compute $\cos^{-1}[(3 + \sqrt{37})/8]$ with a calculator, we would have received an error message. ▰▰

EXERCISE 4.5

In Problems 1–6, find all solutions of the given trigonometric equation if t represents an angle measured in radians.

1. $\sin t = \sqrt{3}/2$ **2.** $\cos t = -\sqrt{2}/2$
3. $\sec t = \sqrt{2}$ **4.** $\tan t = -1$
5. $\cot t = -\sqrt{3}$ **6.** $\csc t = 2$

In Problems 7–12, find all solutions of the given trigonometric equation if x represents a real number.

7. $\cos x = -1$ **8.** $2 \sin x = -1$
9. $\tan x = 0$ **10.** $\sqrt{3} \sec x = 2$
11. $-\csc x = 1$ **12.** $\sqrt{3} \cot x = 1$

In Problems 13–18, find all solutions of the given trigonometric equation if θ represents an angle measured in degrees.

13. $\csc \theta = 2\sqrt{3}/3$ **14.** $2 \sin \theta = \sqrt{2}$
15. $\cot \theta + 1 = 0$ **16.** $\sqrt{3} \sin \theta = \cos \theta$
17. $\sec \theta = -2$ **18.** $2 \cos \theta + \sqrt{2} = 0$

In Problems 19–46, find all solutions of the given trigonometric equation if x is a real number and θ is an angle measured in degrees.

19. $\cos^2 x - 1 = 0$
20. $2 \sin^2 x - 3 \sin x + 1 = 0$
21. $3 \sec^2 x = \sec x$
22. $\tan^2 x + (\sqrt{3} - 1) \tan x - \sqrt{3} = 0$
23. $2 \cos^2 \theta - 3 \cos \theta - 2 = 0$
24. $2 \sin^2 \theta - \sin \theta - 1 = 0$
25. $\cot^2 \theta + \cot \theta = 0$
26. $2 \sin^2 \theta + (2 - \sqrt{3}) \sin \theta - \sqrt{3} = 0$
27. $\cos 2x = -1$ **28.** $\sec 2x = 2$
29. $2 \sin 3\theta = 1$ **30.** $\tan 4\theta = -1$
31. $\cot (x/2) = 1$ **32.** $\csc (\theta/3) = -1$
33. $\sin 2x + \sin x = 0$ **34.** $\cos 2x + \sin^2 x = 1$
35. $\cos 2\theta = \sin \theta$
36. $\sin 2\theta + 2 \sin \theta - 2 \cos \theta = 2$
37. $\sin^4 x - 2 \sin^2 x + 1 = 0$

38. $\tan^4 \theta - 2 \sec^2 \theta + 3 = 0$

39. $\sec x \sin^2 x = \tan x$

40. $\dfrac{1 + \cos \theta}{\cos \theta} = 2$

41. $1 + \cot \theta = \csc \theta$

42. $\sin x + \cos x = 0$

43. $\sqrt{\dfrac{1 + 2 \sin x}{2}} = 1$

44. $\sin x + \sqrt{\sin x} = 0$

45. $\cos \theta - \sqrt{\cos \theta} = 0$

46. $\cos \theta \sqrt{1 + \tan^2 \theta} = 1$

In Problems 47–54, find the first three positive *t*-intercepts of the graph of the given function.

47. $f(t) = -5 \sin (3t + \pi)$

48. $f(t) = 2 \cos \left(t + \dfrac{\pi}{4} \right)$

49. $f(t) = 2 - \sec \dfrac{\pi}{2} t$

50. $f(t) = 1 + \cos \pi t$

51. $f(t) = \sin t + \tan t$

52. $f(t) = 1 - 2 \cos \left(t + \dfrac{\pi}{3} \right)$

53. $f(t) = \sin t - \sin 2t$

54. $f(t) = \cos t + \cos 3t$ [*Hint*: Use a sum formula from Section 4.4.]

In Problems 55–60, find the solutions of the given equation in the indicated interval. (Give the answers to the nearest hundredth.)

55. $20 \cos^2 x + \cos x - 1 = 0$, $[0, \pi]$

56. $3 \sin^2 x - 8 \sin x + 4 = 0$, $[-\pi/2, \pi/2]$

57. $\tan^2 x + \tan x - 1 = 0$, $(-\pi/2, \pi/2)$

58. $3 \sin 2x + \cos x = 0$, $[-\pi/2, \pi/2]$

59. $5 \cos^3 x - 3 \cos^2 x - \cos x = 0$, $[0, \pi]$

60. $\tan^4 x - 3 \tan^2 x + 1 = 0$, $(-\pi/2, \pi/2)$

In Problems 61–64, determine by graphing whether the given equation has any solutions.

61. $\tan x = x$ [*Hint*: Graph $y = \tan x$ and $y = x$ on the same set of axes.]

62. $\sin x = x$

63. $\cot x - x = 0$

64. $\cos x + x + 1 = 0$

65. From Problem 52 in Section 4.3, we have that the area of the isosceles triangle with vertex angle θ as shown in Figure 18 is $(x^2/2) \sin \theta$. If the length of x is 4, what value of θ will give a triangle with area 4?

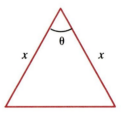

FIGURE 18

66. An object travels in a circular path centered at the origin with constant angular speed. The y-coordinate of the object at any time t seconds is given by

$$y = 8 \cos \left(\pi t - \dfrac{\pi}{12} \right).$$

At what time(s) t does the object cross the x-axis?

67. What is the vertex angle of the cone of sound waves made by an airplane flying at Mach 2? [*Hint*: See Problem 51 in Section 4.3.]

68. An electric generator produces a 60-cycle alternating current given by

$$i(t) = 30 \sin 120\pi(t - \tfrac{7}{36}),$$

where $i(t)$ is the current in amperes at t seconds. Find the smallest positive value of t for which the current is 15 amperes.

69. If the voltage given by

$$V = V_0 \sin (\omega t + \alpha)$$

is impressed on a series circuit, an alternating current is produced. If $V_0 = 110$ volts, $\omega = 120\pi$ radians per second, and $\alpha = -\pi/6$, when is the voltage equal to zero?

70. Consider a ray of light passing from one medium (such as air) into another medium (such as a crystal). Let ϕ be the angle of incidence and θ the angle of refraction. As shown in Figure 19, these angles are measured from a vertical line. According to Snell's law, there is a constant c, depending on the two mediums, such that

$$\dfrac{\sin \phi}{\sin \theta} = c.$$

Assume that for light passing from air into a crystal, $c = 1.437$. Find ϕ and θ such that the angle of incidence is twice the angle of refraction.

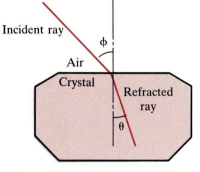

Incident ray

ϕ

Air

Crystal

Refracted ray

θ

FIGURE 19

71. On the basis of data collected from 1966 to 1980, the extent of snow cover S, measured in millions of square kilometers, in the northern hemisphere as a function of time can be approximated by the function

$$S(w) = 25 + 21 \cos (2\pi(w - 5)/52),$$

where w is the number of weeks past January 1.

(a) How much snow cover does this formula predict for April Fool's Day? (Round w to the nearest integer.)

(b) In which week does the formula predict the least amount of snow cover?

(c) What month does this fall in?

IMPORTANT CONCEPTS

Identities	Special formulas	Cofunction formulas
Pythagorean	addition	Reduction of $a_1 \sin bt + a_2 \cos bt$
reciprocal	subtraction	Trigonometric equations
quotient	double-angle	
even–odd	half-angle	
	product and sum	

CHAPTER 4 REVIEW EXERCISE

In Problems 1–10, answer true or false.

1. The equation $\sin x \cos x \tan x = 1 - \cos^2 x$ is an identity. ___

2. The equation $1 + \tan \alpha = \sec \alpha$ is an identity. ___

3. $\cos (u + v) = \cos u + \cos v$. ___

4. The real number $x = \pi/2$ is a solution of $\cos 2x + \sin x = 0$. ___

5. For any real numbers a_1, a_2, b, and t, it is possible to determine A and ϕ such that $a_1 \sin bt + a_2 \cos bt = A \sin (bt + \phi)$. ___

6. $\sin (\pi/5 - \pi/4) =$ $\sin (\pi/5) \cos (\pi/4) - \cos (\pi/5) \sin (\pi/4)$ ___

7. $|\sin (t/2)| = \sqrt{\tfrac{1}{2}(1 - \cos t)}$ ___

8. An x-intercept of the graph of $f(x) = \sin 2x \cos x$ is $x = 5\pi/2$. ___

9. The value $x = \sin^{-1}\left(\dfrac{3 + \sqrt{41}}{8}\right)$ is a solution of $4 \sin^2 x - 3 \sin x - 2 = 0$. ___

10. $2 \sin 4\theta \cos 4\theta = \sin 8\theta$ ___

In Problems 11–18, verify the given identity.

11. $(\sin \alpha + \cos \alpha)^2 + (\sin \alpha - \cos \alpha)^2 = 2$

12. $\dfrac{\sec (-\theta)}{\csc (-\theta)} = -\tan \theta$

13. $\dfrac{\tan x}{\sec x + 1} = \dfrac{\sec x - 1}{\tan x}$ 14. $\dfrac{\tan^3 t - 1}{\tan t - 1} = \sec^2 t + \tan t$

15. $\sin^2 \theta \sec \theta = \sec \theta - \cos \theta$

16. $(\sin x + \cos x)^2 - 1 = \sin 2x$

17. $\dfrac{\tan^2 \beta}{\sec \beta - 1} = 1 + \sec \beta$ 18. $\dfrac{\sin t + \tan t}{1 + \sec t} = \sin t$

In Problems 19 and 20, show that the given trigonometric equation is not an identity.

19. $\sin \theta + \cos \theta = 1$ 20. $\sqrt{1 + \sec^2 x} = \tan x$

In Problems 21–30, find the exact value of the given expression.

21. $\cos (3\pi/8)$ 22. $\tan 75°$
23. $\cos (-7\pi/12)$ 24. $\sin (7\pi/8)$
25. $\tan (7\pi/8)$ 26. $\sin (13\pi/12)$
27. $\sin 15° \sin 45°$ 28. $\sin 45° \cos 15°$
29. $\cos 75° + \cos 15°$ 30. $\sin 15° - \sin 75°$

31. Given that $\cos t = \sqrt{2}/3$, where $3\pi/2 < t < 2\pi$, find $\sin \tfrac{1}{2}t$, $\cos \tfrac{1}{2}t$, $\tan \tfrac{1}{2}t$, $\sin 2t$, $\cos 2t$, and $\tan 2t$.

32. If $\sin u = \tfrac{2}{5}$ and $\cos v = \tfrac{5}{13}$, find $\sin (u + v)$ and $\cos (u - v)$, where $0 < u < \pi/2$ and $3\pi/2 < v < 2\pi$.

In Problems 33–36, find all solutions of the given trigonometric equation if x is a real number and θ is an angle measured in degrees.

33. $\cos 2\theta + \sin \theta = 1$ 34. $\tan^2 \theta = 2 \sec \theta - 2$
35. $2 \cos^2 x - \cos x - 3 = 0$ 36. $\sin x - \sqrt{\sin x} = 0$

In Problems 37 and 38, find the solutions (to the nearest hundredth) of the given equation in the indicated interval.

37. $3 \cos 2x + \sin x = 0$, $[-\pi/2, \pi/2]$
38. $\tan^4 x + \tan^2 x - 1 = 0$, $(-\pi/2, \pi/2)$
39. Show that $\arcsin \tfrac{3}{5} + \arcsin \tfrac{5}{13} = \arcsin \tfrac{56}{65}$. [*Hint:* Use the addition formula for the sine function.]
40. Under certain conditions the vertical displacement y (measured in feet from the equilibrium position) of a weight on the end of a spring after t seconds is given by

$$y = -\tfrac{2}{3} \cos 10t + \tfrac{1}{2} \sin 10t$$

(see Figure 37 in Chapter 3).
  (a) Show that $y = \tfrac{5}{6} \sin (10t - 0.927)$.
  (b) At what times does the weight pass through the equilibrium position?

41. Under certain conditions it can be shown that the displacement angle θ (measured in radians) of a plane pendulum at time t seconds is given by

$$\theta = c_1 \cos \sqrt{g/L}t + c_2 \sin \sqrt{g/L}t,$$

where L is the length of the pendulum, g is the acceleration due to gravity, and c_1 and c_2 are constants that depend on how the pendulum is set in motion (see Figure 20). If $c_1 = \sqrt{3}$ and $c_2 = 1$, determine A, b, and ϕ are such that $\theta = A \sin (bt + \phi)$.

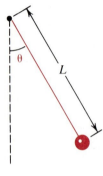

FIGURE 20

42. From a spacecraft S the point P on earth is observed, as shown in Figure 21. To locate P we must determine the angle β formed by the line from P to the center C of the earth and the line from S to C. Let A be a point on the horizon circle (see Problem 86 in the Chapter 2 Review Exercise) such that $AS \perp AC$, as shown in the figure. Measurements of the angle α formed by PS and CS and the angle θ formed by AS and CS can be made from the spacecraft.
  (a) Show that

$$\tan \alpha = \frac{\sin \theta \sin \beta}{1 - \sin \theta \cos \beta}.$$

  [*Hint:* Drop a perpendicular from P to CS.]
  (b) If the measurements $\alpha = 27°$ and $\theta = 30°$ are obtained, find β.

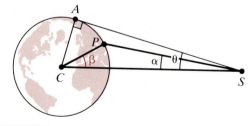

FIGURE 21

5

Vectors and Complex Numbers

Gauss

In the quest to find the solutions or roots of equations, mathematicians of antiquity often encountered square roots of negative numbers. The Italian physician and mathematician Geronimo Cardano (1501–1576) called expressions such as $\sqrt{-5}$ and $1 + \sqrt{-1}$ "numerificti." "Fictitious" or "imaginary" numbers like these were ignored and discarded even though they were, in turn, recognized as formal solutions of the equations $x^2 + 5 = 0$ and $x^2 - 2x + 2 = 0$. Over the next 200 years, mathematicians' understanding and acceptance of the usefulness of "imaginary numbers" slowly evolved; even the name "imaginary number," with its implicit connotation of nonexistence, gave way to the more acceptable but less colorful name "complex number." In 1748 the Swiss mathematician Leonhard Euler (1707–1783) showed how the trigonometric functions sine and cosine could be defined using $\sqrt{-1}$. It was Euler who first introduced the symbol i to denote $\sqrt{-1}$. Still a mathematical curiosity, the complex number system finally attained respectability through the work of the German mathematician Carl Friedrich Gauss (1777–1855). Gauss, the historically recognized "prince of mathematicians," was one of the first to use the complex number system as a working tool; he was also one of the first mathematicians to represent a complex number geometrically. But it was through the prolific efforts of Augustin-Louis Cauchy (1789–1857) that the field of *complex variables* became an established and sophisticated branch of mathematics.

We will see that the two topics of study in this chapter, vectors in the plane and complex numbers, share an important property. A vector and a complex number are uniquely determined by an ordered pair of real numbers (a, b).

5.1 Vectors

In order to describe certain physical quantities accurately, we must have two pieces of information: a magnitude and a direction. For example, when we discuss the flight of an airplane, both its speed and its heading are important. Quantities that involve both magnitude and direction are represented by **vectors.** As we see in Figure 1, a vector is frequently depicted as a directed line segment, that is, a line segment with a direction specified by an arrowhead. We use bold-faced letters, such as **v** or **w,** to denote vectors. If a vector extends from point A to point B with an arrowhead at B, then A is called the **initial point** and B is called the **terminal point.** The vector is denoted $\mathbf{v} = \overrightarrow{AB}$.

$\mathbf{v} = \overrightarrow{AB}$

B Terminal point

A Initial point

FIGURE 1

MAGNITUDE AND DIRECTION

The length of the directed line segment is called the **magnitude** of the vector $\overrightarrow{AB}$ and is denoted by $|\overrightarrow{AB}|$. Two vectors $\overrightarrow{AB}$ and $\overrightarrow{CD}$ are said to be **equal,** written

$$\overrightarrow{AB} = \overrightarrow{CD},$$

if they have both the same magnitude and the same direction, as shown in Figure 2. Thus vectors can be translated from one position to another so long as neither the magnitude nor the direction is changed.

In this section we will consider only vectors that lie in the same coordinate plane. Since we can move a vector provided its magnitude and its direction are unchanged, we can place the initial point at the origin. Then, as shown in Figure 3, the terminal point P will have rectangular coordinates (x, y). Conversely, every ordered pair of real numbers (x, y) determines a vector $\overrightarrow{OP}$, where P has rectangular coordinates (x, y). Thus we have a one-to-one correspondence between vectors and ordered pairs of real numbers. We say that $\mathbf{v} = \overrightarrow{OP}$ is the **position vector** of the point $P(x, y)$. The numbers x and y are called the **components** of $\mathbf{v} = \overrightarrow{OP}$, and we write

$$\mathbf{v} = \langle x, y \rangle.$$

To avoid confusing the vector $\langle x, y \rangle$ with the point $P(x, y)$, we will use the special brackets $\langle\ \rangle$ to denote the vector. Since the magnitude of $\langle a, b \rangle$ is the distance from (a, b) to the origin, we have the following result.

B

D

C

A

$$\overrightarrow{AB} = \overrightarrow{CD}$$

FIGURE 2

y

$P(x, y)$

$\mathbf{v} = \overrightarrow{OP} = \langle x, y \rangle$

O x

FIGURE 3

Magnitude

The **magnitude** $|\mathbf{v}|$ of the vector $\mathbf{v} = \langle a, b \rangle$ is

$$|\mathbf{v}| = \sqrt{a^2 + b^2}.$$

The **zero vector,** denoted by $\mathbf{0}$, is defined by $\mathbf{0} = \langle 0, 0 \rangle$. Thus the magnitude of the zero vector is zero.

Let $\mathbf{v} = \langle x, y \rangle$ be a nonzero vector. Let θ be an angle in standard position formed by $\mathbf{v}$ and the positive x-axis (see Figure 4). Then θ is called a **direction angle** for $\mathbf{v}$. We note that any angle coterminal with θ is also a direction angle for $\mathbf{v}$. Thus a vector $\mathbf{v}$ can be specified by giving either its components $\mathbf{v} = \langle x, y \rangle$ or its magnitude $|\mathbf{v}|$ and a direction angle. From trigonometry, we obtain the following relationships.

Direction Angle

For any nonzero vector $\mathbf{v} = \langle x, y \rangle$ with **direction angle** θ:

$$\cos \theta = \frac{x}{|\mathbf{v}|}, \quad \text{or} \quad x = |\mathbf{v}| \cos \theta,$$

$$\sin \theta = \frac{y}{|\mathbf{v}|}, \quad \text{or} \quad y = |\mathbf{v}| \sin \theta,$$

$$\tan \theta = \frac{y}{x}, \quad \text{if } x \neq 0,$$

where $|\mathbf{v}| = \sqrt{x^2 + y^2}$.

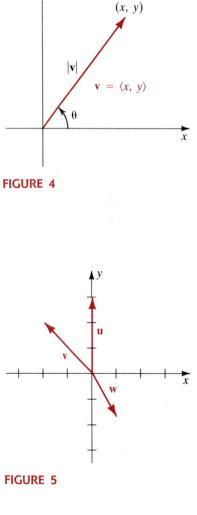

FIGURE 4

The zero vector is not assigned any direction.

EXAMPLE 1

Sketch each of the following vectors. Find the magnitude and the smallest positive direction angle θ of each vector.

(a) $\mathbf{v} = \langle -2, 2 \rangle$ **(b)** $\mathbf{u} = \langle 0, 3 \rangle$ **(c)** $\mathbf{w} = \langle 1, -\sqrt{3} \rangle$

Solution The vectors are sketched in Figure 5. This will assist us in finding the desired direction angles.

FIGURE 5

(a) From the preceding discussion we have

$$|\mathbf{v}| = \sqrt{(-2)^2 + 2^2} = \sqrt{8} = 2\sqrt{2}$$

and $$\tan \theta = \frac{2}{-2} = -1.$$

Since θ is a quadrant II angle (Figure 5), we conclude that $\theta = 3\pi/4$.

(b) The magnitude of $\mathbf{u}$ is $|\mathbf{u}| = \sqrt{0^2 + 3^2} = 3$ and from Figure 5 we see that $\theta = \pi/2$.

(c) The magnitude of $\mathbf{w}$ is $|\mathbf{w}| = \sqrt{1^2 + (-\sqrt{3})^2} = \sqrt{4} = 2$. Since $\tan \theta = -\sqrt{3}$ and θ is a quadrant IV angle, we select $\theta = 5\pi/3$.

OPERATIONS ON VECTORS

There are a variety of operations on vectors. We will define the **sum, difference,** and **scalar multiple.** A constant k is referred to as a **scalar.**

Operations on Vectors

Let $\mathbf{u} = \langle a, b \rangle$ and $\mathbf{v} = \langle c, d \rangle$ be vectors, and let k be any real number. Then we define the:

Sum: $\mathbf{u} + \mathbf{v} = \langle a + c, b + d \rangle$;

Difference: $\mathbf{u} - \mathbf{v} = \langle a - c, b - d \rangle$;

Scalar multiple: $k\mathbf{u} = \langle ka, kb \rangle$.

These operations satisfy the following properties.

Properties of Vectors

(i) $\mathbf{u} + \mathbf{v} = \mathbf{v} + \mathbf{u}$

(ii) $\mathbf{u} + (\mathbf{v} + \mathbf{w}) = (\mathbf{u} + \mathbf{v}) + \mathbf{w}$

(iii) $k(\mathbf{u} + \mathbf{v}) = k\mathbf{u} + k\mathbf{v}$

(iv) $\mathbf{u} - \mathbf{v} = \mathbf{u} + (-1)\mathbf{v}$

(v) $\mathbf{0} + \mathbf{u} = \mathbf{u} = \mathbf{u} + \mathbf{0}$

(vi) $0\mathbf{u} = \mathbf{0}, \quad 1\mathbf{u} = \mathbf{u}$

■ EXAMPLE 2

If $\mathbf{u} = \langle 2, 1 \rangle$ and $\mathbf{v} = \langle -1, 5 \rangle$ find $4\mathbf{u}$, $\mathbf{u} + \mathbf{v}$, and $3\mathbf{u} - 2\mathbf{v}$.

Solution From the definitions of addition, subtraction, and scalar multiples of vectors, we find

$$4\mathbf{u} = 4\langle 2, 1 \rangle = \langle 8, 4 \rangle,$$
$$\mathbf{u} + \mathbf{v} = \langle 2, 1 \rangle + \langle -1, 5 \rangle = \langle 1, 6 \rangle,$$
$$3\mathbf{u} - 2\mathbf{v} = 3\langle 2, 1 \rangle - 2\langle -1, 5 \rangle = \langle 6, 3 \rangle - \langle -2, 10 \rangle = \langle 8, -7 \rangle.$$

GEOMETRIC INTERPRETATIONS

The sum $\mathbf{u} + \mathbf{v}$ of two vectors can be interpreted geometrically as follows. If $\mathbf{u} = \langle a, b \rangle$ and $\mathbf{v} = \langle c, d \rangle$, then $\mathbf{u}, \mathbf{v},$ and $\mathbf{u} + \mathbf{v}$ can be represented by directed line segments from the origin to the points $A(a, b)$, $B(c, d)$, and $C(a + c, b + d)$, respectively. As shown in Figure 6, if the vector $\mathbf{v}$ is translated so that its initial point is A, then its terminal point will be C. Thus a geometric representation of the sum $\mathbf{u} + \mathbf{v}$ can be obtained by placing the initial point of $\mathbf{v}$ on the terminal point of $\mathbf{u}$ and drawing the vector from the initial point of $\mathbf{u}$ to the terminal point of $\mathbf{v}$. By examining the coordinates of the quadrilateral $OACB$ in Figure 6, we see that it is a parallelogram formed by the vectors $\mathbf{u}$ and $\mathbf{v}$, with $\mathbf{u} + \mathbf{v}$ as one of its diagonals.

We now consider scalar multiples of the vector $\mathbf{v} = \langle x, y \rangle$. Let k be any real number; then

$$|k\mathbf{v}| = \sqrt{(kx)^2 + (ky)^2} = \sqrt{k^2(x^2 + y^2)}$$
$$= \sqrt{k^2}\sqrt{x^2 + y^2} = |k|\sqrt{x^2 + y^2} = |k||\mathbf{v}|.$$

We have derived the following property of scalar multiplication:

$$|k\mathbf{v}| = |k||\mathbf{v}|.$$

This property states that in the scalar multiplication of a vector $\mathbf{v}$ by a real number k, the magnitude of $\mathbf{v}$ is multiplied by $|k|$. As shown in Figure 7, if $k > 0$, the direction of $\mathbf{v}$ does not change; but if $k < 0$, the direction of $\mathbf{v}$ is reversed. In particular, $\mathbf{v}$ and $-\mathbf{v}$ have the same length but opposite direction.

The geometric interpretation of the difference $\mathbf{u} - \mathbf{v}$ of two vectors is obtained by observing that

$$\mathbf{u} = \mathbf{v} + (\mathbf{u} - \mathbf{v}).$$

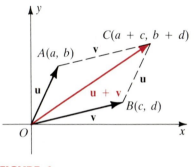

FIGURE 6

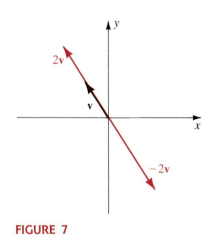

FIGURE 7

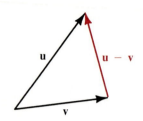

FIGURE 8

Thus, **u** − **v** is the vector that when added to **v** yields **u.** As we see in Figure 8, the initial point of **u** − **v** will be at the terminal point of **v,** and the terminal point of **u** − **v** coincides with the terminal point of **u.** Hence the vector **u** − **v** is one diagonal of the parallelogram determined by **u** and **v,** with **u** + **v** being the other diagonal (see Figure 9).

▮ EXAMPLE 3

Let **u** = ⟨−1, 1⟩ and let **v** = ⟨3, 2⟩. Sketch the geometric interpretations of **u** + **v** and **u** − **v.**

Solution We form the parallelogram determined by the vectors **u** and **v** and identify the diagonals **u** + **v** and **u** − **v,** as shown in Figure 10.

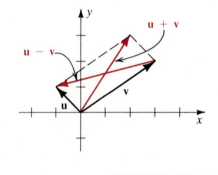

FIGURE 10

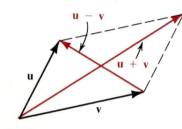

FIGURE 9

▮ EXAMPLE 4

Let **u** = ⟨1, 2⟩. The geometric interpretations of 2**u** and −**u** are shown in Figure 11.

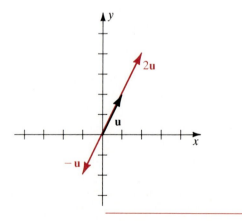

FIGURE 11

UNIT VECTORS

Any vector with magnitude 1 is called a **unit vector.** The unit vectors in the direction of the positive x- and y-axes are denoted

$$\mathbf{i} = \langle 1, 0 \rangle \quad \text{and} \quad \mathbf{j} = \langle 0, 1 \rangle.$$

See Figure 12.

The unit vectors **i** and **j** provide us with another way of denoting vectors. If $\mathbf{u} = \langle a, b \rangle$, then

$$\mathbf{u} = \langle a, 0 \rangle + \langle 0, b \rangle = a\langle 1, 0 \rangle + b\langle 0, 1 \rangle,$$

or

$$\mathbf{u} = \langle a, b \rangle = a\mathbf{i} + b\mathbf{j}.$$

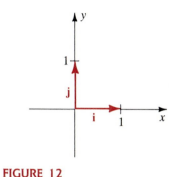

FIGURE 12

As shown in Figure 13, since **i** and **j** are unit vectors, the vectors $a\mathbf{i}$ and $b\mathbf{j}$ are horizontal and vertical vectors of length $|a|$ and $|b|$, respectively. For this reason, a is called the **horizontal component** of **u,** and b is called the **vertical component.** The vector $a\mathbf{i} + b\mathbf{j}$ is often referred to as a **linear combination** of **i** and **j.** Using this notation for the vectors $\mathbf{u} = a\mathbf{i} + b\mathbf{j}$ and $\mathbf{v} = c\mathbf{i} + d\mathbf{j}$, we can write the definitions of the sum, difference, and scalar multiples of **u** and **v** as follows:

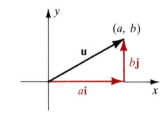

FIGURE 13

Sum: $\qquad (a\mathbf{i} + b\mathbf{j}) + (c\mathbf{i} + d\mathbf{j}) = (a + c)\mathbf{i} + (b + d)\mathbf{j},$

Difference: $\qquad (a\mathbf{i} + b\mathbf{j}) - (c\mathbf{i} + d\mathbf{j}) = (a - c)\mathbf{i} + (b - d)\mathbf{j},$

Scalar multiple: $\qquad k(a\mathbf{i} + b\mathbf{j}) = (ka)\mathbf{i} + (kb)\mathbf{j}.$

EXAMPLE 5

If $\mathbf{u} = 3\mathbf{i} + \mathbf{j}$ and $\mathbf{v} = 5\mathbf{i} - 2\mathbf{j}$, find $4\mathbf{u} - 2\mathbf{v}$.

Solution We have

$$\begin{aligned}
4\mathbf{u} - 2\mathbf{v} &= 4(3\mathbf{i} + \mathbf{j}) - 2(5\mathbf{i} - 2\mathbf{j}) \\
&= (12\mathbf{i} + 4\mathbf{j}) - (10\mathbf{i} - 4\mathbf{j}) \\
&= 2\mathbf{i} + 8\mathbf{j}.
\end{aligned}$$

THE POLAR FORM OF A VECTOR

There is yet another way of representing vectors. For a nonzero vector $\mathbf{v} = \langle x, y \rangle$ with direction angle θ, we recall that $x = |\mathbf{v}| \cos \theta$ and $y = |\mathbf{v}| \sin \theta$. Thus,

$$\mathbf{v} = x\mathbf{i} + y\mathbf{j} = |\mathbf{v}| \cos \theta \mathbf{i} + |\mathbf{v}| \sin \theta \mathbf{j},$$

or

$$\mathbf{v} = |\mathbf{v}|(\cos \theta \mathbf{i} + \sin \theta \mathbf{j}).$$

This latter representation is called the **polar form** of **v,** or **trigonometric form** of the vector **v.**

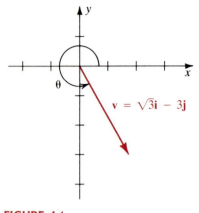

FIGURE 14

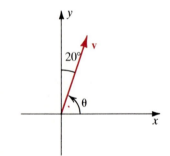

FIGURE 15

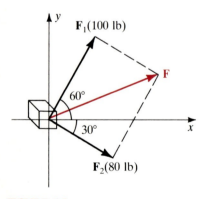

FIGURE 16

▦ EXAMPLE 6

Express $\mathbf{v} = \sqrt{3}\mathbf{i} - 3\mathbf{j}$ in polar form.

Solution To write $\mathbf{v}$ in polar form, we must find the magnitude $|\mathbf{v}|$ and the direction angle θ. We find

$$|\mathbf{v}| = \sqrt{(\sqrt{3})^2 + (-3)^2} = \sqrt{12} = 2\sqrt{3}$$

and $\tan \theta = -3/\sqrt{3} = -\sqrt{3}.$

To determine θ, we sketch $\mathbf{v}$ to find that θ is a quadrant IV angle (see Figure 14). Thus, $\theta = 5\pi/3$, and in polar form,

$$\mathbf{v} = 2\sqrt{3}\left(\cos \frac{5\pi}{3}\mathbf{i} + \sin \frac{5\pi}{3}\mathbf{j}\right).$$

■

▦ EXAMPLE 7

Given that an airplane is flying at 200 mph on a bearing of N 20° E, express its velocity as a vector.

Solution The desired vector $\mathbf{v}$ is shown in Figure 15. Since $\theta = 90° - 20° = 70°$ and $|\mathbf{v}| = 200$, we have

$$\mathbf{v} = 200(\cos 70°\mathbf{i} + \sin 70°\mathbf{j}) \approx 68.4\mathbf{i} + 187.9\mathbf{j}.$$

■

In physics it is shown that when two forces act simultaneously at the same point P on an object, the object reacts as though a single force equal to the vector sum of the two forces is acting on the object at P. This single force is called the **resultant force.**

▦ EXAMPLE 8

Two people push on a crate with forces $\mathbf{F}_1$ and $\mathbf{F}_2$, whose magnitudes and directions are shown in Figure 16. Find the magnitude and the direction of the resultant force.

Solution From Figure 16, we see that the direction angles for the two forces $\mathbf{F}_1$ and $\mathbf{F}_2$ are $\theta_1 = 60°$ and $\theta_2 = 330°$, respectively. Thus,

$$\mathbf{F}_1 = 100(\cos 60°\mathbf{i} + \sin 60°\mathbf{j}) = 50\mathbf{i} + 50\sqrt{3}\mathbf{j}$$

and $\mathbf{F}_2 = 80(\cos 330°\mathbf{i} + \sin 330°\mathbf{j}) = 40\sqrt{3}\mathbf{i} - 40\mathbf{j}.$

The resultant force $\mathbf{F}$ can then be found by vector addition:

$$\mathbf{F} = \mathbf{F}_1 + \mathbf{F}_2 = (50\mathbf{i} + 50\sqrt{3}\mathbf{j}) + (40\sqrt{3}\mathbf{i} - 40\mathbf{j})$$
$$= (50 + 40\sqrt{3})\mathbf{i} + (50\sqrt{3} - 40)\mathbf{j}.$$

Thus the magnitude $|\mathbf{F}|$ of the resultant force is

$$|\mathbf{F}| = \sqrt{(50 + 40\sqrt{3})^2 + (50\sqrt{3} - 40)^2} \approx 128.06.$$

If θ is a direction angle for $\mathbf{F}$, then

$$\tan \theta = (50\sqrt{3} - 40)/(50 + 40\sqrt{3}) \approx 0.3907.$$

Since θ is a quadrant I angle, we find that

$$\theta \approx 21.34°.$$

EXERCISE 5.1

In Problems 1–8, sketch the given vector. Find the magnitude and the smallest positive direction angle of each vector.

1. $\langle \sqrt{3}, -1 \rangle$
2. $\langle 4, -4 \rangle$
3. $\langle 5, 0 \rangle$
4. $\langle -2, 2\sqrt{3} \rangle$
5. $-4\mathbf{i} + 4\sqrt{3}\mathbf{j}$
6. $\mathbf{i} - \mathbf{j}$
7. $-10\mathbf{i} + 10\mathbf{j}$
8. $-3\mathbf{j}$

In Problems 9–14, find $\mathbf{u} + \mathbf{v}$, $\mathbf{u} - \mathbf{v}$, $-3\mathbf{u}$, and $3\mathbf{u} - 4\mathbf{v}$.

9. $\mathbf{u} = \langle 2, 3 \rangle$, $\mathbf{v} = \langle 1, -1 \rangle$
10. $\mathbf{u} = \langle 4, -2 \rangle$, $\mathbf{v} = \langle 10, 2 \rangle$
11. $\mathbf{u} = \langle -4, 2 \rangle$, $\mathbf{v} = \langle 4, 1 \rangle$
12. $\mathbf{u} = \langle -1, -5 \rangle$, $\mathbf{v} = \langle 8, 7 \rangle$
13. $\mathbf{u} = \langle -5, -7 \rangle$, $\mathbf{v} = \langle \frac{1}{2}, -\frac{1}{4} \rangle$
14. $\mathbf{u} = \langle 0.1, 0.2 \rangle$, $\mathbf{v} = \langle -0.3, 0.4 \rangle$

In Problems 15–20, find $\mathbf{u} - 4\mathbf{v}$ and $2\mathbf{u} + 5\mathbf{v}$.

15. $\mathbf{u} = \mathbf{i} - 2\mathbf{j}$, $\mathbf{v} = 8\mathbf{i} + 3\mathbf{j}$
16. $\mathbf{u} = \mathbf{j}$, $\mathbf{v} = 4\mathbf{i} - \mathbf{j}$
17. $\mathbf{u} = \frac{1}{2}\mathbf{i} - \frac{3}{2}\mathbf{j}$, $\mathbf{v} = 2\mathbf{i}$
18. $\mathbf{u} = 2\mathbf{i} - 3\mathbf{j}$, $\mathbf{v} = 3\mathbf{i} - 2\mathbf{j}$
19. $\mathbf{u} = 0.2\mathbf{i} + 0.1\mathbf{j}$, $\mathbf{v} = -1.4\mathbf{i} - 2.1\mathbf{j}$
20. $\mathbf{u} = 5\mathbf{i} - 10\mathbf{j}$, $\mathbf{v} = -10\mathbf{i}$

In Problems 21–24, sketch the geometric interpretations of $\mathbf{u} + \mathbf{v}$ and $\mathbf{u} - \mathbf{v}$.

21. $\mathbf{u} = 2\mathbf{i} + 3\mathbf{j}$, $\mathbf{v} = -\mathbf{i} + 2\mathbf{j}$
22. $\mathbf{u} = -4\mathbf{i} + \mathbf{j}$, $\mathbf{v} = 2\mathbf{i} + 2\mathbf{j}$
23. $\mathbf{u} = 5\mathbf{i} - \mathbf{j}$, $\mathbf{v} = 4\mathbf{i} - 3\mathbf{j}$
24. $\mathbf{u} = 2\mathbf{i} - 7\mathbf{j}$, $\mathbf{v} = -7\mathbf{i} - 3\mathbf{j}$

In Problems 25–28, sketch the geometric interpretations of $2\mathbf{u}$ and $-2\mathbf{u}$.

25. $\mathbf{u} = \langle -2, 1 \rangle$
26. $\mathbf{u} = \langle 4, 7 \rangle$
27. $\mathbf{u} = 3\mathbf{i} - 5\mathbf{j}$
28. $\mathbf{u} = -\frac{1}{2}\mathbf{i} + \frac{3}{2}\mathbf{j}$

In Problems 29–32, if $\mathbf{u} = 3\mathbf{i} - \mathbf{j}$ and $\mathbf{v} = 2\mathbf{i} + 4\mathbf{j}$, find the horizontal and the vertical components of the indicated vector.

29. $2\mathbf{u} - \mathbf{v}$
30. $3(\mathbf{u} + \mathbf{v})$
31. $\mathbf{v} - 4\mathbf{u}$
32. $4(\mathbf{u} + 3\mathbf{v})$

In Problems 33–36, express the given vector (a) in polar form and (b) as a linear combination of the unit vectors $\mathbf{i}$ and $\mathbf{j}$.

33. $\langle -\sqrt{2}, \sqrt{2} \rangle$
34. $\langle 7, 7\sqrt{3} \rangle$
35. $\langle -3\sqrt{3}, 3 \rangle$
36. $\langle -4, -4 \rangle$

37. Two forces $\mathbf{F}_1$ and $\mathbf{F}_2$ of magnitudes 4 kg and 7 kg, respectively, act on a point. If the angle between the forces is 47°, find the magnitude of the resultant force $\mathbf{F}$ and the angle between $\mathbf{F}_1$ and $\mathbf{F}$.

38. The resultant **F** of two forces **F₁** and **F₂** has a magnitude of 100 lb and direction as shown in Figure 17. If $\mathbf{F_1} = -200\mathbf{i}$, find the horizontal and the vertical components of **F₂**.

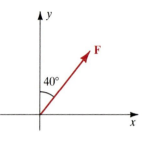

FIGURE 17

39. A 10-lb weight is hanging from a rope. A 2-lb force is applied horizontally to the weight, moving the rope from its horizontal position (see Figure 18). Find the resultant of this force and the force due to gravity.

FIGURE 18

40. A small boat is pulled along a canal by two tow ropes on opposite sides of the canal. The angle between the ropes is 50°. If one rope is pulled with a force of 250 lb and the other with a force of 400 lb, find the magnitude of the resultant force and the angle it makes with the 250-lb force.

41. The current in a river that is 0.5 mi across is 6 mph. A swimmer heads out from shore perpendicular to the current at 2 mph. In what direction is the swimmer actually going?

42. As a freight train, traveling at 10 mph, passes a landing, a mail sack is tossed out perpendicular to the train with a velocity of 15 feet per second. In what direction does it slide on the landing?

43. In order for an airplane to fly due north at 300 mph, it must set a course 10° west of north (N 10° W) because of a strong wind blowing due east. What is the speed of the wind?

44. A hiker walks 1.0 mi to the northeast, then 1.5 mi to the east, and then 2.0 mi to the southeast. What are the hiker's distance and bearing from the starting point? [*Hint:* Each part of the journey can be represented by a vector. Find the vector sum.]

5.2   The Dot Product

In Section 5.1 we saw that two vectors **u** and **v** can be combined by means of addition and subtraction. We can also combine vectors using a special product called the **dot,** or **inner, product.** This product is not another vector but is a real number defined in terms of the components of the vectors.

DEFINITION 1

The **dot product u · v** of vectors $\mathbf{u} = \langle a_1, a_2 \rangle = a_1\mathbf{i} + a_2\mathbf{j}$ and $\mathbf{v} = \langle b_1, b_2 \rangle = b_1\mathbf{i} + b_2\mathbf{j}$ is the real number

$$\mathbf{u} \cdot \mathbf{v} = a_1 b_1 + a_2 b_2. \tag{1}$$

EXAMPLE 1

Find the dot product of each of the following pairs of vectors.

(a) $\mathbf{u} = 5\mathbf{i} - 7\mathbf{j}, \mathbf{v} = -10\mathbf{i} - 6\mathbf{j}$
(b) $\mathbf{u} = \langle 4, 9 \rangle, \mathbf{v} = \langle 3, 2 \rangle$
(c) $\mathbf{u} = \mathbf{i}, \mathbf{v} = -6\mathbf{i}$

Solution

(a) With the identifications $a_1 = 5$, $a_2 = -7$, $b_1 = -10$, and $b_2 = -6$, it follows from (1) that

$$\mathbf{u} \cdot \mathbf{v} = (5)(-10) + (-7)(-6) = -50 + 42 = -12.$$

(b) $\mathbf{u} \cdot \mathbf{v} = (4)(3) + (9)(2) = 12 + 18 = 30$
(c) In this case, we identify $a_1 = 1$, $a_2 = 0$, $b_1 = -6$, and $b_2 = 0$. Thus, $\mathbf{u} \cdot \mathbf{v} = (1)(-6) = -6$.

In Section 5.1 we saw that two nonzero vectors **u** and **v** can be translated in the plane so that they have a common initial point O. Suppose that $\mathbf{u} = \overrightarrow{OA}$ and $\mathbf{v} = \overrightarrow{OB}$. We define **the angle θ between u and v** as the angle AOB satisfying $0 \le \theta \le \pi$ (see Figure 19(a)). We say that two nonzero vectors **u** and **v** are **parallel** if either $\theta = 0$ or $\theta = \pi$. The vectors point in the same direction if $\theta = 0$ and point in the opposite direction if $\theta = \pi$ (see Figures 19(b) and (c)). Vectors **u** and **v** are said to be **orthogonal** or **perpendicular** if $\theta = \pi/2$ (see Figure 19(d)).

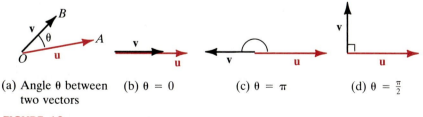

(a) Angle θ between two vectors   (b) θ = 0   (c) θ = π   (d) θ = $\frac{\pi}{2}$

FIGURE 19

Using the law of cosines, we can obtain an alternative formula for the dot product in terms of the angle between the two vectors **u** and **v**. If the vectors

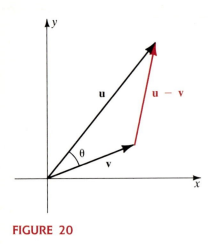

$\mathbf{u} = a_1\mathbf{i} + a_2\mathbf{j}$ and $\mathbf{v} = b_1\mathbf{i} + b_2\mathbf{j}$ form two sides of a triangle, as shown in Figure 20, then

$$|\mathbf{u} - \mathbf{v}|^2 = |\mathbf{u}|^2 + |\mathbf{v}|^2 - 2|\mathbf{u}||\mathbf{v}| \cos \theta,$$

or $\quad (a_1 - b_1)^2 + (a_2 - b_2)^2 = a_1^2 + a_2^2 + b_1^2 + b_2^2 - 2|\mathbf{u}||\mathbf{v}| \cos \theta.$

After multiplying out and simplifying, we see that the last equation becomes

$$-2a_1b_1 - 2a_2b_2 = -2|\mathbf{u}||\mathbf{v}| \cos \theta, \quad \text{or} \quad a_1b_1 + a_2b_2 = |\mathbf{u}||\mathbf{v}| \cos \theta.$$

Since $\mathbf{u} \cdot \mathbf{v} = a_1b_1 + a_2b_2$, we have proved the next result.

THEOREM 1

For nonzero vectors $\mathbf{u}$ and $\mathbf{v}$,

$$\mathbf{u} \cdot \mathbf{v} = |\mathbf{u}||\mathbf{v}| \cos \theta, \tag{2}$$

where $0 \le \theta \le \pi$.

We can now determine the angle between two nonzero vectors by using (2) in the form

$$\cos \theta = \frac{\mathbf{u} \cdot \mathbf{v}}{|\mathbf{u}||\mathbf{v}|}. \tag{3}$$

EXAMPLE 2

Find the angle between $\mathbf{u} = -2\mathbf{i} + 5\mathbf{j}$ and $\mathbf{v} = 6\mathbf{i} + \mathbf{j}.$

Solution   We have

$$\mathbf{u} \cdot \mathbf{v} = (-2)(6) + (5)(1) = -7, \qquad |\mathbf{u}| = \sqrt{29}, \qquad |\mathbf{v}| = \sqrt{37},$$

and so from (3), $\qquad \cos \theta = \dfrac{-7}{\sqrt{29}\sqrt{37}}.$

Since $0 \le \theta \le \pi$, we can write

$$\theta = \arccos \left(-7/\sqrt{29}\sqrt{37}\right) \approx 1.79 \text{ radians.}$$

That is, $\theta \approx 102.34°$. The vectors and the angle between them are illustrated in Figure 21.

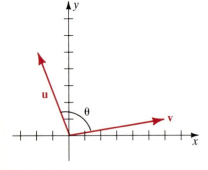

It is apparent from (2) that if $\theta = \pi/2$, then $\mathbf{u} \cdot \mathbf{v} = 0$ and, conversely, that if $\mathbf{u} \cdot \mathbf{v} = 0$, then $\theta = \pi/2$. We summarize these facts in the next theorem.

THEOREM 2

Two nonzero vectors $\mathbf{u}$ and $\mathbf{v}$ are **orthogonal** if and only if $\mathbf{u} \cdot \mathbf{v} = 0$.

EXAMPLE 3

Verify that each of the following pairs of vectors is orthogonal:
(a) $\mathbf{u} = 3\mathbf{i} - 9\mathbf{j}$, $\mathbf{v} = 15\mathbf{i} + 5\mathbf{j}$ and **(b)** $\mathbf{i}$, $\mathbf{j}$.

Solution

(a) From (1) we have

$$\mathbf{u} \cdot \mathbf{v} = 3(15) + (-9)(5) = 45 - 45 = 0.$$

It follows from Theorem 2 that $\mathbf{u}$ and $\mathbf{v}$ are orthogonal.

(b) Since $\mathbf{i} = \langle 1, 0 \rangle$ and $\mathbf{j} = \langle 0, 1 \rangle$, we have

$$\mathbf{i} \cdot \mathbf{j} = (1)(0) + (0)(1) = 0.$$

Thus, $\mathbf{i}$ and $\mathbf{j}$ are orthogonal.

The dot product has the following properties.

Properties of the Dot Product

Let $\mathbf{u}$, $\mathbf{v}$, and $\mathbf{w}$ be vectors, and let k be a real number.

(i) $\mathbf{u} \cdot \mathbf{v} = \mathbf{v} \cdot \mathbf{u}$

(ii) $\mathbf{u} \cdot (\mathbf{v} + \mathbf{w}) = \mathbf{u} \cdot \mathbf{v} + \mathbf{u} \cdot \mathbf{w}$

(iii) $k(\mathbf{u} \cdot \mathbf{v}) = (k\mathbf{u} \cdot \mathbf{v}) = \mathbf{u} \cdot (k\mathbf{v})$

(iv) $\mathbf{u} \cdot \mathbf{u} = |\mathbf{u}|^2$

We will prove (iii) and (iv) and leave the proofs of (i) and (ii) as exercises. Suppose that $\mathbf{u} = a_1\mathbf{i} + a_2\mathbf{j}$ and $\mathbf{v} = b_1\mathbf{i} + b_2\mathbf{j}$. Then by definition,

$$k(\mathbf{u} \cdot \mathbf{v}) = k(a_1 b_1 + a_2 b_2) = (ka_1)b_1 + (ka_2)b_2 = (k\mathbf{u}) \cdot \mathbf{v}.$$

Now, $\quad (k\mathbf{u}) \cdot \mathbf{v} = (ka_1)b_1 + (ka_2)b_2 = a_1(kb_1) + a_2(kb_2) = \mathbf{u} \cdot (k\mathbf{v})$.

This proves (iii).

To prove (iv), we observe that

$$\mathbf{u} \cdot \mathbf{u} = a_1^2 + a_2^2 = |\mathbf{u}|^2.$$

Property (iv) can be used to express the magnitude of a vector **u** in terms of the dot product:

$$|\mathbf{u}| = \sqrt{\mathbf{u} \cdot \mathbf{u}}.$$

COMPONENT OF A VECTOR ALONG ANOTHER VECTOR

The components a_1 and a_2 of a vector $\mathbf{v} = a_1\mathbf{i} + a_2\mathbf{j}$ can be expressed in terms of the dot product. Using the fact that the unit vectors **i** and **j** satisfy $\mathbf{i} \cdot \mathbf{i} = 1$ and $\mathbf{i} \cdot \mathbf{j} = 0$, we see that $a_1 = \mathbf{v} \cdot \mathbf{i}$ and $a_2 = \mathbf{v} \cdot \mathbf{j}$. The numbers a_1 and a_2 are also called the **components of v along the vectors i and j**, respectively, and can be denoted as

$$\text{comp}_{\mathbf{i}}\,\mathbf{v} = \mathbf{v} \cdot \mathbf{i} \quad \text{and} \quad \text{comp}_{\mathbf{j}}\,\mathbf{v} = \mathbf{v} \cdot \mathbf{j}. \tag{4}$$

In general, for nonzero vectors **u** and **v**, we define the **component of v along u** to be the number $|\mathbf{v}| \cos \theta$ (see Figure 22). Dividing $\mathbf{u} \cdot \mathbf{v} = |\mathbf{u}||\mathbf{v}| \cos \theta$ by $|\mathbf{u}|$ gives $|\mathbf{v}| \cos \theta = \mathbf{u} \cdot \mathbf{v}/|\mathbf{u}|$ and so we have proved the following result.

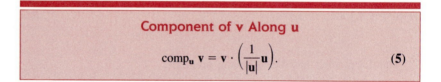

Component of v Along u

$$\text{comp}_{\mathbf{u}}\,\mathbf{v} = \mathbf{v} \cdot \left(\frac{1}{|\mathbf{u}|}\mathbf{u}\right). \tag{5}$$

Note that multiplying a nonzero vector **u** by the reciprocal of its magnitude $|\mathbf{u}|$ yields a unit vector. (Verify this.) Hence formula (5) is a generalization of (4):

To find the component of a vector **v** along a vector **u,** we dot **v** with a unit vector $(1/|\mathbf{u}|)\mathbf{u}$ that has the same direction as **u.**

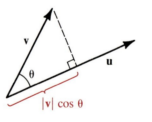

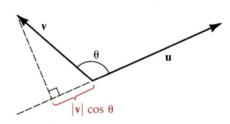

(a) $\text{comp}_{\mathbf{u}}\,\mathbf{v} > 0$ for $0 \leqslant \theta \leqslant \frac{\pi}{2}$ (b) $\text{comp}_{\mathbf{u}}\,\mathbf{v} < 0$ for $\frac{\pi}{2} < \theta \leqslant \pi$

FIGURE 22

▬ EXAMPLE 4

If $\mathbf{u} = 4\mathbf{i} + 5\mathbf{j}$ and $\mathbf{v} = 9\mathbf{i} + 3\mathbf{j}$, find **(a)** $\text{comp}_{\mathbf{u}}\ \mathbf{v}$ and **(b)** $\text{comp}_{\mathbf{v}}\ \mathbf{u}$.

Solution

(a) We first find the magnitude of **u**,

$$|\mathbf{u}| = \sqrt{4^2 + 5^2} = \sqrt{41},$$

and then construct a unit vector in the direction of **u**:

$$\left(\frac{1}{|\mathbf{u}|}\right)\mathbf{u} = \frac{1}{\sqrt{41}}(4\mathbf{i} + 5\mathbf{j}).$$

Hence
$$\text{comp}_{\mathbf{u}}\ \mathbf{v} = (9\mathbf{i} + 3\mathbf{j}) \cdot \left(\frac{1}{\sqrt{41}}(4\mathbf{i} + 5\mathbf{j})\right)$$

$$= \frac{51}{\sqrt{41}} \approx 7.96.$$

(b) Proceeding as in part (a), we form

$$|\mathbf{v}| = \sqrt{90} = 3\sqrt{10}, \qquad \left(\frac{1}{|\mathbf{v}|}\right)\mathbf{v} = \frac{1}{3\sqrt{10}}(9\mathbf{i} + 3\mathbf{j}),$$

and so
$$\text{comp}_{\mathbf{v}}\ \mathbf{u} = \mathbf{u} \cdot \left(\frac{1}{|\mathbf{v}|}\mathbf{v}\right)$$

$$= (4\mathbf{i} + 5\mathbf{j}) \cdot \left(\frac{1}{3\sqrt{10}}(9\mathbf{i} + 3\mathbf{j})\right)$$

$$= \frac{51}{3\sqrt{10}} \approx 5.38.$$

The components found in this example are shown in Figure 23.

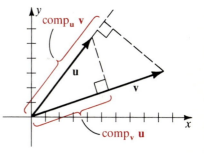

FIGURE 23

WORK

Suppose that a constant force **F** of magnitude F is applied to an object. If the object moves a distance d, either vertically or horizontally, in the same direction as the force, then the **work** W done is defined to be

$$W = \text{force} \times \text{distance} = Fd \qquad \textbf{(6)}$$

(see Figure 24). If a force is measured in pounds and the distance is measured in feet, then the unit of work is the foot-pound (ft-lb).

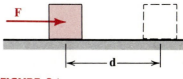

FIGURE 24

▬ EXAMPLE 5

If a person pushes against an unmovable object, such as the tree shown in Figure 25(a), with a constant force **F,** then no work is done since the force does not act through a distance. On the other hand, if the same person lifts a 16-lb bowling ball vertically a distance of 5 ft (see Figure 25(b)), then the work done is

$$W = 16 \times 5 = 80 \text{ ft-lb.}$$

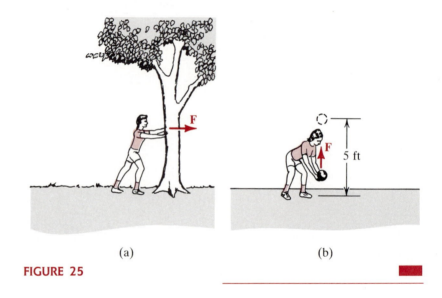

(a)   (b)

FIGURE 25   ▬

In metric units, force is measured in dynes (cgs system) or in newtons (mks system). Correspondingly, the units of work are the dyne-centimeter, or **erg,** and the newton-meter, or **joule.** Thus if a force of 40 newtons is needed to push a wheelbarrow over level ground through a distance of 3 meters, then the work done is $W = 40 \times 3 = 120$ joules. To convert from one system of units to another, we can use

$$1 \text{ newton} = 10^5 \text{ dynes} = 0.224 \text{ lb,}$$
$$1 \text{ ft-lb} = 1.356 \text{ joules} = 1.356 \times 10^7 \text{ ergs.}$$

Thus $80 \text{ ft-lb} = 80 \times 1.356 = 108.48$ joules

and $120 \text{ joules} = 120 \div 1.356 = 88.5$ ft-lb.

WORK AS A DOT PRODUCT

Now if a constant force **F** is applied to an object at an angle θ to the direction of motion, then the work done is given by

W = (component of **F** in the direction of displacement) × distance.   **(7)**

As we see from Figure 26, (7) becomes

$$W = (|\mathbf{F}| \cos \theta) \, |\mathbf{d}|$$
$$= |\mathbf{F}||\mathbf{d}| \cos \theta. \tag{8}$$

Comparing (8) with (2), we see that the work is given by a dot product:

$$W = \mathbf{F} \cdot \mathbf{d}. \tag{9}$$

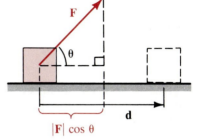

FIGURE 26

EXAMPLE 6

Find the work done by the constant force **F** = 3**i** + 6**j** if its point of application moves on the line segment from the origin to the point $P(5, 2)$. Assume that $|\mathbf{F}|$ is measured in newtons and $|\mathbf{d}|$ in meters.

Solution   We take the vector **d** to be the position vector $\overrightarrow{OP}$ = 5**i** + 2**j**. It follows from (9) that the work done is given by

$$W = (3\mathbf{i} + 6\mathbf{j}) \cdot (5\mathbf{i} + 2\mathbf{j})$$
$$= (3)(5) + (6)(2) = 27 \text{ joules}$$

(see Figure 27).

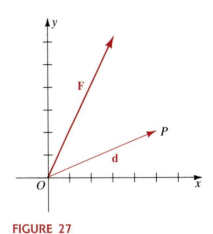

FIGURE 27

EXERCISE 5.2

In Problems 1–8, find the dot product **u** · **v** of the given vectors.

1. **u** = ⟨4, 2⟩, **v** = ⟨3, −1⟩   **2.** **u** = ⟨1, −2⟩, **v** = ⟨4, 0⟩
3. **u** = ⟨6, 0⟩, **v** = ⟨7, 9⟩   **4.** **u** = ⟨−10, 5⟩, **v** = ⟨−2, −11⟩
5. **u** = 3**i** − 2**j**, **v** = **i** + **j**   **6.** **u** = −9**i** − 3**j**, **v** = 8**i** + 15**j**
7. **u** = 4**i**, **v** = −3**j**   **8.** **u** = 5**j**, **v** = −**j**

In Problems 9–14, find the angle (to the nearest hundredth of a degree) between the given pair of vectors.

9. **u** = ⟨1, 4⟩, **v** = ⟨2, −1⟩   **10.** **u** = ⟨3, 5⟩, **v** = ⟨−4, −2⟩
11. **u** = **i** − **j**, **v** = 3**i** + **j**   **12.** **u** = 2**i** − **j**, **v** = 4**i** + **j**
13. **u** = ⟨3, 3⟩, **v** = ⟨5, 0⟩   **14.** **u** = 3**i** + 4**j**, **v** = 5**i** + 12**j**

In Problems 15–18, show that the given pair of vectors is orthogonal.

15. **u** = ⟨−3, 6⟩, **v** = ⟨20, 10⟩ **16.** **u** = ⟨4, −6⟩, **v** = ⟨−9, −6⟩
17. **u** = **i** + **j**, **v** = 5**i** − 5**j**   **18.** **u** = 6**i**, **v** = −100**j**

In Problems 19 and 20, find a constant k such that the given vectors are orthogonal.

19. **u** = 2k**i** − 6**j**, **v** = **i** + 5**j** **20.** **u** = ⟨9k, 4⟩, **v** = ⟨−k, 16⟩

In Problems 21–36, let **u** = 2**i** + 3**j**, **v** = −**i** − 2**j**, and **w** = 3**i** + 4**j**. Find the indicated number or vector.

21. **u** · **v**   **22.** **v** · **w**   **23.** **u** · **w**
24. **v** · (3**u**)   **25.** (2**w**) · (5**u**)   **26.** 6(−4**v**) · (2**w**)

27. $\mathbf{u} \cdot (\mathbf{u} + \mathbf{v} + \mathbf{w})$ **28.** $4\mathbf{v} \cdot (\mathbf{u} - 3\mathbf{w})$ **29.** $(\mathbf{u} - \mathbf{v}) \cdot (\mathbf{u} + \mathbf{v})$

30. $((3\mathbf{w}) \cdot \mathbf{v})(4\mathbf{u})$ **31.** $(\mathbf{u} \cdot \mathbf{u})\mathbf{v}$ **32.** $\left(\dfrac{\mathbf{u} \cdot \mathbf{v}}{\mathbf{v} \cdot \mathbf{v}}\right)\mathbf{v}$

33. $\text{comp}_\mathbf{u}\, \mathbf{v}$ **34.** $\text{comp}_\mathbf{w}\, \mathbf{v}$ **35.** $\text{comp}_\mathbf{w}\, \mathbf{u}$
36. $\text{comp}_\mathbf{u}\, (\mathbf{v} + \mathbf{w})$

In Problems 37–40 find the work done by the constant force F if its point of application moves on the straight-line segment from P to Q. Assume that $|\mathbf{F}|$ is measured in pounds and distance in feet.

37. $\mathbf{F} = 4\mathbf{i} + 5\mathbf{j}$, $P(0, 0)$, $Q(5, 7)$
38. $\mathbf{F} = 10\mathbf{i} - 2\mathbf{j}$, $P(0, 0)$, $Q(6, -2)$
39. $\mathbf{F} = 8\mathbf{i} + \mathbf{j}$, $P(1, 1)$, $Q(6, 9)$
40. $\mathbf{F} = 5\mathbf{i} + 3\mathbf{j}$, $P(-2, 3)$, $Q(-6, 12)$

In Problems 41 and 42 find the work done in moving the object the indicated distance. Assume that $|\mathbf{F}|$ is measured in newtons and distance in meters.

41.

FIGURE 28

42.

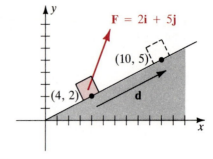

FIGURE 29

43. Prove property (i) of the dot product.
44. Prove property (ii) of the dot product.
45. If the vectors $\mathbf{u}$ and $\mathbf{v}$ are parallel, prove that $\mathbf{v} \cdot \mathbf{u} = \pm|\mathbf{u}||\mathbf{v}|$.

5.3 Complex Numbers

Some equations have no real solutions. For example, the quadratic equation $x^2 + 1 = 0$ has no real roots because there is no real number x such that $x^2 = -1$. In this section we will study the set of **complex numbers,** which contains solutions to equations such as $x^2 + 1 = 0$. The set of complex numbers C contains the set of real numbers R as well as numbers whose squares are negative.

To obtain the complex numbers C, we begin by defining the **imaginary unit,** denoted by the letter i, as the number that satisfies

$$i^2 = -1.$$

It is common practice to write

$$i = \sqrt{-1}.$$

With i we are able to define the principal square root of a negative number, as follows.

DEFINITION 2

If c is a positive real number, then the **principal square root of** $-c$, denoted $\sqrt{-c}$, is defined by

$$\sqrt{-c} = \sqrt{(-1)c} = \sqrt{-1}\sqrt{c} = i\sqrt{c} = \sqrt{c}\,i.$$

EXAMPLE 1

Find the principal square root of **(a)** $\sqrt{-4}$ and **(b)** $\sqrt{-5}$.

Solution

(a) $\sqrt{-4} = \sqrt{(-1)(4)} = \sqrt{-1}\sqrt{4} = i(2) = 2i$
(b) $\sqrt{-5} = \sqrt{(-1)(5)} = \sqrt{-1}\sqrt{5} = i\sqrt{5} = \sqrt{5}\,i$

THE COMPLEX NUMBER SYSTEM

The **complex number system** contains the imaginary unit i, all real numbers, products such as bi, b real, and sums such as $a + bi$, where a and b are real numbers. In particular, a **complex number** is defined to be any expression of the form

$$z = a + bi, \tag{10}$$

where a and b are real numbers and $i^2 = -1$. The numbers a and b are called the **real** and **imaginary parts** of z, respectively. A complex number of the form $0 + bi$ is said to be a **pure imaginary number.** Note that by choosing $b = 0$ in (10), we obtain a real number. Thus the set of real numbers is a subset of the set of complex numbers.

EXAMPLE 2

(a) The complex number $z = 4 + (-5)i$ can be written as $z = 4 - 5i$. Its real part is 4 and its imaginary part is -5.

(b) $z = 10i$ is a pure imaginary number.
(c) $z = 6 + 0i$ is a real number.

▌ EXAMPLE 3

Express each of the following in the form $a + bi$.

(a) $-3 + \sqrt{-7}$ **(b)** $2 - \sqrt{-25}$

Solution

(a) $-3 + \sqrt{-7} = -3 + i\sqrt{7} = -3 + \sqrt{7}i$
(b) $2 - \sqrt{-25} = 2 - i\sqrt{25} = 2 - 5i$

   In order to solve certain equations involving complex numbers, it is necessary to specify when two complex numbers are equal.

DEFINITION 3 ▐

> Two complex numbers are **equal** if and only if their real parts are equal and their imaginary parts are equal. That is, if
>
> $$z_1 = a_1 + b_1 i \quad \text{and} \quad z_2 = a_2 + b_2 i,$$
>
> then
>
> $$z_1 = z_2 \quad \text{if and only if} \quad a_1 = a_2 \text{ and } b_1 = b_2.$$

▌ EXAMPLE 4

Solve for x and y:

$$(2x + 1) + (-2y + 3)i = 2 - 4i.$$

Solution   By Definition 3 we must have

$$2x + 1 = 2 \quad \text{and} \quad -2y + 3 = -4.$$

These equations yield $x = \frac{1}{2}$ and $y = \frac{7}{2}$.

Addition and multiplication for complex numbers are defined as follows.

DEFINITION 4

If $z_1 = a_1 + b_1 i$ and $z_2 = a_2 + b_2 i$, then:

(i) their **sum** is given by $z_1 + z_2 = (a_1 + a_2) + (b_1 + b_2)i$
(ii) and their **product** is given by
$$z_1 z_2 = (a_1 a_2 - b_1 b_2) + (a_1 b_2 + b_1 a_2)i.$$

PROPERTIES OF THE COMPLEX NUMBER SYSTEM

Using Definition 4, we can show that the associative, commutative, and distributive laws hold for complex numbers. We further observe that in Definition 4(i) the sum of two complex numbers is obtained by adding their corresponding real and imaginary parts.

Also, rather than memorizing the definition of the product, we can multiply two complex numbers by using the familiar associative, commutative, and distributive properties and the fact that $i^2 = -1$. Applying this approach, we find that

$$
\begin{aligned}
(a + bi)(c + di) &= (a + bi)c + (a + bi)di \\
&= ac + (bc)i + (ad)i + (bd)i^2 \\
&= ac + (bc)i + (ad)i + (bd)(-1) \\
&= ac + (bd)(-1) + (bc)i + (ad)i \\
&= (ac - bd) + (bc + ad)i.
\end{aligned}
$$

This is the same result as the product given by Definition 4(ii).

These techniques are illustrated in the following example.

EXAMPLE 5

If $z_1 = 5 - 6i$ and $z_2 = 2 + 4i$, find **(a)** $z_1 + z_2$ and **(b)** $z_1 z_2$.

Solution

(a) Combining like terms, we have

$$(5 - 6i) + (2 + 4i) = (5 + 2) + (-6 + 4)i = 7 - 2i.$$

(b) Using the distributive law, we can write the product $(5 - 6i)(2 + 4i)$ as

$$
\begin{aligned}
(5 - 6i)(2 + 4i) &= (5 - 6i)2 + (5 - 6i)4i \\
&= 10 - 12i + 20i - 24i^2 \\
&= 10 - 24(-1) + (-12 + 20)i \\
&= 34 + 8i.
\end{aligned}
$$

Note of Caution: Not all the properties of the real number system hold for complex numbers. In particular, the property of radicals $\sqrt{a}\sqrt{b} = \sqrt{ab}$ is *not* true when both a and b are negative. To see this, consider that

$$\sqrt{-1}\sqrt{-1} = ii = i^2 = -1,$$

whereas

$$\sqrt{(-1)(-1)} = \sqrt{1} = 1.$$

Thus, $\sqrt{-1}\sqrt{-1} \neq \sqrt{(-1)(-1)}$. However, if *only one of a or b is negative,* then we do have $\sqrt{a}\sqrt{b} = \sqrt{ab}$.

In the set of complex numbers, the **additive identity** is the number $0 = 0 + 0i$, and the **multiplicative identity** is the number $1 = 1 + 0i$.

The **additive inverse** of a complex number $z = a + bi$ is $-z = -a - bi$. The additive inverse is used to define **subtraction** of complex numbers:

$$(a + bi) - (c + di) = (a + bi) + [-(c + di)]$$
$$= (a + bi) + (-c - di)$$
$$= (a - c) + (b - d)i.$$

In other words, to subtract complex numbers, we subtract their corresponding real and imaginary parts.

In order to obtain the **multiplicative inverse** of $z = a + bi$, we introduce the concept of the **conjugate** of a complex number. If $z = a + bi$ is a complex number, then $\bar{z} = a - bi$ is called its **conjugate.** For example, the conjugate of $8 + 13i$ is $8 - 13i$, and the conjugate of $-5 - 2i$ is $-5 + 2i$.

The following computations show that both the sum and the product of a complex number z and its conjugate $\bar{z}$ are real numbers:

$$z + \bar{z} = (a + bi) + (a - bi) = 2a,$$
$$z\bar{z} = (a + bi)(a - bi) = a^2 - b^2i^2 = a^2 + b^2. \qquad \textbf{(11)}$$

The latter property makes conjugates very useful in finding the multiplicative inverse $1/z$ or in dividing two complex numbers, as the next example shows.

■ EXAMPLE 6

For $z_1 = 3 - 2i$ and $z_2 = 4 + 5i$, express each of the following in the form $a + bi$.

(a) $\dfrac{1}{z_1}$ **(b)** $\dfrac{z_1}{z_2}$

Solution   We multiply both the numerator and the denominator by the conjugate of the denominator, use (11), and simplify.

(a) $\dfrac{1}{z_1} = \dfrac{1}{3 - 2i} = \dfrac{1}{3 - 2i} \cdot \dfrac{3 + 2i}{3 + 2i} = \dfrac{3 + 2i}{9 + 4} = \dfrac{3}{13} + \dfrac{2}{13}i$

(b) $\dfrac{z_1}{z_2} = \dfrac{3 - 2i}{4 + 5i} = \dfrac{3 - 2i}{4 + 5i} \cdot \dfrac{4 - 5i}{4 - 5i} = \dfrac{12 - 8i - 15i + 10i^2}{16 + 25}$

$\qquad = \dfrac{2 - 23i}{41} = \dfrac{2}{41} - \dfrac{23}{41}i$

Complex numbers make it possible to solve quadratic equations $ax^2 + bx + c = 0$ when the discriminant $b^2 - 4ac$ is negative. We now see that the solutions from the quadratic formula

$$x = \dfrac{-b + \sqrt{b^2 - 4ac}}{2a} \quad \text{and} \quad x = \dfrac{-b - \sqrt{b^2 - 4ac}}{2a}$$

represent complex numbers. In fact, the solutions are conjugates. As Example 7 shows these solutions can be written in the standard form (10).

■ EXAMPLE 7

Solve $x^2 - 8x + 25 = 0$.

Solution   From the quadratic formula, we obtain

$$x = \dfrac{-(-8) \pm \sqrt{(-8)^2 - 4(1)(25)}}{2},$$

that is, $x = \dfrac{8 \pm \sqrt{-36}}{2} = \dfrac{8 \pm 6i}{2} = 4 \pm 3i.$

Thus the solutions are the conjugates $4 + 3i$ and $4 - 3i$.

■ EXAMPLE 8

Solve $x^2 + 16 = 0$.

Solution   The given equation can be written

$$x^2 = -16$$

and so $x = \pm\sqrt{-16} = \pm 4i.$

Therefore, the solutions are the conjugates $4i$ and $-4i$.

We can now obtain solutions to any quadratic equation.

EXERCISE 5.3

In Problems 1–46, perform the indicated operation. Write the answer in the form $a + bi$.

1. $\sqrt{-100}$

2. $-\sqrt{-8}$

3. $-3 - \sqrt{-3}$

4. $\sqrt{-5} - \sqrt{-125} + 5$

5. $(3 + i) - (4 - 3i)$

6. $(5 + 6i) + (-7 + 2i)$

7. $2(4 - 5i) + 3(-2 - i)$

8. $-2(6 + 4i) + 5(4 - 8i)$

9. $i(-10 + 9i) - 5i$

10. $i(4 + 13i) - i(1 - 9i)$

11. $3i(1 + i) - 4(2 - i)$

12. $i + i(1 - 2i) + i(4 + 3i)$

13. $(3 - 2i)(1 - i)$

14. $(4 + 6i)(-3 + 4i)$

15. $(7 + 14i)(2 + i)$

16. $(-5 - \sqrt{3}i)(2 - \sqrt{3}i)$

17. $(4 + 5i) - (2 - i)(1 + i)$

18. $(-3 + 6i) + (2 + 4i)(-3 + 2i)$

19. $i(1 - 2i)(2 + 5i)$

20. $i(\sqrt{2} - i)(1 - \sqrt{2}i)$

21. $(1 + i)(1 + 2i)(1 + 3i)$

22. $(2 + i)(2 - i)(4 - 2i)$

23. $(1 - i)[2(2 - i) - 5(1 + 3i)]$

24. $(4 + i)[i(1 + 3i) - 2(-5 + 3i)]$

25. i^8

26. i^{11}

27. i^{-2}

28. i^{-3}

29. $\dfrac{1}{4 - 3i}$

30. $\dfrac{5}{3 + i}$

31. $\dfrac{4}{5 + 4i}$

32. $\dfrac{1}{-1 + 2i}$

33. $\dfrac{i}{1 + i}$

34. $\dfrac{i}{4 - i}$

35. $\dfrac{4 + 6i}{i}$

36. $\dfrac{3 - 5i}{i}$

37. $\dfrac{1 + i}{1 - i}$

38. $\dfrac{2 - 3i}{1 + 2i}$

39. $\dfrac{4 + 2i}{2 - 7i}$

40. $\dfrac{\frac{1}{2} - \frac{7}{2}i}{4 + 2i}$

41. $i\left(\dfrac{10 - i}{1 + i}\right)$

42. $i\left(\dfrac{1 - 2\sqrt{3}i}{1 + \sqrt{3}i}\right)$

43. $(1 + i)\dfrac{2i}{1 - 5i}$

44. $(5 - 3i)\dfrac{1 - i}{2 - i}$

45. $(4 - 9i) + \dfrac{25i}{2 + i}$

46. $i(-6 + \dfrac{11}{5}i) + \dfrac{2 + i}{2 - i}$

In Problems 47–52, solve for x and y.

47. $2(x + yi) = i(3 - 4i)$

48. $(x + yi) + 4(1 - i) = 5 - 7i$

49. $i(x + yi) = (1 - 6i)(2 + 3i)$

50. $10 + 6yi = 5x + 24i$

51. $(1 + i)(x - yi) = i(14 + 7i) - (2 + 13i)$

52. $i^2(1 - i)(1 + i) = 3x + yi + i(y + xi)$

In Problems 53–64, solve the given equation.

53. $x^2 + 9 = 0$

54. $x^2 + 8 = 0$

55. $2x^2 = -5$

56. $3x^2 = -1$

57. $2x^2 - x + 1 = 0$

58. $x^2 - 2x + 10 = 0$

59. $x^2 + 8x + 52 = 0$

60. $3x^2 + 2x + 5 = 0$

61. $4x^2 - x + 2 = 0$

62. $x^2 + x + 2 = 0$

63. $x^4 + 3x^2 + 2 = 0$

64. $2x^4 + 9x^2 + 4 = 0$

65. Find a complex number $z = x + yi$ such that $z^2 = i$.

66. Find a complex number $z = x + yi$ such that $z^2 = -3 + 4i$.

Let $z_1 = a + bi$ and $z_2 = c + di$ be any two complex numbers. In Problems 67–70, prove the given properties involving the conjugates of z_1 and z_2, $\bar{z}_1 = a - bi$ and $\bar{z}_2 = c - di$.

67. $\bar{z}_1 = z_1$ if and only if z_1 is a real number.

68. $\overline{z_1 + z_2} = \bar{z}_1 + \bar{z}_2$

69. $\overline{z_1 \cdot z_2} = \bar{z}_1 \cdot \bar{z}_2$

70. $\overline{z_1^2} = \bar{z}_1^2$

71. When light crosses from one medium to another—for example, from air to water—it is refracted (or bent) and some of it is absorbed. These phenomena can be described by an index of refraction n and an absorption coefficient k. Engineers have found it useful to combine these into a complex index of refraction,

$$m = n - ik.$$

In theoretical treatments of backscattering of radar by rain droplets, the expression

$$K = \dfrac{m^2 - 1}{m^2 + 2}$$

is important. Find K if the complex index of refraction m is $5 - 3i$.

Trigonometric Form of a Complex Number 5.4

We now develop a trigonometric form for representing complex numbers.

GEOMETRIC REPRESENTATION

We begin by considering a **geometric representation** for complex numbers. A complex number $z = a + bi$ is uniquely determined by the pair of real numbers (a, b). Thus we can identify the first and second elements of an ordered pair with the real and imaginary parts of z, respectively. For example, $(2, -5)$ corresponds to $z = 2 - 5i$. When each point in a coordinate plane is identified with a complex number in this way, the plane is called the **complex plane.*** As Figure 30 shows, the vertical or y-axis is called the **imaginary axis,** and the horizontal or x-axis is designated the **real axis.**

FIGURE 30

■ EXAMPLE 1

Graph the complex numbers

$$z_1 = 5 + 4i, \quad z_2 = -2i, \quad z_3 = -2 - 3i, \quad \text{and} \quad z_4 = -4 + 2i.$$

Solution As shown in Figure 31, z_1, z_2, z_3, and z_4 are identified with the points $(5, 4)$, $(0, -2)$, $(-2, -3)$, and $(-4, 2)$, respectively.

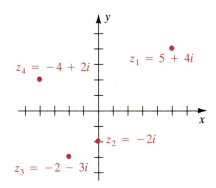

FIGURE 31

TRIGONOMETRIC FORM

If $z = a + bi$ is a nonzero complex number and $P(a, b)$ is its geometric representation, as shown in Figure 32, then the distance from P to the origin is given by $r = \sqrt{a^2 + b^2}$. This distance is called the **modulus, or absolute value,** of z and is denoted by $|z|$. Let θ be the angle in standard position whose terminal side passes through $P(a, b)$. Then $\cos \theta = a/r$ and $\sin \theta = b/r$, from which we obtain

$$a = r \cos \theta \quad \text{and} \quad b = r \sin \theta.$$

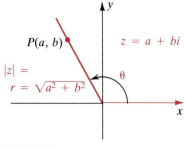

FIGURE 32

---

*The plane is sometimes referred to as the **Argand plane,** after the French mathematician Jean R. Argand (1768–1822).

Substituting these expressions for a and b in $z = a + bi$, we obtain

$$z = a + bi = (r \cos \theta) + (r \sin \theta)i,$$

or $\qquad\qquad\qquad z = r[\cos \theta + i \sin \theta],$ $\qquad\qquad$ **(12)**

where $r = \sqrt{a^2 + b^2}$.

We say that (12) is a **trigonometric form,** or **polar form,** of the complex number z. The angle θ is called the **argument,** or **amplitude,** of z and satisfies $\tan \theta = b/a$. However, θ is not necessarily arctan (b/a), since θ is not restricted to $(-\pi/2, \pi/2)$. (See Examples 2 and 3.) Also, the argument θ is not uniquely determined, since $\cos \theta = \cos (\theta + 2k\pi)$ and $\sin \theta = \sin (\theta + 2k\pi)$ for any integer k. If $z = a + bi = 0$, then $a = b = 0$. In this case, $r = 0$ and we can take any angle θ as the argument. Indeed, we know that if $a_1 + b_1i = a_2 + b_2i$, then $a_1 = a_2$ and $b_1 = b_2$. However, if $r_1[\cos \theta_1 + i \sin \theta_1] = r_2[\cos \theta_2 + i \sin \theta_2]$, then $r_1 = r_2$, but θ_1 need not equal θ_2.

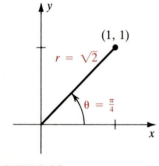

FIGURE 33

■■■ **EXAMPLE 2**

Write in trigonometric form **(a)** $1 + i$ and **(b)** $1 - \sqrt{3}i$.

Solution

(a) We identify $a = 1$ and $b = 1$. Then

$$r = |1 + i| = \sqrt{(1)^2 + (1)^2} = \sqrt{2}.$$

Since $\tan \theta = b/a = 1$ and $(1, 1)$ lies in the first quadrant, we take $\theta = \pi/4$, as shown in Figure 33. Thus,

$$z = \sqrt{2}\left[\cos \frac{\pi}{4} + i \sin \frac{\pi}{4}\right].$$

(b) In this case,

$$r = |1 - \sqrt{3}i| = \sqrt{1^2 + (-\sqrt{3})^2} = 2.$$

From $\tan \theta = -\sqrt{3}/1 = -\sqrt{3}$ and the fact that $(1, -\sqrt{3})$ lies in the fourth quadrant, we take $\theta = 5\pi/3$, as shown in Figure 34. Thus,

$$z = 2\left[\cos \frac{5\pi}{3} + i \sin \frac{5\pi}{3}\right].$$

Alternatively, we could take $\theta = -\pi/3$ and write

$$z = 2\left[\cos \left(-\frac{\pi}{3}\right) + i \sin \left(-\frac{\pi}{3}\right)\right].$$

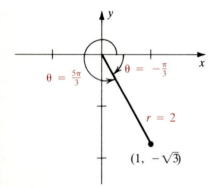

FIGURE 34

EXAMPLE 3

Find the trigonometric form of $z = -4 + 5i$.

Solution The modulus is

$$r = |-4 + 5i| = \sqrt{16 + 25} = \sqrt{41}.$$

Since $\tan \theta = -\frac{5}{4}$, we note that θ is not related to one of the special angles 0, $\pi/6$, $\pi/4$, $\pi/3$, and $\pi/2$. Since $(-4, 5)$ lies in the second quadrant (Figure 35), we must take care to adjust the value of θ obtained from a calculator or tables so that our final answer is a quadrant II angle. One approach is to use a calculator set in radian mode to obtain the reference angle $\theta' = \tan^{-1} \frac{5}{4} \approx$ 0.8961 radian. The desired quadrant II angle is then

$$\theta = \pi - \theta' \approx 2.2455.$$

Thus, $\qquad z \approx \sqrt{41}[\cos 2.2455 + i \sin 2.2455].$

Alternatively, with the calculator set in degree mode, we would obtain $\theta' \approx$ 51.34° and $\theta = 180° - \theta' \approx 128.66°$, from which it follows that

$$z \approx \sqrt{41}[\cos 128.66° + i \sin 128.66°].$$

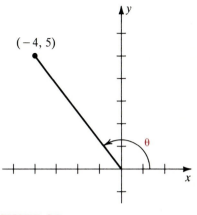

FIGURE 35

EXAMPLE 4

Find the modulus and the argument of $z_1 z_2$, where $z_1 = 2i$ and $z_2 = 1 + i$.

Solution The product is

$$z_1 z_2 = 2i(1 + i) = -2 + 2i,$$

and hence

$$r = |z_1 z_2| = |-2 + 2i| = \sqrt{8} = 2\sqrt{2}.$$

Identifying $a = -2$ and $b = 2$, we have $\tan \theta = -1$. Since θ is a quadrant II angle, we conclude that $\theta = 3\pi/4$ (see Figure 36).

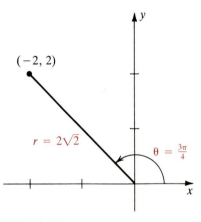

FIGURE 36

MULTIPLICATION AND DIVISION IN TRIGONOMETRIC FORM

In the preceding example note that the modulus $r = 2\sqrt{2}$ of the product $z_1 z_2$ is the product of the modulus $r_1 = 2$ of z_1 and the modulus $r_2 = \sqrt{2}$ of z_2. Also, the argument $\theta = 3\pi/4$ of $z_1 z_2$ is the sum of the arguments $\theta_1 = \pi/2$ and $\theta_2 = \pi/4$ of z_1 and z_2, respectively. We have illustrated a particular example of the following theorem, which describes how to multiply and divide complex numbers when they are written in trigonometric form.

THEOREM 3 ▰▰▰▰▰▰▰▰▰▰▰▰▰▰▰▰▰▰▰▰▰▰▰▰▰▰▰▰

If $z_1 = r_1[\cos \theta_1 + i \sin \theta_1]$ and $z_2 = r_2[\cos \theta_2 + i \sin \theta_2]$, then:

(a) $z_1 z_2 = r_1 r_2[\cos (\theta_1 + \theta_2) + i \sin (\theta_1 + \theta_2)]$;

(b) $\dfrac{z_1}{z_2} = \dfrac{r_1}{r_2}[\cos (\theta_1 - \theta_2) + i \sin (\theta_1 - \theta_2)]$, $r_2 \neq 0$.

PROOF   We will prove (b) and leave (a) as an exercise (see Problem 55). If we multiply the numerator and the denominator of

$$\frac{z_1}{z_2} = \frac{r_1[\cos \theta_1 + i \sin \theta_1]}{r_2[\cos \theta_2 + i \sin \theta_2]}$$

by $\cos \theta_2 - i \sin \theta_2$, we obtain

$$\frac{z_1}{z_2} = \frac{r_1}{r_2} \frac{[\cos \theta_1 + i \sin \theta_1][\cos \theta_2 - i \sin \theta_2]}{\cos^2 \theta_2 + \sin^2 \theta_2}$$

$$= \frac{r_1}{r_2}[\cos \theta_1 + i \sin \theta_1][\cos \theta_2 - i \sin \theta_2].$$

Performing the multiplication and then using the subtraction formulas from Section 4.2, we have

$$\frac{z_1}{z_2} = \frac{r_1}{r_2}[(\cos \theta_1 \cos \theta_2 + \sin \theta_1 \sin \theta_2)$$

$$+ i(\sin \theta_1 \cos \theta_2 - \cos \theta_1 \sin \theta_2)]$$

$$= \frac{r_1}{r_2}[\cos (\theta_1 - \theta_2) + i \sin (\theta_1 - \theta_2)].$$

▱

▰▰ **EXAMPLE 5**

If $z_1 = 4[\cos 75° + i \sin 75°]$ and $z_2 = \dfrac{1}{2}[\cos 45° + i \sin 45°]$, find $z_1 z_2$ and z_1/z_2. Express each answer in the form $a + bi$.

Solution   From Theorem 3 we have

$$z_1 z_2 = 4 \cdot \frac{1}{2}[\cos (75° + 45°) + i \sin (75° + 45°)]$$

$$= 2[\cos 120° + i \sin 120°] = 2\left[-\frac{1}{2} + \frac{\sqrt{3}}{2}i\right] = -1 + \sqrt{3}i$$

and

$$\frac{z_1}{z_2} = \frac{4}{1/2}[\cos(75° - 45°) + i\sin(75° - 45°)]$$

$$= 8[\cos 30° + i\sin 30°] = 8\left[\frac{\sqrt{3}}{2} + \frac{1}{2}i\right] = 4\sqrt{3} + 4i.$$

EXERCISE 5.4

In Problems 1–10, graph z_1 and z_2. Evaluate and graph the indicated complex number.

1. $z_1 = 2 + 5i; \bar{z_1}$

2. $z_1 = -8 - 4i; \frac{1}{4}\bar{z_1}$

3. $z_1 = 1 + i, z_2 = 2 - 2i; z_1 + z_2$

4. $z_1 = 4i, z_2 = -4 + i; z_1 - z_2$

5. $z_1 = 6 - 3i, z_2 = -i; \bar{z_1} + z_2$

6. $z_1 = 5 + 2i, z_2 = -1 + 2i; z_1 + \bar{z_2}$

7. $z_1 = -2i, z_2 = 1 - i; z_1 z_2$

8. $z_1 = 1 + i, z_2 = 2 - i; z_1 z_2$

9. $z_1 = 2\sqrt{3} + 2i, z_2 = 1 - \sqrt{3}i; \frac{z_1}{z_2}$

10. $z_1 = i, z_2 = 1 - i; \frac{z_1}{z_2}$

In Problems 11–22, find the modulus and an argument of the given complex number.

11. $z = \frac{1}{2} - \frac{\sqrt{3}}{2}i$ **12.** $z = 4 + 3i$

13. $z = \sqrt{2} - 4i$ **14.** $z = -5 + 2i$

15. $z = \frac{3}{4} - \frac{1}{4}i$ **16.** $z = -8 - 2i$

17. $z = 3 + 3i$ **18.** $z = -1 - i$

19. $z = \sqrt{3} + i$ **20.** $z = 2 - 2\sqrt{3}i$

21. $z = 2 - i$ **22.** $z = 4 + 8i$

In Problems 23–32, write the given complex number in trigonometric form.

23. $z = -4i$ **24.** $z = 15i$

25. $z = 5\sqrt{3} + 5i$ **26.** $z = 3 + i$

27. $z = -2 + 5i$ **28.** $z = 2 + 2\sqrt{3}i$

29. $z = 3 - 5i$ **30.** $z = -10 + 6i$

31. $z = -2 - 2i$ **32.** $z = 1 - i$

In Problems 33–42, write the given complex number in the form $z = a + bi$. Do not use a calculator.

33. $z = \sqrt{2}\left[\cos\frac{\pi}{4} + i\sin\frac{\pi}{4}\right]$

34. $z = 6\left[\cos\left(-\frac{\pi}{4}\right) + i\sin\left(-\frac{\pi}{4}\right)\right]$

35. $z = 10[\cos 210° + i\sin 210°]$

36. $z = \sqrt{5}[\cos 420° + i\sin 420°]$

37. $z = 2[\cos 30° + i\sin 30°]$

38. $z = 7\left[\cos\frac{7\pi}{12} + i\sin\frac{7\pi}{12}\right]$

39. $z = \cos\left(-\frac{5\pi}{3}\right) + i\sin\left(-\frac{5\pi}{3}\right)$

40. $z = \frac{3}{2}\left[\cos\left(-\frac{3\pi}{4}\right) + i\sin\left(-\frac{3\pi}{4}\right)\right]$

41. $z = 4[\cos(\tan^{-1}2) + i\sin(\tan^{-1}2)]$

42. $z = 20\left[\cos\left(\tan^{-1}\frac{3}{5}\right) + i\sin\left(\tan^{-1}\frac{3}{5}\right)\right]$

In Problems 43–48, find $z_1 z_2$ and z_1/z_2 in trigonometric form by first writing z_1 and z_2 in trigonometric form.

43. $z_1 = 3i, z_2 = 6 + 6i$

44. $z_1 = 1 + i, z_2 = -1 + i$

45. $z_1 = 1 + \sqrt{3}i, z_2 = 2\sqrt{3} + 2i$

46. $z_1 = 5i, z_2 = -10i$

47. $z_1 = \sqrt{3} + i, z_2 = 5 - 5i$

48. $z_1 = -\sqrt{2} + \sqrt{2}i, z_2 = \frac{5\sqrt{2}}{2} + \frac{5\sqrt{2}}{2}i$

In Problems 49–52, find z_1z_2 and z_1/z_2. Write the answer in the form $a + bi$.

49. $z_1 = \sqrt{6}\left[\cos\dfrac{\pi}{3} + i\sin\dfrac{\pi}{3}\right]$,

$z_2 = \sqrt{2}\left[\cos\left(-\dfrac{\pi}{4}\right) + i\sin\left(-\dfrac{\pi}{4}\right)\right]$

50. $z_1 = 10\left[\cos\dfrac{7\pi}{6} + i\sin\dfrac{7\pi}{6}\right]$,

$z_2 = \dfrac{1}{2}\left[\cos\dfrac{\pi}{6} + i\sin\dfrac{\pi}{6}\right]$

51. $z_1 = 3\left[\cos\dfrac{\pi}{4} + i\sin\dfrac{\pi}{4}\right]$,

$z_2 = 4\left[\cos\left(-\dfrac{\pi}{8}\right) + i\sin\left(-\dfrac{\pi}{8}\right)\right]$

52. $z_1 = \cos 57° + i\sin 57°$,

$z_2 = 7[\cos 73° + i\sin 73°]$

53. Show that $|z| = \sqrt{z\bar{z}}$.

54. If $z = r[\cos\theta + i\sin\theta]$, write $1/z$ in trigonometric form. What is the trigonometric form of $\bar{z}$?

55. Prove part (a) of Theorem 3.

5.5

DeMoivre's Theorem and nth Roots

The trigonometric form of a complex number can be used to compute powers and roots of the number. For example, suppose we wish to compute z_1^3, where

$$z_1 = 1 + i = \sqrt{2}\left[\cos\frac{\pi}{4} + i\sin\frac{\pi}{4}\right].$$

From Theorem 3(a), with $z_1 = z_2$, we obtain the square of z_1:

$$z_1^2 = z_1 \cdot z_1 = (\sqrt{2})(\sqrt{2})\left[\cos\left(\frac{\pi}{4} + \frac{\pi}{4}\right) + i\sin\left(\frac{\pi}{4} + \frac{\pi}{4}\right)\right]$$

$$= (\sqrt{2})^2\left[\cos 2\left(\frac{\pi}{4}\right) + i\sin 2\left(\frac{\pi}{4}\right)\right].$$

Then

$$z_1^3 = z_1^2 \cdot z_1 = (\sqrt{2})^2(\sqrt{2})\left[\cos\left(\frac{2\pi}{4} + \frac{\pi}{4}\right) + i\sin\left(\frac{2\pi}{4} + \frac{\pi}{4}\right)\right]$$

$$= (\sqrt{2})^3\left[\cos 3\left(\frac{\pi}{4}\right) + i\sin 3\left(\frac{\pi}{4}\right)\right].$$

This illustrates a particular case of the next result, known as **DeMoivre's theorem.***

*Named after the French mathematician Abraham DeMoivre (1667–1754).

THEOREM 4

DeMoivre's Theorem

If $z = r[\cos \theta + i \sin \theta]$ and n is a positive integer, then

$$z^n = r^n[\cos n\theta + i \sin n\theta].$$

EXAMPLE 1

Evaluate $(\sqrt{3} + i)^8$.

Solution First, we calculate

$$r = \sqrt{(\sqrt{3})^2 + 1^2} = 2 \quad \text{and} \quad \tan \theta = \frac{1}{\sqrt{3}},$$

so that $\theta = \pi/6$ since $(\sqrt{3}, 1)$ lies in quadrant I. It then follows from DeMoivre's theorem that

$$(\sqrt{3} + i)^8 = 2^8\left[\cos 8\left(\frac{\pi}{6}\right) + i \sin 8\left(\frac{\pi}{6}\right)\right] = 256\left[\cos \frac{4\pi}{3} + i \sin \frac{4\pi}{3}\right]$$

$$= 256\left[-\frac{1}{2} - \frac{\sqrt{3}}{2}i\right] = -128 - 128\sqrt{3}i.$$

ROOTS OF z

We say that w is an nth **root** of a complex number z if $w^n = z$. In the following discussion we will consider a method for expressing an nth root of a complex number when it is written in trigonometric form.

Let the modulus and the argument of w be ρ and ϕ, respectively, so that $w = \rho[\cos \phi + i \sin \phi]$. If w is an nth root of $z = r[\cos \theta + i \sin \theta]$, then by DeMoivre's theorem

$$w^n = z$$

can be written as

$$\rho^n[\cos n\phi + i \sin n\phi] = r[\cos \theta + i \sin \theta].$$

When two complex numbers are equal, their moduli are necessarily equal. Thus we have

$$\rho^n = r \quad \text{or} \quad \rho = r^{1/n}$$

and $\qquad\qquad \cos n\phi + i \sin n\phi = \cos \theta + i \sin \theta.$

Equating the real and imaginary parts in this equation gives

$$\cos n\phi = \cos \theta, \qquad \sin n\phi = \sin \theta,$$

from which it follows that $n\phi = \theta + 2k\pi$, or

$$\phi = \frac{\theta + 2k\pi}{n},$$

where k is any integer. As k takes on the successive integer values 0, 1, 2, . . . , $n - 1$, we obtain n distinct roots of z. For $k \geq n$, the values of $\sin \phi$ and $\cos \phi$ repeat the values obtained by letting $k = 0, 1, 2, \ldots, n - 1$. To see this, suppose that $k = n + m$, where $m = 0, 1, 2, \ldots$. Then

$$\phi = \frac{\theta + 2(n + m)\pi}{n} = \frac{\theta + 2m\pi}{n} + 2\pi.$$

Since the sine and cosine each have period 2π, we have

$$\sin \phi = \sin \left(\frac{\theta + 2m\pi}{n} \right) \quad \text{and} \quad \cos \phi = \cos \left(\frac{\theta + 2m\pi}{n} \right),$$

and so no new roots are obtained for $k \geq n$. These results are summarized in the following theorem.

THEOREM 5

The nth roots of a nonzero complex number $z = r[\cos \theta + i \sin \theta]$ are given by

$$w = r^{1/n} \left[\cos \left(\frac{\theta + 2k\pi}{n} \right) + i \sin \left(\frac{\theta + 2k\pi}{n} \right) \right],$$

where $k = 0, 1, 2, \ldots, n - 1$, and θ is measured in radians.

We denote the n values of w by $z_0, z_1, \ldots, z_{n-1}$ (corresponding to $k = 0$, 1, . . . , $n - 1$, respectively).

EXAMPLE 2

Find the three cube roots of i.

Solution   In the trigonometric form for i, $r = 1$ and $\theta = \pi/2$, so that

$$i = \cos \frac{\pi}{2} + i \sin \frac{\pi}{2}.$$

With $n = 3$ we find from Theorem 5 that

$$w = 1^{1/3}\left[\cos\left(\frac{\pi/2 + 2k\pi}{3}\right) + i\sin\left(\frac{\pi/2 + 2k\pi}{3}\right)\right], \quad k = 0, 1, 2.$$

Now for

$$k = 0, \quad z_0 = \cos\frac{\pi}{6} + i\sin\frac{\pi}{6},$$

$$k = 1, \quad z_1 = \cos\left(\frac{\pi}{6} + \frac{2\pi}{3}\right) + i\sin\left(\frac{\pi}{6} + \frac{2\pi}{3}\right)$$

$$= \cos\frac{5\pi}{6} + i\sin\frac{5\pi}{6},$$

$$k = 2, \quad z_2 = \cos\left(\frac{\pi}{6} + \frac{4\pi}{3}\right) + i\sin\left(\frac{\pi}{6} + \frac{4\pi}{3}\right)$$

$$= \cos\frac{3\pi}{2} + i\sin\frac{3\pi}{2}.$$

Therefore,

$$z_0 = \frac{\sqrt{3}}{2} + \frac{1}{2}i, \quad z_1 = -\frac{\sqrt{3}}{2} + \frac{1}{2}i, \quad z_2 = -i. \quad \blacksquare$$

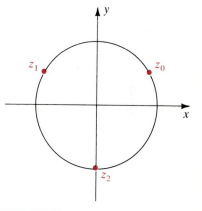

The three cube roots of i found in Example 2 are plotted in Figure 37. We note that they are equally spaced around a circle of radius 1 centered at the origin. In general, the nth roots of a nonzero complex number z are equally spaced on the circumference of the circle of radius $|z|^{1/n}$ with center at the origin (see Problem 32).

EXAMPLE 3

Find the five fifth roots of $1 + i$.

FIGURE 37

Solution The modulus and argument of $1 + i$ are $\sqrt{2}$ and $45°$, respectively, so that

$$1 + i = \sqrt{2}(\cos 45° + i\sin 45°).$$

We have chosen to express the argument in degrees because $45°$ is divisible by 5. With $n = 5$, $r^{1/n} = (\sqrt{2})^{1/5} = 2^{1/10}$. Thus, from Theorem 3, we obtain

$$k = 0, \quad z_0 = 2^{1/10}[\cos 9° + i\sin 9°],$$

$$k = 1, \quad z_1 = 2^{1/10}[\cos 81° + i\sin 81°],$$

$$k = 2, \quad z_2 = 2^{1/10}[\cos 153° + i\sin 153°],$$

$$k = 3, \quad z_3 = 2^{1/10}[\cos 225° + i \sin 225°],$$
$$k = 4, \quad z_4 = 2^{1/10}[\cos 297° + i \sin 297°].$$

Using a calculator, we can obtain these approximations:

$$z_0 \approx 1.0586 + 0.1677i,$$
$$z_1 \approx 0.1677 + 1.0586i,$$
$$z_2 \approx -0.9550 + 0.4866i,$$
$$z_3 \approx -0.7579 - 0.7579i,$$
$$z_4 \approx 0.4866 - 0.9550i.$$

EXERCISE 5.5

In Problems 1–14, use DeMoivre's theorem to calculate the given power.

1. i^{30}

2. $(-i)^{15}$

3. $(1 + i)^6$

4. $(1 - i)^9$

5. $(-2 + 2i)^4$

6. $(-4 - 4i)^3$

7. $(\sqrt{3} + i)^5$

8. $(-\sqrt{3} + i)^{10}$

9. $\left(\dfrac{\sqrt{2}}{2} - \dfrac{\sqrt{6}}{2}i\right)^9$

10. $\left(\dfrac{\sqrt{3}}{6} + \dfrac{1}{2}i\right)^8$

11. $(1 + 2i)^4$

12. $[2(\cos 67° + i \sin 67°)]^3$

13. $\left[\cos \dfrac{5\pi}{8} + i \sin \dfrac{5\pi}{8}\right]^{24}$

14. $\left[\sqrt{3}\left(\cos \dfrac{\pi}{24} + i \sin \dfrac{\pi}{24}\right)\right]^6$

In Problems 15–24, evaluate the indicated roots.

15. The three cube roots of -8

16. The three cube roots of 1

17. The four fourth roots of i

18. The two square roots of i

19. The four fourth roots of $-1 - \sqrt{3}i$

20. The two square roots of $-1 + \sqrt{3}i$

21. The two square roots of $1 + i$

22. The three cube roots of $-2\sqrt{3} + 2i$

23. The six sixth roots of $64[\cos 54° + i \sin 54°]$

24. The two square roots of $81\left[\cos \dfrac{5\pi}{3} + i \sin \dfrac{5\pi}{3}\right]$

25. Factor the polynomial $x^2 + 8 + 8\sqrt{3}i$.

26. Factor the polynomial $x^3 - 125i$.

27. For what positive integers n will $(\sqrt{2}/2 + \sqrt{2}i/2)^n$ be equal to 1? equal to i? equal to $-\sqrt{2}/2 - \sqrt{2}i/2$? equal to $\sqrt{2}/2 + \sqrt{2}i/2$?

28. Evaluate

$$\frac{\left[\cos \dfrac{\pi}{9} + i \sin \dfrac{\pi}{9}\right]^{12}}{\left[\dfrac{1}{2}\left(\cos \dfrac{\pi}{6} + i \sin \dfrac{\pi}{6}\right)\right]^5}.$$

29. Evaluate

$$\frac{\left(2\left[\cos \dfrac{\pi}{16} + i \sin \dfrac{\pi}{16}\right]\right)^{10}}{\left(4\left[\cos \dfrac{3\pi}{8} + i \sin \dfrac{3\pi}{8}\right]\right)^3}.$$

30. Prove that DeMoivre's theorem holds for negative integers n. [*Hint*: Use $z^{-n} = 1/z^n$ and apply DeMoivre's theorem to the denominator.]

31. DeMoivre's theorem implies

$$[\cos \theta + i \sin \theta]^2 = \cos 2\theta + i \sin 2\theta$$

and

$$[\cos \theta + i \sin \theta]^3 = \cos 3\theta + i \sin 3\theta.$$

Use this information to derive trigonometric identities for $\cos 2\theta$, $\sin 2\theta$, $\cos 3\theta$, and $\sin 3\theta$ by multiplying out the powers and equating real and imaginary parts.

32. (a) If the nonzero complex number z has modulus r, then show that any nth root of z lies on the circle of radius $\sqrt[n]{r}$ centered at the origin. [*Hint:* Find the modulus of $z^{1/n}$].

(b) Show that the nth roots of z are equally spaced on this circle. [*Hint:* Find the difference between the arguments of any two successive nth roots of z.]

33. Show that the n nth roots of a complex number z are given by $u, uw_1, uw_2, \ldots, uw_{n-1}$, where u satisfies $u^n = z$ and 1, $w_1, w_2, \ldots, w_{n-1}$ are the n nth roots of 1.

IMPORTANT CONCEPTS

Vectors	linear combination	product
components	Resultant force	additive identity
magnitude	Dot product	multiplicative identity
direction angle	Angle between vectors	multiplicative inverse
initial point	Component of a vector along	conjugate
terminal point	another vector	Complex plane
position vector	Work	imaginary axis
zero vector	Complex numbers	real axis
sum	imaginary unit	Modulus
difference	principal square root	Trigonometric form
scalar multiple	pure imaginary number	Argument
scalar	equality of complex numbers	DeMoivre's theorem
unit vector	sum	Roots of a complex number

CHAPTER 5 REVIEW EXERCISE

In Problems 1–10, fill in the blanks or answer true or false.

1. The vectors $\mathbf{u} = \langle 2, 6 \rangle$ and $\mathbf{v} = \langle -1, -3 \rangle$ are parallel. ___

2. The vectors $\mathbf{u} = 4\mathbf{i} - 8\mathbf{j}$ and $\mathbf{v} = -20\mathbf{i} + 10\mathbf{j}$ are orthogonal. ___

3. If $\mathbf{u} = k\mathbf{i} - 2\mathbf{j}$ and $\mathbf{v} = 4\mathbf{i} + (k - 8)\mathbf{j}$, then $\mathbf{u} \cdot \mathbf{v} = 0$ if $k = $ ___.

4. If $\mathbf{u} = \mathbf{i} + \mathbf{j}$ and $\mathbf{v} = 4\mathbf{i} - 3\mathbf{j}$, the magnitude of $\mathbf{u} - \mathbf{v}$ is ___.

5. The vectors $\mathbf{u} + \mathbf{v}$ and $\mathbf{u} - \mathbf{v}$ are the diagonals of the parallelogram determined by the nonzero vectors $\mathbf{u}$ and $\mathbf{v}$. ___

6. $\mathbf{u} = \frac{12}{13}\mathbf{i} - \frac{5}{13}\mathbf{j}$ is a unit vector. ___

7. The conjugate of the complex number $4 - 9i$ is ___.

8. Suppose that $z_1 = (x - 2) + (2y + 5)i$ and $z_2 = -4 + 7i$. If $z_1 = z_2$, then $x = $ ___ and $y = $ ___.

9. The modulus of the complex number $3i(4 + 3i)$ is ___.

10. $i(2i^2)(3i^3)(4i^4)(5i^5) = $ ___.

In Problems 11–20, find the indicated number or vector if $\mathbf{u} = \langle 2, -5 \rangle$ and $\mathbf{v} = \langle -4, 3 \rangle$.

11. $-\mathbf{u}$

12. $3\mathbf{u} + \mathbf{v}$

13. $\mathbf{v} - \mathbf{u}$

14. $2\mathbf{v} - 4\mathbf{u}$

15. $(6\mathbf{u}) \cdot \mathbf{v}$

16. $(-4\mathbf{u}) \cdot (-2\mathbf{v})$

17. $\mathbf{u} \cdot (\mathbf{u} + 5\mathbf{v})$

18. $(2\mathbf{v}) \cdot (4\mathbf{u} + 9\mathbf{v})$

19. $\dfrac{\mathbf{u} \cdot \mathbf{v}}{\mathbf{u} \cdot \mathbf{u}}$

20. $\dfrac{1}{\sqrt{\mathbf{v} \cdot \mathbf{v}}} \mathbf{v}$

6.2

Logarithmic Functions

Since the exponential function $y = b^x$ ($b > 0$, $b \neq 1$) is one-to-one, it follows that it has an inverse function. To find the inverse, we use the method discussed in Section 1.6: We interchange the variables x and y to obtain $x = b^y$. This formula defines y as a function of x:

$$y \text{ is that exponent of the base } b \text{ that produces } x. \qquad (4)$$

By replacing the word "exponent" with the word "logarithm," we can rephrase (4) as

$$y \text{ is the logarithm to the base } b \text{ of } x$$

and abbreviate it using the formula $y = \log_b x$. That is,

$$y = \log_b x \quad \text{is equivalent to} \quad x = b^y. \qquad (5)$$

We have the following definition.

DEFINITION 2

> The **logarithmic function with base b,**
>
> $$f(x) = \log_b x$$
>
> is the inverse of the exponential function with base b.

Recall from Section 1.6 that the graph of an inverse function can be obtained by reflecting the graph of the original function in the line $y = x$. This technique is used in Figure 11(a) to obtain the graph of $y = \log_b x$ for $b > 1$. The graph of $y = \log_b x$ for $0 < b < 1$ is shown in Figure 11(b).

PROPERTIES OF THE LOGARITHMIC FUNCTION

As Figure 11 illustrates, the logarithmic function f with base b has the following properties:

- The domain of f is the set of positive real numbers.
- The range of f is the set of real numbers.
- The x-intercept for the graph of f is 1. The graph of f has no y-intercept.
- The y-axis is a vertical asymptote for the graph of f.
- The function f is increasing on the interval $(0, \infty)$ if $b > 1$ and decreasing on the interval $(0, \infty)$ if $0 < b < 1$.
- The function f is one-to-one.

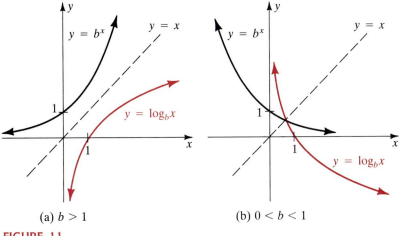

(a) $b > 1$ (b) $0 < b < 1$

Since the two equations $y = \log_b x$ and $b^y = x$ are equivalent, we can use whichever is more convenient. The following table lists several examples of equivalent exponential and logarithmic statements.

LOGARITHMIC FORM	EQUIVALENT EXPONENTIAL FORM
$\log_3 9 = 2$	$9 = 3^2$
$\log_{10} 0.0001 = -4$	$0.0001 = 10^{-4}$
$\log_8 4 = \frac{2}{3}$	$4 = 8^{2/3}$

From (5) it follows that

$$\log_b b = 1 \qquad\qquad (6)$$

and

$$\log_b 1 = 0 \qquad\qquad (7)$$

since $b^1 = b$ and $b^0 = 1$, respectively. It should also be noted that $\log_b x$ is meaningless for $x \leq 0$ since there is no exponent y for which $b^y \leq 0$. The result in (7) confirms that 1 is the x-intercept for the graph of a logarithmic function $f(x) = \log_b x$.

EXAMPLE 1

Solve each of the following for the unknown.

(a) $\log_2 8 = y$ **(b)** $\log_4 x = -\frac{1}{2}$ **(c)** $\log_b 25 = 2$

Solution In each case we use the equivalent exponential form given in (5).

(a) $\log_2 8 = y$ is equivalent to

$$2^y = 8$$
$$= 2^3.$$

Thus we conclude that $y = 3$.

(b) $\log_4 x = -\frac{1}{2}$ is equivalent to

$$4^{-1/2} = x$$

so that $x = 1/4^{1/2} = \frac{1}{2}$.

(c) $\log_b 25 = 2$ is equivalent to

$$b^2 = 25$$
$$= 5^2,$$

and so we find that $b = 5$.

By substituting $y = \log_b x$ into the equivalent equation $x = b^y$, we obtain the important identity

$$x = b^{\log_b x}. \tag{8}$$

EXAMPLE 2

(a) $3^{\log_3 7} = 7$ (b) $10^{\log_{10} 5^2} = 5^2$

THE LAWS OF LOGARITHMS

The following three properties, or **laws of logarithms,** are simply a reformulation of the laws of exponents.

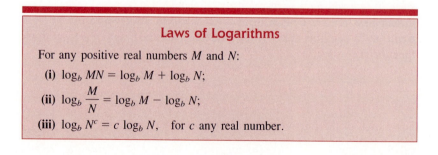

Laws of Logarithms

For any positive real numbers M and N:

(i) $\log_b MN = \log_b M + \log_b N$;

(ii) $\log_b \dfrac{M}{N} = \log_b M - \log_b N$;

(iii) $\log_b N^c = c \log_b N$,   for c any real number.

To verify these laws we use identity (8) to write any two positive numbers M and N as

$$M = b^{\log_b M} \quad \text{and} \quad N = b^{\log_b N}$$

so that

$$MN = b^{\log_b M} \cdot b^{\log_b N}$$

or

$$MN = b^{\log_b M + \log_b N}.$$

From (5) the last exponential statement is equivalent to the logarithmic statement

$$\log_b MN = \log_b M + \log_b N,$$

which is law (i).

Similarly,

$$\frac{M}{N} = \frac{b^{\log_b M}}{b^{\log_b N}}$$

$$= b^{\log_b M - \log_b N}$$

and

$$N^c = (b^{\log_b N})^c$$

$$= b^{c \, \log_b N}$$

are equivalent to laws (ii) and (iii), respectively.

■ EXAMPLE 3

Simplify $\frac{1}{2} \log_9 36 + 2 \log_9 4 - \log_9 4$.

Solution There are several ways to approach this problem. Note, for example, that the second and third terms can be combined as

$$2 \log_9 4 - \log_9 4 = \log_9 4.$$

Alternatively, we can use law (iii) followed by law (ii) to combine these terms:

$$2 \log_9 4 - \log_9 4 = \log_9 4^2 - \log_9 4$$

$$= \log_9 16 - \log_9 4$$

$$= \log_9 \tfrac{16}{4}$$

$$= \log_9 4.$$

Hence,
$$\tfrac{1}{2} \log_9 36 + 2 \log_9 4 - \log_9 4 = \log_9 (36)^{1/2} + \log_9 4 \quad \Leftarrow \boxed{\textbf{By (iii)}}$$

$$= \log_9 6 + \log_9 4$$

$$= \log_9 24. \quad \Leftarrow \boxed{\textbf{By (i)}}$$

EXAMPLE 4

If $\log_b 2 = 0.3010$ and $\log_b 3 = 0.4771$, find each of the following.

(a) $\log_b 6$ **(b)** $\log_b \dfrac{2}{3}$ **(c)** $\dfrac{\log_b 2}{\log_b 3}$

(d) $\log_b 64$ **(e)** $\log_b \sqrt[3]{18}$

Solution We use laws (i)–(iii) to write each of the given logarithms in terms of $\log_b 2$ and $\log_b 3$.

(a) $\log_b 6 = \log_b (3 \cdot 2)$

$\qquad\quad = \log_b 3 + \log_b 2$ ⬅ **By (i)**

$\qquad\quad = 0.4771 + 0.3010$

$\qquad\quad = 0.7781$

(b) $\log_b \tfrac{2}{3} = \log_b 2 - \log_b 3$ ⬅ **By (ii)**

$\qquad\quad = 0.3010 - 0.4771$

$\qquad\quad = -0.1761$

(c) By division,

$$\frac{\log_b 2}{\log_b 3} = \frac{0.3010}{0.4771}$$

$$\approx 0.6309.$$

(d) $\log_b 64 = \log_b 2^6$

$\qquad\quad = 6 \log_b 2$ ⬅ **By (iii)**

$\qquad\quad = 6(0.3010)$

$\qquad\quad = 1.8060$

(e) $\log_b \sqrt[3]{18} = \log_b (18)^{1/3}$

$\qquad\quad = \tfrac{1}{3} \log_b 18$ ⬅ **By (iii)**

$\qquad\quad = \tfrac{1}{3} \log_b (2 \cdot 3^2)$

$\qquad\quad = \tfrac{1}{3}[\log_b 2 + \log_b 3^2]$ ⬅ **By (i)**

$\qquad\quad = \tfrac{1}{3}[\log_b 2 + 2 \log_b 3]$ ⬅ **By (iii)**

$\qquad\quad = \tfrac{1}{3}[0.3010 + 2(0.4771)]$

$\qquad\quad = 0.4184$

Note of Caution: Observe that law (ii) of logarithms is *not* applicable in part (c) of Example 4. In other words, *a quotient of logarithms is not the* *difference of logarithms*. You should also keep in mind that

$$\log_b (M + N) \neq \log_b M + \log_b N.$$

In general, there is no way of expressing $\log_b (M + N)$ in terms of $\log_b M$ and $\log_b N$.

GRAPHS

In the next three examples we will examine the graphs of some functions involving logarithms with base 10.

EXAMPLE 5

Graph $f(x) = \log_{10} x$.

Solution The following table shows the corresponding values of y for selected values of x. Using the points $(1, 0)$, $(10, 1)$, and $(100, 2)$ and the knowledge of the basic shape of a logarithmic graph from Figure 11(a), we obtain the graph shown in Figure 12.

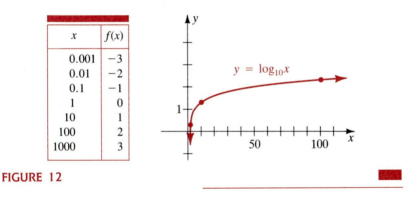

x	$f(x)$
0.001	-3
0.01	-2
0.1	-1
1	0
10	1
100	2
1000	3

FIGURE 12

EXAMPLE 6

Graph $f(x) = \log_{10} (x + 10)$.

Solution The domain of this function is determined by the requirement that $x + 10 > 0$, or $x > -10$. Also, from Section 1.6 we know that the graph of the function is the graph in Figure 12 shifted 10 units to the left. With this information and the accompanying table, we obtain the graph shown in Figure 13.

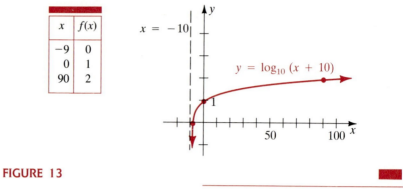

x	f(x)
−9	0
0	1
90	2

FIGURE 13

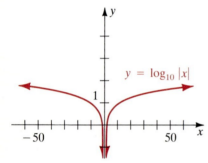

FIGURE 14

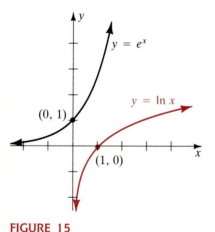

FIGURE 15

EXAMPLE 7

Graph $f(x) = \log_{10} |x|$.

Solution Since $|x| > 0$ for $x \neq 0$, the absolute value extends the domain of the given logarithmic function to all real numbers except $x = 0$. Furthermore, since

$$f(-x) = \log_{10} |-x| = \log_{10} |x| = f(x),$$

we see that f is an even function; thus its graph is symmetric with respect to the y-axis. Now the portion of the graph of f for $x > 0$ is identical to the graph in Figure 12. We obtain the portion of the graph of f for $x < 0$ by symmetry (see Figure 14).

COMMON AND NATURAL LOGARITHMS

As we saw in Definition 2, the base b of a logarithmic function can be any positive real number other than 1. In practice, however, two of the most important bases are $b = 10$ and $b = e = 2.718281828459. \ldots$. Logarithms with $b = 10$ are known as **common logarithms** and logarithms with $b = e$ are called **natural logarithms.** Furthermore, it is customary to write the natural logarithm

$$\log_e x \quad \text{as} \quad \ln x.$$

The symbol "ln x" is usually read phonetically as "ell-en of x." Since $b = e > 1$, the graph of $f(x) = \ln x$ has the logarithmic shape shown in Figure 11(a) (see Figure 15). For base e, (5) becomes

$$y = \ln x \quad \text{is equivalent to} \quad x = e^y.$$

Also, (8) then becomes

$$x = e^{\ln x}.$$

To compute $f(x) = \log_{10} x$ and $f(x) = \ln x$, we use the keys marked $\boxed{\texttt{log}}$ and $\boxed{\texttt{ln}}$, respectively, on a scientific calculator. Consult your user's manual if your calculator does not have these keys.

████ **EXAMPLE 8**

Find the values of **(a)** $\log_{10} 647$ and **(b)** $\ln 123$.

Solution

(a) After entering 647, we press the $\boxed{\texttt{log}}$ key to obtain

$$\log_{10} 647 \approx 2.8109.$$

(b) Using the $\boxed{\texttt{ln}}$ key, we see that

$$\ln 123 \approx 4.8122.$$ ████

THE CHANGE OF BASE FORMULA

It is possible to express a logarithm with base a in terms of logarithms with base b. To do so, let us suppose that

$$y = \log_a x \quad \text{so that} \quad x = a^y.$$

By taking the logarithm with base b of both sides of the last equation, we see that

$$\log_b x = \log_b a^y$$
$$= y \log_b a,$$

or

$$y = \frac{\log_b x}{\log_b a}.$$

But since $y = \log_a x$, we obtain the **change of base formula:**

$$\log_a x = \frac{\log_b x}{\log_b a} \tag{9}$$

████ **EXAMPLE 9**

Find the value of $\log_2 50$.

Solution   We can use the change of base formula to convert the given logarithm to either base 10 or base e. If we choose base 10, (9) gives us

$$\log_2 50 = \frac{\log_{10} 50}{\log_{10} 2}.$$

Using the $\boxed{\text{log}}$ key to calculate the two logarithms and then dividing yields the approximation

$$\log_2 50 \approx 5.6439.$$

Alternative Solution   If we choose base e, (9) gives

$$\log_2 50 = \frac{\ln 50}{\ln 2}.$$

Now using the $\boxed{\text{ln}}$ key, we find

$$\log_2 50 \approx 5.6439.$$

We can check the answer in Example 9 on a calculator by using the $\boxed{y^x}$ key. We enter $y = 2$ and $x = 5.6439$ to get

$$2^{5.6439} \approx 50.$$

EXERCISE 6.2

In Problems 1–12, rewrite the given exponential statement in the equivalent logarithmic form.

1. $4^{-1/2} = \frac{1}{2}$

2. $10^{0.3010} = 2$

3. $9^0 = 1$

4. $(3^2)^{-2} = \frac{1}{81}$

5. $10^y = x$

6. $t^{-s} = v$

7. $\left(\frac{1}{64}\right)^{-1/2} = 8$

8. $(a + b)^2 = a^2 + 2ab + b^2$

9. $36^{-3/2} = \frac{1}{216}$

10. $10^{-3} = 0.001$

11. $8^{2/3} = 4$

12. $e^1 = e$

In Problems 13–24, rewrite the given logarithmic statement in the equivalent exponential form.

13. $\log_3 81 = 4$

14. $\log_2 32 = 5$

15. $\log_{10} 10 = 1$

16. $\log_{17} 17^5 = 5$

17. $\log_5 \frac{1}{25} = -2$

18. $\log_{\sqrt{2}} 2 = 2$

19. $\log_{16} 2 = \frac{1}{4}$

20. $\log_9 \frac{1}{3} = -\frac{1}{2}$

21. $\log_b b^2 = 2$

22. $\log_b u = v$

23. $\ln 1 = 0$

24. $\ln (1/e) = -1$

In Problems 25–36, find the value of the given logarithm without using a calculator.

25. $\log_{10} (0.000001)$

26. $\log_4 \frac{1}{64}$

27. $\log_2 (2^2 + 2^2)$

28. $\log_7 \sqrt[3]{49}$

29. $\log_{64} \frac{1}{32}$

30. $\log_{\sqrt{3}} 9$

31. $\log_{1/2} 16$

32. $\log_8 \frac{1}{4}$

33. $\log_{5/2} \frac{8}{125}$

34. $\log_6 216$

35. $\ln e^e$

36. $\ln (ee^2 e^3)$

In Problems 37–48, solve for the unknown.

37. $\log_b 125 = 3$

38. $\log_{10} N = -2$

39. $\log_7 343 = x$

40. $\log_5 25^c = 4$

41. $\log_2 (1/N) = 5$

42. $\log_b 6 = -1$

43. $2 \log_9 N = 1$

44. $\log_2 4^{-3} = x$

45. $\log_3 \frac{1}{27} = x$

46. $\log_{10} \left(\frac{1}{1000}\right)^c = 1$

47. $\sqrt{\ln N} = 3$

48. $\ln (N/2) = 4$

In Problems 49–52, find the given number without using a calculator.

49. $10^{\log_{10} 6^2}$

50. $25^{\log_5 8}$

51. $e^{-\ln 7}$

52. $(\sqrt{e})^{\ln 9}$

In Problems 53–64, use $\log_b 4 = 0.6021$ and $\log_b 5 = 0.6990$ to evaluate the given logarithm.

53. $\log_b 2$

54. $\log_b 20$

55. $\log_b 64$

56. $\log_b 625$

57. $\log_b \sqrt{5}$

58. $\log_b \frac{5}{4}$

59. $\log_b \sqrt[3]{4}$ **60.** $\log_b 80$
61. $\log_b 0.8$ **62.** $\log_b 3.2$
63. $\log_4 b$ **64.** $\log_5 5b$

In Problems 65–70, simplify and write as one logarithm.

65. $\log_{10} 2 + \log_{10} 5$
66. $\frac{1}{2} \log_5 49 - \frac{1}{3} \log_5 8 + 13 \log_5 1$
67. $\log_{10} (x^4 - 4) - \log_{10} (x^2 + 2)$
68. $\log_{10} \left(\frac{x}{y} \right) - 2 \log_{10} x^3 + \log_{10} y^{-4}$
69. $\log_2 5 + \log_2 5^2 + \log_2 5^3 - \log_2 5^6$
70. $5 \ln 2 + 2 \ln 3 - 3 \ln 4$

In Problems 71–84, graph the given function and state its domain.

71. $f(x) = \log_2 x$ **72.** $f(x) = \log_4 x$

73. $f(x) = \log_2 |x|$ **74.** $f(x) = \log_2 \frac{1}{x}$

75. $f(x) = \log_2 (x - 3)$ **76.** $f(x) = \log_2 (3 - x)$
77. $f(x) = \log_4 (x + 1)$ **78.** $f(x) = \log_4 |x + 1|$
79. $f(x) = \log_2 2x$ **80.** $f(x) = \log_2 \sqrt{x}$
81. $f(x) = -1 + \log_2 x$ **82.** $f(x) = 1 - \log_2 x$
83. $f(x) = \log_{1/2} x$ **84.** $f(x) = \log_{1/4} 2x$

In Problems 85–88, use the change of base formula (9) and the $\boxed{\log}$ **key to find the value of the given logarithm.**

85. $\log_7 23$ **86.** $\log_3 1000$
87. $2 \ln 6$ **88.** $-\log_{1/2} 18$

In Problems 89–92, use the change of base formula (9) and the $\boxed{\ln}$ **key to find the value of the given logarithm.**

89. $\log_5 16$ **90.** $\log_4 537$
91. $\log_{10} 265$ **92.** $\log_{2/3} 41$

93. The logarithm developed by John Napier was actually

$$10^7 \log_{1/e} \left(\frac{x}{10^7} \right).$$

Use the change of base formula (9) to express this logarithm in terms of the natural logarithm.
94. Show that $(\log_a b)(\log_b a) = 1$.
95. Which of the following equations determines the graph shown in Figure 16?
  (a) $y = 2x + 1$ **(b)** $y = 10 + x^2$
  (c) $y = 10x^2$ **(d)** $x^2 y = 10$

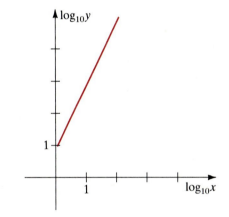

FIGURE 16

96. Verify the identity

$$\ln |\sec \theta - \tan \theta| = -\ln |\sec \theta + \tan \theta|.$$

Exponential and Logarithmic Equations   *6.3*

EXPONENTIAL EQUATIONS

Consider the function $P(t) = 1000(\frac{3}{2})^t$. When $t = 1$, we see that $P(1) = 1500$. Suppose we now turn the problem around; that is, for a given value of P, find the corresponding value of t. For example, if $P = 2250$, then we must solve the exponential equation

$$2250 = 1000(\tfrac{3}{2})^t, \quad \text{or} \quad (\tfrac{3}{2})^t = 2.25$$

for the unknown value of t. By rewriting the last equation in an equivalent logarithmic form, we find

$$t = \log_{3/2} 2.25$$
$$= \log_{3/2} \tfrac{9}{4}$$
$$= \log_{3/2} (\tfrac{3}{2})^2$$
$$= 2 \log_{3/2} (\tfrac{3}{2})$$
$$= 2.$$

▮▮ EXAMPLE 1

Solve $4^{x^2-1} = \tfrac{1}{2}$.

Solution   If we write the given equation in logarithmic form, we find that

$$x^2 - 1 = \log_4 \tfrac{1}{2}.$$

But $\log_4 \tfrac{1}{2} = -\tfrac{1}{2}$, since $4^{-1/2} = \tfrac{1}{2}$. Therefore,

$$x^2 - 1 = -\tfrac{1}{2}$$
$$x^2 = \tfrac{1}{2}$$
$$x = \pm\sqrt{\tfrac{1}{2}}.$$

Check: $4^{(\sqrt{1/2})^2 - 1} = 4^{-1/2} = \tfrac{1}{2}$ and $4^{(-\sqrt{1/2})^2 - 1} = 4^{-1/2} = \tfrac{1}{2}.$ ▮

▮▮ EXAMPLE 2

Solve $4^x = 69$.

Solution   Written in logarithmic form, the equation has the solution

$$x = \log_4 69.$$

Since scientific calculators use only common and natural logarithms, we are not able to give an immediate numerical value for $\log_4 69$. However, we can get out of this small dilemma by using base 10 logarithms and the change of base formula (9):

$$x = \log_4 69 = \frac{\log_{10} 69}{\log_{10} 4}.$$

Now we can use a calculator to get $x \approx 3.0543$.

Alternative Solution   Taking the base 10 logarithm of both sides of the equation $4^x = 69$, we see that

$$\log_{10} 4^x = \log_{10} 69$$
$$x \log_{10} 4 = \log_{10} 69$$
$$x = \frac{\log_{10} 69}{\log_{10} 4} \approx 3.0543.$$

▦ EXAMPLE 3

Solve $5^x - 5^{-x} = 2$.

Solution   Since $5^{-x} = 1/5^x$, the equation is

$$5^x - \frac{1}{5^x} = 2.$$

Multiplying both sides by 5^x then gives

$$(5^x)^2 - 1 = 2(5^x)$$

or

$$(5^x)^2 - 2(5^x) - 1 = 0.$$

This last equation can be interpreted as a quadratic equation in 5^x. Using the quadratic formula to solve for 5^x yields

$$5^x = \frac{2 \pm \sqrt{4 + 4}}{2}$$
$$= 1 \pm \sqrt{2}.$$

But since 5^x is always positive, we must reject the negative number $1 - \sqrt{2}$. Hence,

$$5^x = 1 + \sqrt{2}. \tag{10}$$

Taking the base 10 logarithm of both sides, we get

$$\log_{10} 5^x = \log_{10} (1 + \sqrt{2})$$
$$x \log_{10} 5 = \log_{10} (1 + \sqrt{2})$$
$$x = \frac{\log_{10} (1 + \sqrt{2})}{\log_{10} 5}$$
$$x \approx 0.5476. \tag{11}$$

In Examples 2 and 3 we could use the natural logarithm rather than the common logarithm to solve for x. For example, taking the natural logarithm of both sides of (10), we obtain

$$x = \frac{\ln (1 + \sqrt{2})}{\ln 5}.$$

You should verify on a calculator that this result is the same as (11).

■ EXAMPLE 4

Solve $e^{4t} = 23$.

Solution   We have
$$4t = \ln 23$$
$$t = \tfrac{1}{4} \ln 23.$$

Using a calculator, we find
$$t \approx 0.7839. \qquad \blacksquare$$

Since an exponential function $f(x) = b^x$ is a one-to-one function, an equation such as

$$b^{x_1} = b^{x_2} \quad \text{implies that} \quad x_1 = x_2. \tag{12}$$

■ EXAMPLE 5

Solve $7^{2(x+1)} = 343$.

Solution   By observing that $343 = 7^3$, we have the same base on both sides of the equation:

$$7^{2(x+1)} = 7^3.$$

Thus by (12) we can equate exponents:

$$2(x + 1) = 3$$
$$2x + 2 = 3$$
$$2x = 1$$
$$x = \tfrac{1}{2}. \qquad \blacksquare$$

LOGARITHMIC EQUATIONS

Using the fact that the statements $b^y = x$ and $y = \log_b x$ are equivalent, we can also solve certain equations involving logarithms.

■ **EXAMPLE 6**

Solve $\log_{10} (2x + 50) = 2$.

Solution The equation can be written equivalently as
$$2x + 50 = 10^2,$$
or
$$2x + 50 = 100$$
$$2x = 50$$
$$x = 25.$$

You should check this solution. ■

■ **EXAMPLE 7**

Solve $\log_2 x + \log_2 (x - 2) = 3$. **(13)**

Solution By law (i) of Section 6.2, we can write the sum of logarithms as one logarithm:
$$\log_2 x(x - 2) = 3.$$
Hence we have
$$x(x - 2) = 2^3$$
$$x^2 - 2x = 8$$
$$x^2 - 2x - 8 = 0$$
$$(x - 4)(x + 2) = 0.$$

We conclude that either $x = 4$ or $x = -2$. However, $x = -2$ must be ruled out since $\log_2 x$ in (13) is not defined for $x \le 0$. Thus the only solution of the equation is $x = 4$.

Check:
$$\log_2 4 + \log_2 2 = \log_2 2^2 + \log_2 2$$
$$= 2 \log_2 2 + \log_2 2$$
$$= 2 + 1 = 3. ■$$

■ **EXAMPLE 8**

Solve $\log_3 (7 - x) - \log_3 (1 - x) = 1$. **(14)**

Solution By law (ii) of Section 6.2, we can write the equation as
$$\log_3 \frac{7 - x}{1 - x} = 1$$

so that
$$\frac{7 - x}{1 - x} = 3^1.$$

We multiply the latter equation by $1 - x$ and solve for x:

$$7 - x = 3(1 - x)$$
$$7 - x = 3 - 3x$$
$$2x = -4$$
$$x = -2.$$

Now at this point we should *not* conclude that $x = -2$ is not a solution of (14) simply because it is negative. Indeed, $x = -2$ *is* a solution since

$$\log_3 (7 - (-2)) - \log_3 (1 - (-2)) = \log_3 9 - \log_3 3$$
$$= \log_3 \tfrac{9}{3}$$
$$= \log_3 3$$
$$= 1. \qquad\blacksquare$$

Recall that $f(x) = \log_b x$ is also a one-to-one function. Thus,

$$\log_b x_1 = \log_b x_2 \quad \text{implies that} \quad x_1 = x_2. \tag{15}$$

■ EXAMPLE 9

Solve $2 \log_2 x + 3 \log_2 2 = 3 \log_2 x - \log_2 \tfrac{1}{32}$.

Solution   First, we write each side of the equation as a single logarithm:

$$\log_2 x^2 + \log_2 2^3 = \log_2 x^3 - \log_2 \tfrac{1}{32}$$
$$\log_2 2^3 x^2 = \log_2 \frac{x^3}{1/32}$$
$$\log_2 8x^2 = \log_2 32x^3.$$

It then follows from (15) that

$$8x^2 = 32x^3$$
$$8x^2 - 32x^3 = 0$$
$$8x^2(1 - 4x) = 0,$$

and so $x = 0$ or $x = \tfrac{1}{4}$. But $\log_2 x$ is not defined for $x = 0$; therefore, the only possible solution is $x = \tfrac{1}{4}$.

Alternative Solution   By algebra the original equation can be written as

$$3 \log_2 2 + \log_2 \tfrac{1}{32} = \log_2 x,$$

since $3 \log_2 x - 2 \log_2 x = \log_2 x$. Continuing, we find

$$\log_2 8 + \log_2 \tfrac{1}{32} = \log_2 x$$
$$\log_2 \tfrac{8}{32} = \log_2 x$$
$$\log_2 \tfrac{1}{4} = \log_2 x.$$

In this case, (15) gives $x = \tfrac{1}{4}$. The point here is: Depending on the method you use, you may introduce extraneous solutions.

EXERCISE 6.3

In Problems 1–22, solve the given exponential equation.

1. $5^{x-2} = 1$

2. $3^x = 27^{x^2}$

3. $10^{-2x} = \dfrac{1}{10,000}$

4. $27^x = \dfrac{9^{2x-1}}{3^x}$

5. $5 - 10^{2w} = 0$

6. $7^{-x} = 9$

7. $e^{5x-2} = 30$

8. $4^{\log_2 x} = 9$

9. $2^x + 2^{-x} = 2$

10. $e^x - e^{-x} = -1$

11. $2^{x^2} = 8^{2x-3}$

12. $\dfrac{1}{4}(10^{-2x}) - 25(10^x) = 0$

13. $\left(\dfrac{1}{2}\right)^{-x+2} = 8(2^{x-1})^3$

14. $6^{2t} + 6^t - 6 = 0$

15. $\dfrac{1}{3} = (2^{|x|-2} - 1)^{-1}$

16. $\left(\dfrac{1}{3}\right)^x = 9^{1-2x}$

17. $5^{|x|-1} = 25$

18. $(e^2)^{x^2} - \dfrac{1}{e^{5x+3}} = 0$

19. $4^x = 5^{2x+1}$

20. $3^{x+4} = 2^{x-16}$

21. $(5^x)^2 - 26(5^x) + 25 = 0$

22. $64^t - 10(8^t) + 16 = 0$

In Problems 23–42, solve the given logarithmic equation.

23. $\log_3 5x = \log_3 160$

24. $\ln (10 + x) = \ln (3 + 4x)$

25. $\log_2 x = \log_2 5 + \log_2 9$

26. $3 \log_8 x = \log_8 36 + \log_8 12 - \log_8 2$

27. $\log_{10} \dfrac{1}{x^2} = 2$

28. $\log_3 \sqrt{x^2 + 17} = 2$

29. $\log_2 (\log_3 x) = 2$

30. $\log_5 |1 - x| = 1$

31. $\log_3 81^x - \log_3 3^{2x} = 3$

32. $\dfrac{\log_2 8^x}{\log_2 \tfrac{1}{4}} = \dfrac{1}{2}$

33. $\log_{10} x = 1 + \log_{10} \sqrt{x}$

34. $\log_2 (x - 3) - \log_2 (2x + 1) = -\log_2 4$

35. $\log_2 x + \log_2 (10 - x) = 4$

36. $\log_8 x + \log_8 x^2 = 1$

37. $\log_6 2x - \log_6 (x + 1) = 0$

38. $\log_{10} 54 - \log_{10} 2 = 2 \log_{10} x - \log_{10} \sqrt{x}$

39. $\log_9 \sqrt{10x + 5} - \dfrac{1}{2} = \log_9 \sqrt{x + 1}$

40. $\log_{10} x^2 + \log_{10} x^3 + \log_{10} x^4 - \log_{10} x^5 = \log_{10} 16$

41. $\log_4 x^2 = (\log_4 x)^2$

42. $(\log_{10} x)^2 + \log_{10} x = 2$

In Problems 43 and 44, solve the given equation.

43. $x^{\ln x} = e^9$

44. $x^{\log_{10} x} = \dfrac{1000}{x^2}$

In Problems 45–50, graph the given functions. Determine the (approximate) x-coordinate of the point(s) of intersection of their graphs.

45. $f(x) = 4e^x$,
$g(x) = 3^{-x}$

46. $f(x) = 2^x$,
$g(x) = 3 - 2^x$

47. $f(x) = 3^{x^2}$,
$g(x) = 2(3^x)$

48. $f(x) = \dfrac{1}{3} \cdot 2^{x^2}$,
$g(x) = 2^{x^2} - 1$

49. $f(x) = \log_{10} \dfrac{10}{x}$, $g(x) = \log_{10} x$

50. $f(x) = \log_{10} \dfrac{x}{2}$, $g(x) = \log_2 x$

51. Medical researchers use the empirical formula

$$\log_{10} A = -2.144 + (0.425) \log_{10} m + (0.725) \log_{10} h$$

to estimate body surface area A (measured in square meters), given a person's mass (in kilograms) and height (in centimeters).

(a) Estimate the body surface area of a person whose mass is 70 kg and who is 175 cm tall.

(b) Determine your mass and height and estimate your own body surface area.

52. An empirical formula found by DeGroot and Gebhard relates the diameter d of the pupil of the eye (measured in millimeters) to the luminance B of a light source (measured in millilamberts, mL):

$$\log_{10} d = 0.8558 - 0.000401(8.1 + \log_{10} B)^3$$

(see Figure 17).

(a) The average luminance of clear sky is approximately 255 mL. Find the corresponding pupil diameter.

(b) The luminance of the sun varies from approximately 190,000 mL at sunrise to 51,000,000 mL at noon. Find the corresponding pupil diameters.

(c) Find the luminance corresponding to a pupil diameter of 7 mm.

FIGURE 17

6.4 Applications

A form of the exponential function given by

$$f(t) = Cb^{kt}, \tag{16}$$

where C, b, and k are constants, plays a significant role in describing many diverse phenomena in science, engineering, and business.

GROWTH

In biology (16) is often used as a mathematical model to describe the **growth** of populations of bacteria, small animals, and, in some circumstances, humans.

■ EXAMPLE 1

The population P in a community after t years is given by

$$P(t) = 1000(\tfrac{3}{2})^t.$$

Does the population increase or decrease for increasing time? What is the initial population? What is the population after 1, 2, and 5 years?

Solution We can think of $P(t)$ as a constant multiple of $(\tfrac{3}{2})^t$. Since $1000 > 0$ and $b = \tfrac{3}{2} > 1$, the population increases for increasing time. The initial population occurs when $t = 0$:

$$P(0) = 1000(\tfrac{3}{2})^0 = 1000.$$

Also,

$$P(1) = 1000(\tfrac{3}{2})^1 = 1500,$$
$$P(2) = 1000(\tfrac{3}{2})^2 = 1000(\tfrac{9}{4}) = 2250,$$
$$P(5) = 1000(\tfrac{3}{2})^5 = 1000(\tfrac{243}{32}) \approx 7594.$$

In the early nineteenth century the English clergyman and economist Thomas R. Malthus (1776–1834) used the function $P(t) = P_0 e^{kt}$, $k > 0$, as a mathematical model to predict world population. For specific values of P_0 and k, the functional values $P(t)$ were actually reasonable approximations to the world population during the eighteenth century. Since $P(t)$ is an increasing function, Malthus predicted that the future population growth would surpass the world's ability to produce food. As a consequence, Malthus also predicted wars and eventual worldwide famine. More a doomsayer than a seer, Malthus failed to foresee the food supply keeping pace with the increased population because of the simultaneous advances in science and technology.

A more realistic model for predicting human populations in small countries was advanced by the Belgian mathematician–biologist P. F. Verhulst in 1840. The so-called **logistic function**

$$P(t) = \frac{aP_0}{bP_0 + (a - bP_0)e^{-at}}, \tag{17}$$

where a, b, and P_0 are constants, has also proved to be quite accurate in describing populations of protozoa, bacteria, fruit flies, water fleas, and animals confined to limited spaces. Although (17) is an increasing function, the graph of P possesses horizontal asymptotes at $P = 0$ and $P = a/b$. Thus in contrast with the Malthusian model, populations predicted by (17) must exhibit bounded growth, that is, a population will not increase beyond a certain number. This inhibited growth could be a result of predators, overcrowding, competition for food, pollution, family planning, and so on. You are asked to graph a special case of (17) in Problem 45.

DECAY

Element 88, better known as radium, is radioactive. This means that radium atoms spontaneously **decay,** or disintegrate, by emitting radiation in the form of alpha particles, beta particles, or gamma rays. When an atom disintegrates in this manner, its nucleus is transmuted into the nucleus of another element. The nucleus of a disintegrating radium atom is transmuted into the nucleus of a radon atom. The exponential function (16) can also serve as a mathematical model for approximating the amount remaining of an element that is decaying through radioactivity.

EXAMPLE 2

Suppose there are 20 grams of radium on hand initially. After t years the amount remaining is given by

$$A(t) = 20e^{-0.000418t}.$$

Find the amount of radium remaining after 100 years. What percent of the 20 grams has decayed after 100 years?

Solution Using a calculator, we find that

$$A(100) = 20e^{-0.000418(100)}$$
$$\approx 19.1812 \text{ g}.$$

In other words, after 100 years only

$$\frac{20 - 19.1812}{20} \times 100\% \approx 4.1\%$$

of the initial quantity has decayed.

HALF-LIFE

The **half-life** of a radioactive element is defined to be the time it takes for one half of a given amount of that element to disintegrate and change into a new element. Half-life is a measure of the stability of the element, that is, the shorter the half-life, the more unstable the element. For example, the half-life of the highly radioactive strontium 90, Sr-90, is 29 days, whereas the half-life of the uranium isotope U-238 is 4,560,000,000 years. The half-life of californium, Cf-244, first discovered in 1950, is only 45 minutes. Polonium, Po-213, has a half-life of 0.000001 second.

EXAMPLE 3

If there are A_0 grams of radium present initially, then after t years the number of grams remaining is

$$A(t) = A_0e^{-0.000418t}.$$

Determine the half-life of radium.

Solution We must find the time when $A = A_0/2$; that is, we must solve the equation

$$\frac{A_0}{2} = A_0e^{-0.000418t}$$

for t. Dividing by A_0 and using (9) of Section 6.2 yields

$$-0.000418t = \ln \frac{1}{2},$$

or

$$t = \frac{\ln \frac{1}{2}}{-0.000418}.$$

Using the $\boxed{\ln}$ key and then dividing we obtain the result from a calculator:

$$t \approx 1658 \text{ years.}$$

Note in Example 3 that the initial amount A_0 of radium played no part in the actual calculation of the half-life. Thus the half-life of 1 gram, 20 grams, or 10,000 grams of radium is the same. It takes about 1700 years for one half of *any* given quantity of radium to change into radon.

CARBON DATING

The approximate age of fossils can be determined by a method known as **carbon dating.** This method, invented by the chemist Willard Libby around 1950, is based on the fact that a living organism absorbs radioactive carbon 14, C-14, through the processes of eating and breathing and ceases to absorb C-14 when it dies. As the next example shows, the carbon dating procedure uses the knowledge that the half-life of C-14 is 5600 years.

Libby won the 1960 Nobel prize in chemistry for this work.

EXAMPLE 4

A fossil is found to contain $\frac{1}{1000}$ of the initial amount of C-14 that the organism contained while it was alive. Determine the approximate age of the fossil.

Solution If there was an initial amount of A_0 grams of C-14 in the organism, then t years after its death there are $A(t) = A_0 e^{kt}$ grams remaining. When $t = 5600$, $A(t) = A_0/2$, and so

$$\frac{A_0}{2} = A_0 e^{5600k}.$$

Solving the last equation for k gives

$$e^{5600k} = \frac{1}{2} \quad \text{and} \quad k = \frac{\ln \frac{1}{2}}{5600} = -0.000124.$$

Thus,

$$A(t) = A_0 e^{-0.000124t}.$$

Finally, to determine the age of the fossil, we solve the last equation for t when $A(t) = A_0/1000$:

$$\frac{A_0}{1000} = A_0 e^{-0.000124t}$$

$$t = -\frac{\ln \frac{1}{1000}}{0.000124} \approx 55{,}708 \text{ years.}$$

CIRCUITS

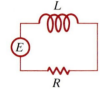

L

E

R

FIGURE 18

The simple **series circuit** shown in Figure 18 consists of a constant voltage E, an inductance of L henries, and a resistance of R ohms. It can be shown that the current I at any time t is given by

$$I = \frac{E}{R}[1 - e^{-(R/L)t}]. \qquad \textbf{(18)}$$

■■■ **EXAMPLE 5**

Solve (18) for t in terms of the current I.

Solution   Using algebra, we find that

$$\frac{IR}{E} = 1 - e^{-(R/L)t}$$

$$e^{-(R/L)t} = 1 - \frac{IR}{E}.$$

It follows that

$$-\left(\frac{R}{L}\right)t = \ln\left(1 - \frac{IR}{E}\right)$$

$$t = -\frac{L}{R}\ln\left(1 - \frac{IR}{E}\right).$$

COMPOUND INTEREST

Investments such as savings accounts pay an annual rate of interest that can be compounded annually, quarterly, monthly, weekly, daily, and so on. In general, if a principal of P dollars is invested at an annual rate r of interest that is compounded n times a year, then the amount S accrued at the end of t years is given by

$$S = P\left(1 + \frac{r}{n}\right)^{nt}. \qquad \textbf{(19)}$$

S is called the **future value** of the principal *P*. Now if the number of compoundings *n* is increased without bound, then interest is said to be **compounded continuously.** To find the future value of *P* in this case, we let $m = n/r$. Then $n = mr$ and

$$\left(1 + \frac{r}{n}\right)^{nt} = \left(1 + \frac{1}{m}\right)^{mrt} = \left[\left(1 + \frac{1}{m}\right)^{m}\right]^{rt}.$$

As *n* increases without bound, necessarily *m* increases without bound, that is, $n \to \infty$ implies that $m \to \infty$. It follows from (2) of Section 6.1 that

$$P\left[\left(1 + \frac{1}{m}\right)^{m}\right]^{rt} \to P[e]^{rt} \quad \text{as} \quad m \to \infty.$$

Thus if an annual rate *r* of interest is compounded continuously, the future value *S* of a principal *P* in *t* years is

$$S = Pe^{rt}. \tag{20}$$

■ EXAMPLE 6

Suppose that a principal of $1000 is deposited in a savings account whose annual rate of interest is 9%. Compare the future value of this principal in 5 years **(a)** if interest is compounded monthly and **(b)** if interest is compounded continuously.

Solution

(a) Since there are 12 months in a year, we identify $n = 12$. Furthermore, with $P = 1000$, $r = 0.09$, and $t = 5$, (19) becomes

$$P = 1000\left(1 + \frac{0.09}{12}\right)^{12(5)}$$

$$= 1000(1.0075)^{60}.$$

Using the $\boxed{y^x}$ key on a calculator, we obtain

$$P \approx \$1565.68.$$

(b) Now from (20), we have

$$P = 1000e^{(0.09)(5)}$$

$$= 1000e^{0.45}$$

$$\approx \$1568.31.$$

By compounding interest continuously rather than monthly, we have gained $2.63 over 5 years.

55. Match the letter of the graph in Figure 19 with the appropriate function.

$f(x) = b^x, b > 2$ ___

$f(x) = b^x, 1 < b < 2$ ___

$f(x) = b^x, \frac{1}{2} < b < 1$ ___

$f(x) = b^x, 0 < b < \frac{1}{2}$ ___

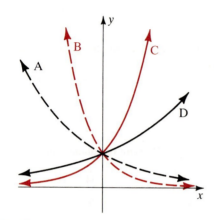

FIGURE 19

56. In Figure 20, fill in the blanks for the coordinates of the points on each graph.

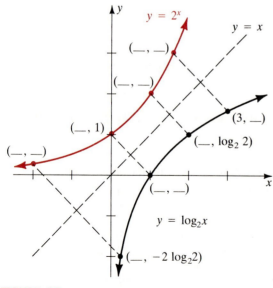

FIGURE 20

57. Use a graph to solve the inequality $\log_{10}(x + 3) > 0$.

58. A principal of $10,000 is deposited in a savings account paying 8% annual interest compounded continuously. What is its future value in 5 years? How long will it take to triple the initial deposit?

59. If an earthquake has a magnitude of 4.2 on the Richter scale, what is the magnitude on the Richter scale of an earthquake that has an intensity 20 times greater? [*Hint:* First solve the equation $10^x = 20$.]

60. The function

$$x(t) = c_1 e^{m_1 t} + c_2 e^{m_2 t}, \ c_1 c_2 < 0, \ m_1 < 0, \ m_2 < 0,$$

represents the vertical displacement at time t of a weight attached to the end of a spring when the entire system is immersed in a heavy fluid. If $c_1 = \frac{3}{2}$, $c_2 = -2$, $m_1 = -1$, and $m_2 = -2$, find the value of t for which $x(t) = 0$.

61. Tritium, an isotope of hydrogen, has a half-life of 12.5 years. How much of an initial quantity of this element remains after 50 years?

62. The amount remaining of a radioactive substance after t years is given by $A(t) = A_0 e^{kt}$, $k > 0$. Show that if $A(t_1) = A_1$ and $A(t_2) = A_2$ for $t_1 < t_2$, then the half-life of the element is

$$t = \frac{(t_2 - t_1) \ln 2}{\ln (A_1/A_2)}.$$

63. Radioactive substances are removed from living substances by two processes: natural physical decay and biological metabolism. Each process contributes to an **effective half-life** E defined by

$$1/E = 1/P + 1/B,$$

where P is the physical half-life of the radioactive substance and B is the biological half-life.

(a) For radioactive iodine, I-131, in the human thyroid gland, $P = 8$ days and $B = 24$ days. Find the effective half-life of I-131 in the thyroid.

(b) Suppose that the amount of I-131 in the thyroid after t days is given by $A(t) = A_0 e^{kt}$, $k > 0$. Use the effective half-life found in part (a) to determine the percentage of radioactive iodine remaining in the human thyroid gland two weeks after its ingestion.

64. An **annuity** is a savings plan where the same amount of money P is deposited into an account at n equally spaced periods (say, years) of time. If the annual rate r of interest is compounded continuously, then the amount accrued in the account immediately after the nth deposit is

$$S = P + Pe^r + Pe^{2r} + \cdots + Pe^{(n-1)r}.$$

What is the value of such an annuity in 15 years if $P = \$3000$ and the annual rate of interest is 7%?

65. In Problem 58 of Exercise 6.1, we saw that the graph of $y = ae^{-be^{-cx}}$ is a Gompertz curve. Solve for x in terms of the other symbols.

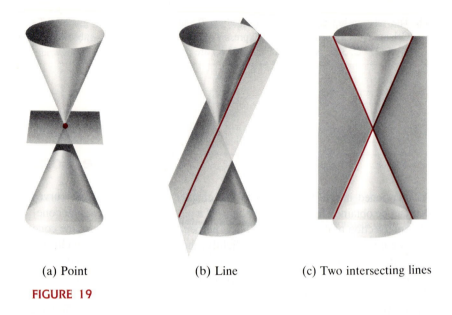

(a) Point   (b) Line   (c) Two intersecting lines

FIGURE 19

In the next three sections we will study the parabola, the ellipse, and the hyperbola by defining each one, in turn, as a set of points satisfying a certain geometric property. From these definitions we will then derive an equation in standard form for each of these conic sections. We have already done this for the circle in Section 1.3. There, from Definition 2 of a circle as the set of all points (x, y) at a fixed distance r from the center $C(h, k)$, we obtained the equation $(x - h)^2 + (y - k)^2 = r^2$.

In Section 7.6 we will see the relationship between the general second-degree equation,

$$Ax^2 + Bxy + Cy^2 + Dx + Ey + F = 0,$$

and the conic sections.

7.3

The Parabola

We begin our study of the parabola with the following geometric definition.

DEFINITION 2

A **parabola** is the set of all points in the plane that are equidistant from a fixed point F, called the **focus,** and a fixed line, called the **directrix.**

AXIS AND VERTEX

A parabola is shown in Figure 20. The line through the focus perpendicular to the directrix is called the **axis.** The point of intersection of the parabola and the axis is called the **vertex,** denoted by V in Figure 20.

EQUATION OF A PARABOLA

We now derive an equation of a parabola with focus $F(0, p)$ and directrix $y = -p$, where $p > 0$. Then we see that the axis of the parabola is along the y-axis, as Figure 21 shows. The origin is necessarily the vertex, since it lies on the axis p units from both the focus and the directrix. If $P(x, y)$ is a point on the parabola, then the distance from P to the directrix is

$$d_1 = y - (-p) = y + p.$$

Using the distance formula, we find the distance from P to the focus:

$$d(P, F) = \sqrt{(x - 0)^2 + (y - p)^2}.$$

From Definition 2, it then follows that

$$\sqrt{(x - 0)^2 + (y - p)^2} = y + p.$$

By squaring both sides and simplifying, we obtain

$$x^2 + (y - p)^2 = (y + p)^2$$

$$x^2 + y^2 - 2py + p^2 = y^2 + 2py + p^2$$

$$x^2 = 4py. \tag{9}$$

This completes the derivation of the equation in standard form of a parabola with focus $(0, p)$ and directrix $y = -p$ for $p > 0$.

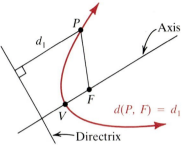

FIGURE 20

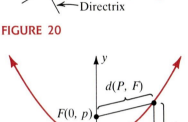

FIGURE 21

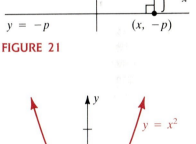

FIGURE 22

--- ## ■ EXAMPLE 1

In Section 1.2 we graphed the curve $y = x^2$. Since this equation has the form of (9) with $p = \frac{1}{4}$, we see now that it is a parabola with vertex at the origin, focus $(0, \frac{1}{4})$, and directrix $y = -\frac{1}{4}$. The graph is shown again in Figure 22. ■

■ EXAMPLE 2

For the parabola with focus $(0, 3)$ and directrix $y = -3$, we have that $p = 3$. Thus from (9) we find the equation of this parabola to be

$$x^2 = 4(3)y,$$

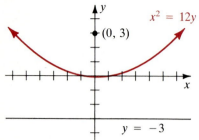

FIGURE 23

or $x^2 = 12y.$

The graph is shown in Figure 23.

Equation (9) does not depend on the assumption that $p > 0$ (see Problem 53). However, the direction in which the parabola opens does depend on the sign of p. Specifically, if $p > 0$, the parabola opens upward; and if $p < 0$, the parabola opens downward. This is illustrated in Figure 24.

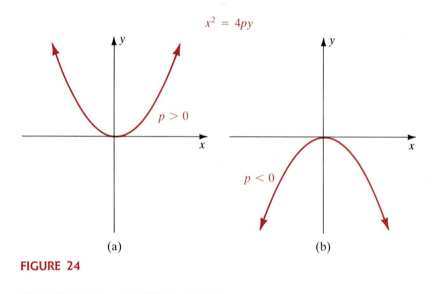

$x^2 = 4py$

(a)   (b)

FIGURE 24

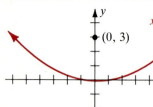

FIGURE 25

■ EXAMPLE 3

Find the equation in standard form of the parabola with directrix $y = 2$ and focus $(0, -2)$. Graph the parabola.

Solution In Figure 25 we have graphed the directrix and the focus. We see from their placement that the form of the equation is

$$x^2 = 4py.$$

Since $p = -2$, the parabola opens downward and the equation must be

$$x^2 = 4(-2)y, \quad \text{or} \quad x^2 = -8y.$$

To graph the parabola, we first plot the vertex at $(0, 0)$ and then locate another pair of points on the parabola. If $y = -2$, then

$$x^2 = -8(-2) = 16, \quad \text{or} \quad x = \pm 4.$$

Thus the points $(4, -2)$ and $(-4, -2)$ lie on the parabola (see Figure 26).

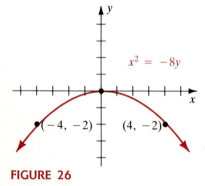

FIGURE 26

In general, for the equation $x^2 = 4py$, the choice of $y = p$ gives $x = \pm 2p$, and thus the points $(\pm 2p, p)$ lie on the parabola (see Problem 56). Also note that the graph of the parabola $x^2 = 4py$ is symmetric with respect to its axis, the y-axis, since substituting $-x$ for x results in an equivalent equation.

If the focus of a parabola lies on the x-axis at $F(p, 0)$ and the directrix is $x = -p$, then the x-axis is the axis of the parabola and the vertex is at $(0, 0)$. As shown in Figure 27, if $p > 0$, the parabola opens to the right; and if $p < 0$, it opens to the left. In these cases, the standard form of the equation is

$$y^2 = 4px \tag{10}$$

(see Problem 54). The graph of this parabola is symmetric with respect to the x-axis.

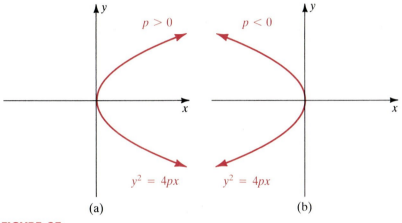

(a)   (b)

FIGURE 27

The following table summarizes information about parabolas with equations (9) and (10).

STANDARD FORMS FOR PARABOLAS

CARTESIAN EQUATION	VERTEX	AXIS	FOCUS	DIRECTRIX	PARABOLA OPENS
$x^2 = 4py$	$(0, 0)$	$x = 0$	$(0, p)$	$y = -p$	Up if $p > 0$, Down if $p < 0$
$y^2 = 4px$	$(0, 0)$	$y = 0$	$(p, 0)$	$x = -p$	Right if $p > 0$, Left if $p < 0$

■ EXAMPLE 4

Find the focus, vertex, directrix, and axis of the parabola $y^2 = -6x$. Graph the parabola and indicate the focus and directrix.

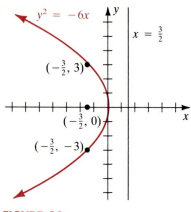

FIGURE 28

Solution The equation has the form $y^2 = 4px$. Thus the vertex is at the origin, the axis is the x-axis, and

$$4p = -6, \quad \text{or} \quad p = -\tfrac{3}{2}.$$

Since $p < 0$, the parabola opens to the left, the focus is $(-\tfrac{3}{2}, 0)$, and the directrix is $x = \tfrac{3}{2}$.

To graph the parabola, we let $x = -\tfrac{3}{2}$. Then

$$y^2 = -6(-\tfrac{3}{2}) = 9, \quad \text{or} \quad y = \pm 3.$$

Thus the points $(-\tfrac{3}{2}, \pm 3)$ lie on the parabola, which is shown in Figure 28.

PARABOLA WITH VERTEX AT (h, k)

Suppose that the parabola is shifted both horizontally and vertically so that its vertex is at the point (h, k) and its axis is the vertical line $x = h$. The standard form of the equation of this parabola is

$$(x - h)^2 = 4p(y - k). \tag{11}$$

(See Section 7.6 for further discussion of this.) Similarly, the standard equation of a parabola with vertex (h, k) and axis the horizontal line $y = k$ is

$$(y - k)^2 = 4p(x - h). \tag{12}$$

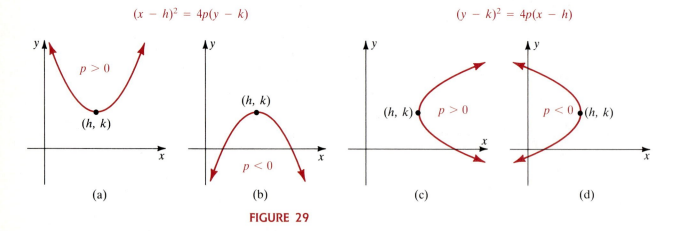

FIGURE 29

As before, the distance from the vertex to the focus in each of these cases is $|p|$, and the distance from the vertex to the directrix is $|p|$. In Figure 29, we illustrate the four types of parabolas that result from taking p positive or negative and the axis either horizontal or vertical. The following table summarizes information about parabolas with equations (11) and (12).

STANDARD FORMS FOR PARABOLAS

CARTESIAN EQUATION	VERTEX	AXIS	FOCUS	DIRECTRIX	PARABOLA OPENS
$(x - h)^2 = 4p(y - k)$	(h, k)	$x = h$	$(h, k + p)$	$y = k - p$	Up if $p > 0$, Down if $p < 0$
$(y - k)^2 = 4p(x - h)$	(h, k)	$y = k$	$(h + p, k)$	$x = h - p$	Right if $p > 0$, Left if $p < 0$

■ **EXAMPLE 5**

Find the equation in standard form of the parabola with vertex $(-3, -1)$ and directrix $y = 2$.

Solution We begin by locating the vertex at $(-3, -1)$ and the directrix $y = 2$ (see Figure 30). Since the vertex lies 3 units below the directrix, it follows that $p = -3$, and the standard form of the equation must be $(x - h)^2 = 4p(y - k)$. Substituting $h = -3$, $k = -1$, and $p = -3$ gives

$$[x - (-3)]^2 = 4(-3)[y - (-1)],$$

or
$$(x + 3)^2 = -12(y + 1).$$

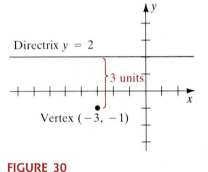

FIGURE 30

■ **EXAMPLE 6**

Find the vertex, focus, axis, and directrix of the parabola $y^2 - 4y - 8x - 28 = 0$. Graph the parabola.

Solution We begin by writing the equation in one of the standard forms. Completing the square in y, we obtain

$$y^2 - 4y + 4 = 8x + 28 + 4$$
$$(y - 2)^2 = 8x + 32,$$

or
$$(y - 2)^2 = 8(x + 4). \tag{13}$$

Comparing this last equation with (12), we conclude that the vertex is $(-4, 2)$. Also, since $4p = 8$, then $p = 2 > 0$ and the parabola opens to the right. We locate the focus two units to the right of the vertex at $(-4 + 2, 2)$, or $(-2, 2)$. The directrix is the vertical line two units to the left of the vertex:

$$x = -4 - 2 = -6.$$

The axis (the horizontal line through the vertex) is $y = 2$. Using $x = -2$ in equation (13), we find that $y = 6$ and $y = -2$. Thus the points $(-2, -2)$ and $(-2, 6)$ lie on the graph of the parabola (see Figure 31).

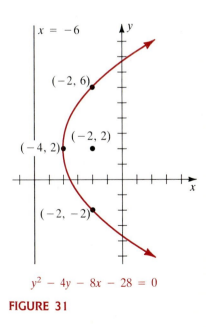

$$y^2 - 4y - 8x - 28 = 0$$

FIGURE 31

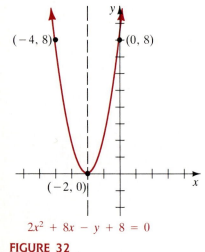

$(-4, 8)$ · $(0, 8)$

$(-2, 0)$|

x

$2x^2 + 8x - y + 8 = 0$

FIGURE 32

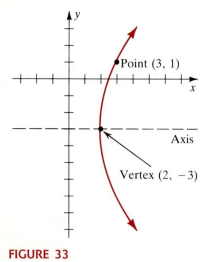

Point (3, 1)

x

Axis

Vertex (2, −3)

FIGURE 33

EXAMPLE 7

Find the vertex and focus of the parabola $2x^2 + 8x - y + 8 = 0$. Graph the parabola.

Solution　In order to complete the square, the coefficient of x^2 must be 1, so we rewrite the equation as

$$2(x^2 + 4x \quad\quad) = y - 8.$$

Then completing the square, we have

$$2(x^2 + 4x + 4) = y - 8 + 8. \tag{14}$$

Note that we added 8 to the right-hand side of (14) because $2(4) = 8$ was added to the left-hand side. Simplifying (14), we find

$$2(x + 2)^2 = y.$$

To obtain one of the standard forms for a parabola, we divide both sides of the last equation by 2 and get

$$(x + 2)^2 = \tfrac{1}{2}y. \tag{15}$$

Thus,　　　　　　　　　$4p = \tfrac{1}{2}, \quad \text{or} \quad p = \tfrac{1}{8}.$

We conclude that the parabola will open up with vertex $(-2, 0)$ and focus $(-2, 0 + \tfrac{1}{8})$, or $(-2, \tfrac{1}{8})$. If we let $x = 0$ in (15), we have

$$(0 + 2)^2 = \tfrac{1}{2}y, \quad \text{or} \quad y = 8.$$

Thus the point $(0, 8)$ is on the parabola. Using the fact that a parabola is symmetric with respect to its axis, we can locate another point on the parabola. Since the axis of this parabola is $x = -2$ and $(0, 8)$ lies on the parabola, we know, by symmetry, that $(-4, 8)$ is also on the parabola. This information is sufficient to sketch the graph, as shown in Figure 32.　■

EXAMPLE 8

Find the equation in standard form of the parabola with vertex $(2, -3)$, axis parallel to the x-axis, and passing through the point $(3, 1)$.

Solution　We use the given information to sketch the parabola shown in Figure 33. The axis of the parabola is horizontal, so the form of the equation must be

$$(y - k)^2 = 4p(x - h).$$

Since the vertex is $(2, -3)$, we conclude that $h = 2$ and $k = -3$. If the point

(3, 1) lies on the graph, its coordinates must satisfy

$$(y + 3)^2 = 4p(x - 2).$$

Thus we can solve

$$(1 + 3)^2 = 4p(3 - 2)$$

for p to obtain

$$16 = 4p, \quad \text{or} \quad p = 4.$$

Therefore, the equation of the parabola is

$$(y + 3)^2 = 16(x - 2).$$

APPLICATIONS OF THE PARABOLA

The parabola has many interesting properties that make it suitable for certain applications. The design of mirrors for telescopes and certain lighting systems is based on an important reflection property of parabolas. As illustrated in Figure 34, a ray of light from a point source located at the focus of a parabola will be reflected along a line parallel to the axis. Thus, the shape of the reflecting surface in most searchlights, automobile headlights, and flashlights is obtained by rotating a parabola about its axis. The light source is placed at the focus. Then, theoretically, the result of this design is a beam of light parallel to the axis. See Figure 35(a). Of course, in reality, some dispersion, or scattering of light, will occur, since there is no point source of light.

Conversely, if an incoming ray of light is parallel to the axis of a parabola, it will be reflected along a line passing through the focus. Reflecting telescopes, satellite dishes, and radar antennae utilize this property by placing the eyepiece of the telescope and the receiving equipment for the antenna at the focus of a parabolic reflector. See Figure 35(b).

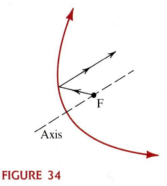

FIGURE 34

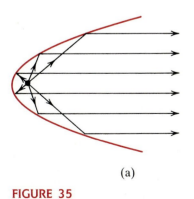
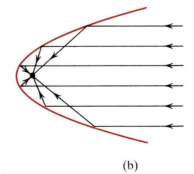

(a)   (b)

FIGURE 35

Parabolas are also important in the design of suspension bridges. The main cables of suspension bridges are usually parabolic in shape, since it can be shown that if the weight of a bridge is distributed uniformly along its length, then a cable in the shape of a parabola will bear the load evenly.

In addition, the path of a projectile will be a parabola if the motion is considered to be in a plane and air resistance is neglected.

■■ EXAMPLE 9

An automobile headlamp is designed so that a cross section through its axis is a parabola and the light source is placed at the focus. If the headlamp is 16 cm across and 6 cm deep, find the location of the light source.

Solution If we superimpose a coordinate system on the parabola, as shown in Figure 36, the parabola will have the equation $y^2 = 4px$ and pass through the point $(6, 8)$. Substituting $x = 6$ and $y = 8$ into the equation, we obtain

$$8^2 = 4p(6), \quad \text{or} \quad p = \tfrac{8}{3}.$$

Thus the focus is at $(\tfrac{8}{3}, 0)$ and the light source is located on the axis of the parabola $\tfrac{8}{3}$ cm to the right of the vertex. ■

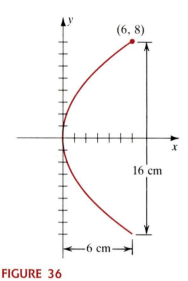

FIGURE 36

The next example shows that applications of the reflection property of parabolas are found in biology as well as physics.

■■ EXAMPLE 10

Tuna, which prey on smaller fish, have been observed swimming in schools of 10 to 20 fish, arrayed approximately in a parabolic shape. One possible explanation for this is that the smaller fish caught in the school of tuna will try to escape by "reflecting" off the parabola. As a result, they are concentrated at the focus and become easy prey for the tuna.* See Figure 37.

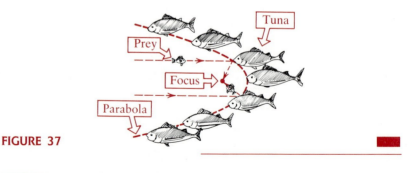

FIGURE 37

*Partridge, "The Structure and Function of Fish Schools," *Scientific American,* **246** (June 1982), 114–123.

For each parabola in Problems 1–24, find the vertex, focus, directrix, and axis. Graph the parabola.

1. $y^2 = 4x$ **2.** $y^2 = \frac{7}{2}x$
3. $y^2 = -\frac{4}{3}x$ **4.** $y^2 = -10x$
5. $x^2 = -16y$ **6.** $x^2 = \frac{1}{10}y$
7. $x^2 = 28y$ **8.** $x^2 = -64y$
9. $(y - 1)^2 = 16x$ **10.** $(y + 3)^2 = -8(x + 2)$
11. $(x + 5)^2 = -4(y + 1)$ **12.** $(x - 2)^2 + y = 0$
13. $y^2 + 12y - 4x + 16 = 0$ **14.** $x^2 + 6x + y + 11 = 0$
15. $x^2 + 5x - \frac{1}{4}y + 6 = 0$ **16.** $x^2 - 2x - 4y + 17 = 0$
17. $y^2 - 8y + 2x + 10 = 0$ **18.** $y^2 - 4y - 4x + 3 = 0$
19. $4x^2 = 2y$ **20.** $3(y - 1)^2 = 9x$
21. $-2x^2 + 12x - 8y - 18 = 0$
22. $4y^2 + 16y - 6x - 2 = 0$
23. $6y^2 - 12y - 24x - 42 = 0$
24. $3x^2 + 30x - 8y + 75 = 0$

In Problems 25–44, find an equation of the parabola satisfying the given conditions.

25. Focus $(0, 7)$, directrix $y = -7$
26. Focus $(0, -5)$, directrix $y = 5$
27. Focus $(-4, 0)$, directrix $x - 4 = 0$
28. Focus $(\frac{3}{2}, 0)$, directrix $x + \frac{3}{2} = 0$
29. Focus $(\frac{5}{2}, 0)$, vertex $(0, 0)$
30. Focus $(0, -10)$, vertex $(0, 0)$
31. Focus $(2, 3)$, directrix $y = -3$
32. Focus $(1, -7)$, directrix $x = -5$
33. Focus $(-1, 4)$, directrix $x = 5$
34. Focus $(-2, 0)$, directrix $y = \frac{3}{2}$
35. Focus $(1, 5)$, vertex $(1, -3)$
36. Focus $(-2, 3)$, vertex $(-2, 5)$
37. Focus $(8, -3)$, vertex $(0, -3)$
38. Focus $(1, 2)$, vertex $(7, 2)$
39. Vertex $(0, 0)$, directrix $y = -\frac{7}{4}$
40. Vertex $(0, 0)$, directrix $x = 6$
41. Vertex $(5, 1)$, directrix $y = 7$
42. Vertex $(-1, 4)$, directrix $x = 0$
43. Vertex $(0, 0)$, axis along the y-axis, through $(-2, 8)$
44. Vertex $(0, 0)$, axis along the x-axis, through $(1, \frac{1}{4})$
45. A large spotlight is designed so that a cross section through its axis is a parabola and the light source is at the focus. Find the position of the light source if the spotlight is 4 ft across at the opening and 2 ft deep.

46. A reflecting telescope has a parabolic mirror that is 20 ft across at the top and 4 ft deep at the center. Where should the eyepiece be located?
47. Suppose that a light ray emanating from the focus of the parabola $y^2 = 4x$ strikes the parabola at $(1, -2)$. What is the equation of the reflected ray?
48. Suppose that two towers of a suspension bridge are 350 ft apart and the vertex of the parabolic cable is tangent to the road midway between the towers. If the cable is 1 ft above the road at a point 20 ft from the vertex, find the height of the towers above the road.
49. Two 75-ft towers of a suspension bridge with a parabolic cable are 250 ft apart. The vertex of the parabola is tangent to the road midway between the towers. Find the height of the cable above the roadway at a point 50 ft from one of the towers.
50. Assume that the water gushing from the end of a horizontal pipe follows a parabolic arc with vertex at the end of the pipe. The pipe is 20 m above the ground. At a point 2 m below the end of the pipe, the horizontal distance from the water to a vertical line through the end of the pipe is 4 m (see Figure 38). Where does the water strike the ground?

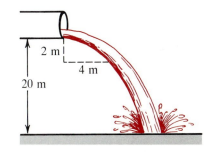

FIGURE 38

51. A dart thrower releases a dart 5 ft above the ground. The dart is thrown horizontally and follows a parabolic path. It hits the ground $10\sqrt{10}$ ft from the dart thrower. At a distance of 10 ft from the dart thrower, how high should a bull's-eye be placed in order for the dart to hit it?
52. The vertical position of a projectile is given by the equation $y = -16t^2$ and the horizontal position by $x = 40t$ for $t \geq 0$. By eliminating t between the two equations, show that the path of the projectile is a parabolic arc. Graph the path of the projectile.
53. Derive equation (9) for $p < 0$.

54. Derive equation (10).

55. Derive the equation of the parabola with vertex (h, k) and focus $(h + p, k)$.

56. The **focal width** of a parabola is the length of the line segment through the focus perpendicular to the axis, with endpoints on the parabola (see Figure 39). This line segment is called the **focal chord,** or **latus rectum.**

   (a) Find the focal width of the parabola $x^2 = 8y$.

   (b) Show that the focal width of the parabolas $x^2 = 4py$ and $y^2 = 4px$ is $4|p|$.

   (c) For the parabolas $x^2 = 4py$ and $y^2 = 4px$, show that the endpoints of the focal chord are $(\pm 2p, p)$ and $(p, \pm 2p)$, respectively.

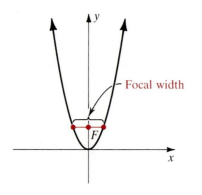

FIGURE 39

57. Prove that on a parabola, the point closest to the focus is the vertex.

58. The orbit of a comet is a parabola with the sun at the focus. When the comet is 50,000,000 km from the sun, the line from the comet to the sun is perpendicular to the axis of the parabola. Use the result of Problem 56(b) to write an equation of the comet's path. (A comet with a parabolic path will not return to earth like comet Halley.)

59. For the comet in Problem 58, use the result of Problem 57 to determine the shortest distance between the sun and the comet.

60. Suppose that two parabolic reflecting surfaces face one another (with foci on a common axis), as shown in Figure 40. Any sound emitted at one focus will be reflected off the parabolas and concentrated at the other focus. The figure shows the paths of two typical sound waves. Using Definition 2, show that all the waves will travel the same distance. [*Note*: This result is important for the following reason. If the sound waves traveled paths of different lengths, then the waves would arrive at the second focus at different times. The result would be interference rather than clear sound.]

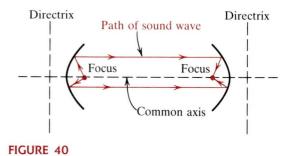

FIGURE 40

7.4 The Ellipse

The ellipse occurs frequently in astronomy. For example, the paths of the planets around the sun are elliptical (with the sun at one focus), as are the orbits of satellites about the earth (with the earth at one focus). In this section we define the ellipse and study some of its properties and applications.

DEFINITION 3

An **ellipse** is the set of all points P in the plane such that the sum of the distances between P and two fixed points F_1 and F_2 is constant. The two fixed points F_1 and F_2 are called **foci** (plural of **focus**). The midpoint of the line segment joining the foci is called the **center.**

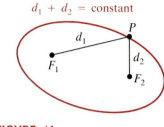

FIGURE 41

An ellipse is shown in Figure 41.

EQUATION OF AN ELLIPSE

We now derive an equation of the ellipse with foci $F_1(-c, 0)$ and $F_2(c, 0)$ on the x-axis. The center of this ellipse is at the origin O (see Figure 42). For convenience, we denote the constant sum of the distances by $2a$, since this choice gives a simple form for the final equation.

For any point $P(x, y)$ on the ellipse, we have from Definition 3

$$d(P, F_1) + d(P, F_2) = 2a.$$

FIGURE 42

We then use the distance formula to obtain

$$\sqrt{(x + c)^2 + (y - 0)^2} + \sqrt{(x - c)^2 + (y - 0)^2} = 2a,$$

or
$$\sqrt{(x + c)^2 + y^2} = 2a - \sqrt{(x - c)^2 + y^2}.$$

We square both sides and simplify:

$$(x + c)^2 + y^2 = 4a^2 - 4a\sqrt{(x - c)^2 + y^2} + (x - c)^2 + y^2$$
$$4a\sqrt{(x - c)^2 + y^2} = 4a^2 + (x - c)^2 + y^2 - (x + c)^2 - y^2$$
$$4a\sqrt{(x - c)^2 + y^2} = 4a^2 + x^2 - 2cx + c^2 - x^2 - 2cx - c^2$$
$$a\sqrt{(x - c)^2 + y^2} = a^2 - cx.$$

Squaring again gives

$$a^2[(x - c)^2 + y^2] = a^4 - 2a^2cx + c^2x^2,$$

or
$$(a^2 - c^2)x^2 + a^2y^2 = a^2(a^2 - c^2).$$

By dividing both sides by $a^2(a^2 - c^2)$, we have

$$\frac{x^2}{a^2} + \frac{y^2}{a^2 - c^2} = 1. \qquad (16)$$

Since the sum of the lengths of two sides of a triangle is greater than the length of the third side, we see from triangle F_1PF_2 in Figure 42 that $d_1 + d_2 =$

$2a > 2c$, or $a > c$. Since a and c are both positive, $a^2 - c^2 > 0$. Thus we can set $b = \sqrt{a^2 - c^2}$ in equation (16) to obtain

$$\frac{x^2}{a^2} + \frac{y^2}{b^2} = 1. \tag{17}$$

This is the standard form of the equation of the ellipse with foci $(-c, 0)$ and $(c, 0)$ and fixed distance $2a$.

MAJOR AND MINOR AXES; VERTICES

The line segment through the foci with endpoints on the ellipse is called the **major axis.** The line segment with endpoints on the ellipse that is perpendicular to the major axis at the center is called the **minor axis.** The endpoints of the axes are called **vertices.**

For an ellipse of the form $x^2/a^2 + y^2/b^2 = 1$, the major and the minor axes lie on the x- and y-axis, respectively. Therefore, to determine the coordinates of the vertices, we simply find the x- and y-intercepts of the ellipse. Letting $y = 0$ for the x-intercepts, we have $x^2/a^2 = 1$, or $x = \pm a$. Similarly, setting $x = 0$, we find that $y = \pm b$. Thus, as shown in Figure 43(a), the vertices of the ellipse are $(\pm a, 0)$ and $(0, \pm b)$. It is easy to see that the length of the major axis is $2a$ and the length of the minor axis is $2b$.

Another important observation is that $a > b$, since $a^2 = b^2 + c^2$. Thus the major axis is longer than the minor axis. The relationship $a^2 = b^2 + c^2$ can be remembered by noting that a is the hypotenuse of the right triangle VOF_2, shown in Figure 43(b).

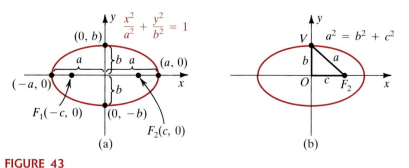

FIGURE 43

■ EXAMPLE 1

For the ellipse

$$\frac{x^2}{25} + \frac{y^2}{16} = 1,$$

since $25 > 16$, we make the identification $a^2 = 25$ and $b^2 = 16$. As Figure 44 shows, the vertices are $(\pm 5, 0)$ and $(0, \pm 4)$. ■

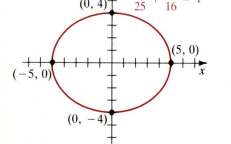

FIGURE 44

If the ellipse is positioned so that the foci are on the y-axis at $F_1(0, -c)$ and $F_2(0, c)$, a derivation similar to that of (17) yields the standard form

$$\frac{x^2}{b^2} + \frac{y^2}{a^2} = 1 \qquad \textbf{(18)}$$

(see Problem 48). Here again, $a > c$, $a^2 = b^2 + c^2$, and $a > b$. The vertices are located at $(0, \pm a)$ and $(\pm b, 0)$, and the center is at the origin. As shown in Figure 45, for an ellipse of this type, the major axis is vertical and the minor axis is horizontal.

The following table summarizes the results obtained above.

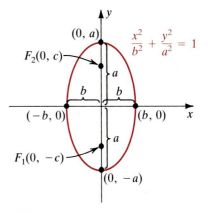

FIGURE 45

STANDARD FORMS FOR ELLIPSES

EQUATION	CENTER	FOCI	MAJOR AXIS	VERTICES ON MAJOR AXIS	VERTICES ON MINOR AXIS
$\dfrac{x^2}{a^2} + \dfrac{y^2}{b^2} = 1$	$(0, 0)$	$(\pm c, 0)$	Horizontal, lies on x-axis	$(\pm a, 0)$	$(0, \pm b)$
$\dfrac{x^2}{b^2} + \dfrac{y^2}{a^2} = 1$	$(0, 0)$	$(0, \pm c)$	Vertical, lies on y-axis	$(0, \pm a)$	$(\pm b, 0)$
Important Relationships: $a > b$, $a^2 = b^2 + c^2$					

◼ EXAMPLE 2

Find the vertices and foci for the ellipse $9x^2 + 3y^2 = 27$. Graph the ellipse.

Solution Dividing by 27, we find that

$$\frac{x^2}{3} + \frac{y^2}{9} = 1,$$

or

$$\frac{x^2}{(\sqrt{3})^2} + \frac{y^2}{3^2} = 1.$$

Since $3 > \sqrt{3}$, we have $a = 3$ and $b = \sqrt{3}$. The denominator of the y^2-term is then a^2, and the major axis is vertical. Thus the vertices are $(0, \pm 3)$ and $(\pm\sqrt{3}, 0)$. Since

$$c^2 = a^2 - b^2 = 9 - 3 = 6,$$

we have $c = \sqrt{6}$. The foci are on the y-axis at $(0, \pm\sqrt{6})$. The ellipse is shown in Figure 46.

Although the foci are indicated, you should be aware that they do *not* lie on the graph of the ellipse (just as the focus of a parabola does not lie on the graph of the parabola).

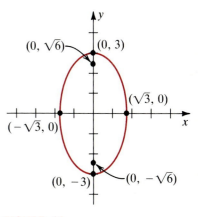

FIGURE 46

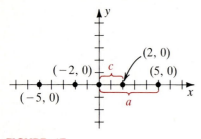

FIGURE 47

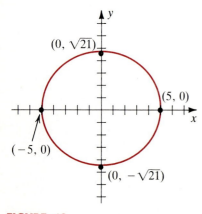

FIGURE 48

Note of Caution: The relationship $a > b$ is key to identifying which of the two standard forms applies to a given ellipse. For $x^2/16 + y^2/4 = 1$, since $16 > 4$, we identify $a^2 = 16$ and $b^2 = 4$. Then we conclude that the major axis is horizontal. For $x^2/4 + y^2/16 = 1$, since $16 > 4$, we again identify $a^2 = 16$ and $b^2 = 4$; for this ellipse, however, the major axis is vertical.

EXAMPLE 3

Find an equation of the ellipse with vertices $(\pm 5, 0)$ and foci $(\pm 2, 0)$. Graph the ellipse.

Solution   In Figure 47 we have plotted the vertices and the foci. We see that $a = 5$ and $c = 2$. Since $b^2 = a^2 - c^2$, we have

$$b^2 = 25 - 4 = 21,$$

and so $b = \sqrt{21}$. Since the foci lie on the x-axis and the center is at the origin, the equation of the ellipse has the form

$$\frac{x^2}{a^2} + \frac{y^2}{b^2} = 1.$$

Substituting the values of a and b gives

$$\frac{x^2}{25} + \frac{y^2}{21} = 1.$$

The graph is shown in Figure 48.

EXAMPLE 4

Find an equation of the ellipse with foci $(0, \pm\sqrt{3})$ such that the length of the major axis is 12.

Solution   Because the foci are on the y-axis, the equation of the ellipse must be of the form

$$\frac{x^2}{b^2} + \frac{y^2}{a^2} = 1.$$

Since the length of the major axis is $2a = 12$, we have $a = 6$. Now, from $c = \sqrt{3}$, it follows that

$$b^2 = a^2 - c^2 = 36 - 3 = 33.$$

Thus the equation of the ellipse is

$$\frac{x^2}{33} + \frac{y^2}{36} = 1.$$

ELLIPSE WITH CENTER AT (h, k)

Suppose that the center of an ellipse is at the point (h, k) and the foci are located at $(h - c, k)$ and $(h + c, k)$. The standard form of the equation of this ellipse is

$$\frac{(x - h)^2}{a^2} + \frac{(y - k)^2}{b^2} = 1. \tag{19}$$

For this ellipse, the major axis is horizontal and the minor axis is vertical.

Similarly, the equation of the ellipse with center (h, k) and foci located at $(h, k - c)$ and $(h, k + c)$ is

$$\frac{(x - h)^2}{b^2} + \frac{(y - k)^2}{a^2} = 1. \tag{20}$$

In this ellipse, the major axis is vertical and the minor axis is horizontal. As before, $a > c$ and $b^2 = a^2 - c^2$ so that $a > b$.

The vertices can be obtained by setting $x = h$ and $y = k$ in (19) and (20) and solving for y and x, respectively. They are labeled in Figure 49. Note too that the length of the major axis is $2a$ and the length of the minor axis is $2b$.

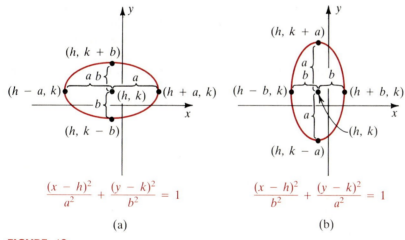

FIGURE 49

It is not a good idea to memorize formulas for the coordinates of the vertices and the foci of ellipses centered at (h, k). It is better to work with the values of a, b, and c. For this reason, no summary table is given for ellipses centered at (h, k). As we see from Figure 49, a is the distance from the center (h, k) to the vertices on the major axis, and b is the distance from the center to the vertices on the minor axis. Also, recall that c is the distance from the center to the foci. This approach is illustrated in the following examples.

or $b = 2$, and the equation of the hyperbola is

$$\frac{x^2}{4^2} - \frac{y^2}{2^2} = 1.$$

HYPERBOLA WITH CENTER AT (h, k)

Suppose that the center of a hyperbola is at the point (h, k) and that the foci are located at c units to the left and the right of the center at $(h - c, k)$ and $(h + c, k)$. The standard form of the equation of this hyperbola is

$$\frac{(x - h)^2}{a^2} - \frac{(y - k)^2}{b^2} = 1. \qquad (27)$$

As Figure 66(a) shows, this hyperbola has a horizontal transverse axis with vertices a units to the left and the right of the center at $(h - a, k)$ and $(h + a, k)$. The asymptotes are the straight lines obtained by solving

$$\frac{(x - h)^2}{a^2} - \frac{(y - k)^2}{b^2} = 0 \qquad (28)$$

for y.

The standard form of the equation of the hyperbola with center (h, k) and foci c units above and below the center $(h, k - c)$ and $(h, k + c)$ is

$$\frac{(y - k)^2}{a^2} - \frac{(x - h)^2}{b^2} = 1. \qquad (29)$$

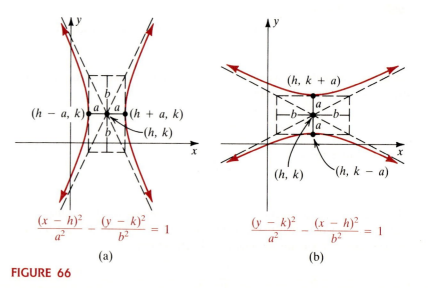

(a)   (b)

FIGURE 66

As we see in Figure 66(b), this hyperbola has a vertical transverse axis and vertices a units above and below the center at $(h, k - a)$ and $(h, k + a)$. The asymptotes are found by solving

$$\frac{(y - k)^2}{a^2} - \frac{(x - h)^2}{b^2} = 0 \tag{30}$$

for y. As before, a^2 is the denominator of the term with the positive coefficient, and we also retain the relationship

$$b^2 = c^2 - a^2.$$

What we stated earlier for the ellipse also holds for the hyperbola, that is, it is not a good idea to memorize formulas for the coordinates of the vertices and the foci of a hyperbola centered at (h, k). As illustrated in the following example, you should work with the values of a, b, and c and the auxiliary rectangle.

■ EXAMPLE 4

Find the center, vertices, foci, and asymptotes of the hyperbola $4x^2 - y^2 - 8x - 4y - 4 = 0$. Graph the hyperbola.

Solution   Before completing the square in x and y, we factor 4 from the x^2- and x-terms and factor -1 from the y^2- and y-terms so that the leading coefficient in each expression is 1. Then we have

$$4(x^2 - 2x \quad) - (y^2 + 4y \quad) = 4$$
$$4(x^2 - 2x + 1) - (y^2 + 4y + 4) = 4 + 4 - 4$$
$$4(x - 1)^2 - (y + 2)^2 = 4$$
$$\frac{(x - 1)^2}{1} - \frac{(y + 2)^2}{4} = 1.$$

We see now that the center is $(1, -2)$. Since the term involving x has the positive coefficient, the transverse axis is horizontal, and we identify $a = 1$ and $b = 2$. As shown in Figure 66(a), the vertices can be located by measuring 1 unit to the left and the right of the center. Thus the vertices are $(2, -2)$ and $(0, -2)$. From $b^2 = c^2 - a^2$, we have

$$c^2 = a^2 + b^2 = 1 + 4 = 5,$$

and so $c = \sqrt{5}$. Thus the foci are located $\sqrt{5}$ units to the left and the right of the center $(1, -2)$ at $(1 - \sqrt{5}, -2)$ and $(1 + \sqrt{5}, -2)$.

To find the asymptotes, we solve

$$\frac{(x - 1)^2}{1} - \frac{(y + 2)^2}{4} = 0$$

$4x^2 - y^2 - 8x - 4y - 4 = 0$

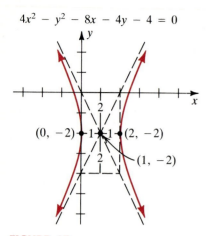

$(0, -2)$ $(2, -2)$ $(1, -2)$

FIGURE 67

for y:

$$\frac{(y + 2)^2}{4} = (x - 1)^2$$

$$\frac{y + 2}{2} = \pm(x - 1)$$

$$y + 2 = \pm(2x - 2)$$

$$y = \pm(2x - 2) - 2.$$

Thus, $y = 2x - 4$ and $y = -2x$. The graph is drawn by locating the center $(1, -2)$ and using the values $a = 1$ and $b = 2$ to draw the auxiliary rectangle that determines the asymptotes (see Figure 67). ■

■ **EXAMPLE 5**

Find an equation of the hyperbola with center $(2, -3)$, passing through the point $(4, 1)$, and having one vertex at $(2, 0)$.

Solution Since the distance from the center to one vertex is a, we have $a = 3$. From the given location of the center and the vertex, it follows that the transverse axis is vertical. Therefore, from (29) we know that the equation is

$$\frac{(y + 3)^2}{3^2} - \frac{(x - 2)^2}{b^2} = 1, \qquad \textbf{(31)}$$

where b^2 is yet to be determined. Since the point $(4, 1)$ is on the graph of the hyperbola, its coordinates must satisfy equation (31). Hence,

$$\frac{(1 + 3)^2}{9} - \frac{(4 - 2)^2}{b^2} = 1$$

$$\frac{16}{9} - \frac{4}{b^2} = 1$$

$$\frac{7}{9} = \frac{4}{b^2},$$

and so

$$b^2 = \frac{36}{7}.$$

We conclude that the desired equation is

$$\frac{(y + 3)^2}{9} - \frac{(x - 2)^2}{\frac{36}{7}} = 1.$$

■

APPLICATIONS OF THE HYPERBOLA

The hyperbola has several important applications involving sounding techniques. In particular, several navigational systems utilize hyperbolas, as follows. Two fixed radio transmitters at a known distance from each other transmit synchronized signals. The difference in reception times by a navigator determines the difference $2a$ of the distances from the navigator to the two transmitters. This information locates the navigator somewhere on the hyperbola with foci at the transmitters and fixed difference in distances from the foci equal to $2a$. By using two sets of signals obtained from a single master station paired with each of two secondary stations, the long-range navigation system LORAN locates a ship or plane at the intersection of two hyperbolas (see Figure 68).

The next example illustrates the use of a hyperbola in another situation involving sounding techniques.

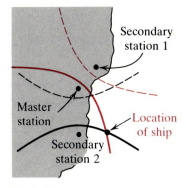

FIGURE 68

◼ EXAMPLE 6

The sound of a dynamite blast is heard at different times at two points A and B. From this it is determined that the blast occurred 1000 m closer to A than to B. If A and B are 2600 m apart, show that the location of the blast lies on a particular branch of a hyperbola, and find the equation of the branch.

Solution In Figure 69, we have placed the points A and B on the x-axis at $(1300, 0)$ and $(-1300, 0)$, respectively. If $P(x, y)$ denotes the location of the blast, then

$$d(P, B) - d(P, A) = 1000.$$

From Definition 4 and the derivation following it, we see that this is the equation for the right branch of a hyperbola with fixed distance difference $2a = 1000$ and $c = 1300$. Thus the equation has the form

$$\frac{x^2}{a^2} - \frac{y^2}{b^2} = 1, \quad \text{where } x \geq 0,$$

or, equivalently,

$$x = a\sqrt{1 + \frac{y^2}{b^2}}.$$

With $a = 500$ and $c = 1300$, we have

$$\begin{aligned} b^2 &= c^2 - a^2 \\ &= (1300)^2 - (500)^2 \\ &= (1200)^2. \end{aligned}$$

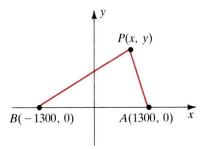

FIGURE 69

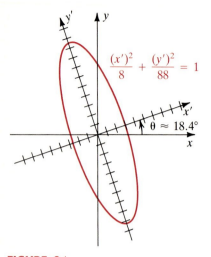

$$\frac{(x')^2}{8} + \frac{(y')^2}{88} = 1$$

$\theta \approx 18.4°$

FIGURE 84

We recognize this to be the equation of an ellipse, and we use the new axes to sketch the graph, as shown in Figure 84. From (40) we have that $\sin \theta = 1/\sqrt{10}$, and so $\theta \approx 18.4°$. This rotation angle is shown in the figure.

EXERCISE 7.6

Let the origin of the *XY*-system be at the *xy*-point (3, 2). In Problems 1–6, find the *XY*-coordinates of the given *xy*-point.

1. (5, 6)   **2.** (−2, 4)   **3.** (−3, −1)
4. (5, −2)   **5.** (3, −1)   **6.** (7, 2)

Let the origin of the *XY*-system be at the *xy*-point (4, −3). In Problems 7–12, find the *xy*-coordinates of the given *XY*-point.

7. (2, 5)   **8.** (−3, 7)   **9.** (−8, −8)
10. (4, −3)   **11.** (0, 0)   **12.** (9, 0)

In Problems 13–20, express the given *xy*-equation in terms of the *XY*-system, where the origin of the *XY*-system is at the indicated *xy*-point.

13. $(x - 2)^2 + (y + 5)^2 = 9$; (2, −5)
14. $(x + 1)^2 + (y + 2)^2 = 4$; (−1, −2)
15. $(x - 4)^2 + (y + 3)^2 = 5$; (4, −3)
16. $x^2 + y^2 - 2x - 8y + 7 = 0$; (1, 4)
17. $x^2 + y^2 + 6x - 2y + 2 = 0$; (−3, 1)
18. $2x - 5y + 7 = 0$; (−1, 1)
19. $3x + 2y - 6 = 0$; (0, 3)
20. $7x - 4y + 15 = 0$; (−1, 2)

In Problems 21–28, express the given *XY*-equation in terms of the *xy*-system, where the origin of the *XY*-system is at the indicated *xy*-point.

21. $X^2 + Y^2 = 25$; (−3, −1)
22. $X^2 + Y^2 = 7$; (−2, 5)
23. $X^2 + Y^2 = 1$; (4, −3)
24. $X^2 + Y^2 = r^2$; (h, k)
25. $2X - 3Y = 0$; (5, 0)
26. $X + 7Y = 0$; (1, 1)
27. $3X + 5Y = 0$; (5, −3)
28. $aX + bY + c = 0$; (h, k)

In Problems 29–44, discuss and sketch the graph of the given equation.

29. $y^2 - 6x + 2y = 0$
30. $4x^2 - y^2 + 16x - 2y + 19 = 0$
31. $2x^2 + 2y^2 - 8x + 4y + 5 = 0$
32. $3x^2 - 4y^2 + 12x - 8y - 16 = 0$
33. $4x^2 + 3y^2 - 8x + 6y + 7 = 0$
34. $x^2 + y^2 + 2x - 10y + 30 = 0$
35. $4x^2 + 6y^2 + 8x - 12y - 2 = 0$
36. $x^2 - y^2 - 10x + 6y + 16 = 0$

37. $3x^2 + y^2 - 6x = 0$
38. $2x^2 + 4x - 3y + 6 = 0$
39. $2x^2 + y^2 + 24x + 12y + 109 = 0$
40. $-x^2 - y^2 + 4x - 10y - 4 = 0$
41. $-x^2 - y^2 + 6x + 4y - 13 = 0$
42. $-x^2 + y^2 - 4x + 14y + 45 = 0$
43. $-2x^2 + 6y^2 + 8x - 12y = 0$
44. $-2x^2 - 4y^2 + 8x - 4y + 3 = 0$
45. Show that any equation of the form

$$xy + ax + by + c = 0$$

can be written in the form $XY = d$ by a suitable translation of axes. Determine d.

In Problems 46–48, refer to the equation

$$Ax^2 + Cy^2 + Dx + Ey + F = 0.$$

46. Show that if A and C have the same signs, then the graph of the equation is an ellipse, a circle, a point, or does not exist.
47. Show that if A and C have opposite signs, then the graph of the equation is either a hyperbola or a pair of intersecting lines.
48. Show that if either $A = 0$ or $C = 0$, then the graph of the equation is either a parabola, two parallel lines, or one line, or does not exist.
49. Derive equation (19) and verify Figure 49(a). [*Hint*: Let XY be a translation of the xy-coordinate system to a new origin at (h, k). Consider the ellipse $X^2/a^2 + Y^2/b^2 = 1$. Find its equation in the xy-coordinate system. Find the vertices and foci of this ellipse in the xy-coordinate system.]
50. Derive equation (27) and verify Figure 66(a). [*Hint*: Let XY be a translation of the xy-coordinate system to a new origin at (h, k). Consider the hyperbola $X^2/a^2 - Y^2/b^2 = 1$. Find its equation in the xy-coordinate system. Find the vertices, foci, and asymptotes of this hyperbola in the xy-coordinate system.]

In Problems 51–56, find the xy-coordinates of the given $x'y'$-point. Use the specified angle of rotation θ.

51. $(2, -8)$, $\theta = 30°$ **52.** $(-5, 7)$, $\theta = 45°$

53. $(0, 4)$, $\theta = \dfrac{\pi}{2}$ **54.** $(3, 0)$, $\theta = \dfrac{\pi}{3}$

55. $(4, 6)$, $\theta = 15°$ **56.** $(1, 1)$, $\theta = 75°$

In Problems 57–62, use the given angle of rotation θ to represent the equation relative to the $x'y'$-system.

57. $7x^2 - 6\sqrt{3}xy + 13y^2 = 16$, $\theta = 30°$

58. $x^2 + 2xy + y^2 + 16\sqrt{2}y = 0$, $\theta = \dfrac{\pi}{4}$

59. $2x^2 + \sqrt{3}xy + y^2 = 2$, $\theta = \dfrac{\pi}{6}$

60. $x^2 + xy + y^2 = 24$, $\theta = 45°$
61. $3xy + 4y^2 + 18 = 0$, $\theta = \arctan 3$
62. $2x^2 + 4xy + 5y^2 = 36$, $\theta = \arctan 2$

In Problems 63–66, eliminate the xy-term in the given equation by performing a suitable rotation of axes.

63. $6x^2 + 3xy + 2y^2 = 468$
64. $x^2 + 4xy + 4y^2 - 4\sqrt{5}x + 2\sqrt{5}y = 0$
65. $x^2 - 2xy + y^2 + 4\sqrt{2}x + 4\sqrt{2}y = 0$
66. $-7x^2 + 5xy + 5y^2 = 330$

In Problems 67–70, for the given xy-equation, perform a suitable rotation of axes so that the resulting $x'y'$-equation has no $x'y'$-term. Sketch the graph.

67. $6x^2 + 6xy + 6y^2 = 24$
68. $8x^2 - 12xy + 17y^2 = 80$
69. $x^2 - 2xy + y^2 - 8x - 8y = 0$
70. $41x^2 - 24xy + 34y^2 - 25 = 0$
71. (a) Using Figure 85, show that

$$x' = r \cos \phi, \qquad y' = r \sin \phi$$

and

$$x = r \cos (\theta + \phi), \qquad y = r \sin (\theta + \phi).$$

(b) Using the results from part (a), derive the rotation equations (37).

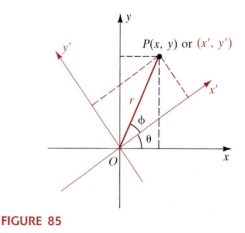

FIGURE 85

IMPORTANT CONCEPTS

Slope
  rise
  run
Equations of lines
  point–slope form
  slope–intercept form
Horizontal line
Vertical line
Angle of inclination
Parallel lines
Perpendicular lines
Conic section
  degenerate conic

Parabola
  focus
  directrix
  axis
  vertex
Ellipse
  foci
  center
  major axis
  minor axis
  vertices

Hyperbola
  foci
  center
  transverse axis
  vertices
  conjugate axis
  asymptotes
  auxiliary rectangle
  branches
Translation of axes
  translation equations
Rotation of axes
  rotation equations

CHAPTER 7 REVIEW EXERCISE

In Problems 1–10, fill in the blank or answer true or false.

1. The x- and y-intercepts and the slope of the line

$$-4x + 3y - 48 = 0$$

are _____, _____, and _____,
respectively.

2. The lines $2x - 5y = 1$ and $kx + 3y + 3 = 0$ are parallel if
$k =$ _____.

3. The graph of $x = -6$ is a _____ line.

4. The angle of inclination of the line $\sqrt{3}x - y + 1 = 0$ is
_____.

5. The graph of $x^2 - 4x + y^2 + 10y + 29 = 0$ is a degenerate
conic. ___

6. The parabola $x^2 = -4y$ has a vertical axis. ___

7. The foci of the ellipse lie on its graph. ___

8. For a hyperbola, the length of the transverse axis is always
greater than the length of the conjugate axis. ___

9. If the origin of the XY-system is at $(2, -3)$ in the xy-system,
then the XY-coordinates of the xy-point $(1, -1)$ are $(3, -4)$.
___

10. If the angle of rotation from the x-axis to the x'-axis is $45°$,
then the xy-coordinates of the $x'y'$-point $(2, 0)$ are
$(\sqrt{2}, \sqrt{2})$. ___

**In Problems 11–14, find an equation of the line satisfying the
given conditions.**

11. Passing through $(1, 4)$ and $(-2, 3)$

12. With x-intercept 2 and y-intercept 4

13. Passing through $(1, 2)$ and perpendicular to the line
$y = 3x - 5$

14. Passing through $(2, 4)$ and parallel to the line through
$(-1, 1)$ and $(4, -3)$

**In Problems 15 and 16, find the slope and the x- and y-inter-
cepts of the given line (if they exist). Graph.**

15. $y = 2(x - 2) + 3$ **16.** $3x - y + 2 = 0$

17. Find the angle of inclination of the line $2x + 2y - 3 = 0$.

18. Find the equation of the line passing through $(4, 3)$ with
angle of inclination $\alpha = 120°$.

19. Assume that there is a linear relationship between the monthly rent r in an apartment complex with 100 units and the number x of units rented.

  (a) Find a formula expressing x in terms of r if all units are rented when the monthly rent is $450, but 6 units are vacant when the rent is $540.

  (b) How many units will be rented if the rent is $585?

20. The relationship between the temperature T in degrees Fahrenheit and the number x of chirps of the black field cricket heard each 14 seconds is approximated by the formula

$$T = \frac{x}{14} + 40.$$

  (a) According to this formula, the black field cricket will begin chirping above what threshold temperature?

  (b) If the number of chirps heard in 14 seconds increases by 7, what is the corresponding increase in the temperature?

In Problems 21–24, find the vertex, focus, directrix, and axis of the given parabola, and sketch the graph.

21. $(y - 3)^2 = -8x$

22. $8(x + 4)^2 = y - 2$

23. $x^2 - 2x + 4y + 1 = 0$

24. $y^2 + 10y + 8x + 41 = 0$

In Problems 25–28, find an equation of the parabola satisfying the given conditions.

25. Focus at $(1, -3)$, directrix $y = -7$

26. Focus at $(3, -1)$, vertex $(0, -1)$

27. Vertex at $(1, 2)$, vertical axis, passing through $(4, 5)$

28. Vertex at $(-1, -4)$, directrix $x = 2$

In Problems 29–32, find the center, vertices, and foci of the given ellipse, and sketch the graph.

29. $\dfrac{x^2}{3} + \dfrac{(y + 5)^2}{25} = 1$

30. $\dfrac{(x - 2)^2}{16} + \dfrac{(y + 5)^2}{4} = 1$

31. $4x^2 + y^2 + 8x - 6y + 9 = 0$

32. $5x^2 + 9y^2 - 20x + 54y + 56 = 0$

In Problems 33–36, find an equation of the ellipse satisfying the given conditions.

33. Vertices at $(0, \pm 4)$, foci at $(\pm 5, 0)$

34. Foci at $(2, -1 \pm \sqrt{2})$, one vertex at $(4, -1)$

35. Vertices at $(\pm 2, -2)$, passing through $(1, -2 + \sqrt{3}/2)$

36. Center at $(2, 4)$, one focus at $(2, 1)$, one vertex at $(2, 0)$

In Problems 37–40, find the center, vertices, foci, and asymptotes of the given hyperbola, and sketch the graph.

37. $(x - 1)(x + 1) = y^2$

38. $y^2 - \dfrac{(x + 3)^2}{4} = 1$

39. $9x^2 - y^2 - 54x - 2y + 71 = 0$

40. $16y^2 - 9x^2 - 64y - 80 = 0$

In Problems 41–44, find an equation of the hyperbola satisfying the given conditions.

41. Center at $(0, 0)$, one vertex at $(6, 0)$, and one focus at $(8, 0)$

42. Foci at $(2, \pm 3)$, one vertex at $(2, -\frac{3}{2})$

43. Foci at $(\pm 2\sqrt{5}, 0)$, asymptotes $y = \pm 2x$

44. Vertices at $(-3, 2)$ and $(-3, 4)$, one focus at $(-3, 3 + \sqrt{2})$

45. Express the equation

$$x^2 + y^2 + 2x - 10y + 6 = 0$$

in terms of the XY-system, where the origin of the XY-system is at the xy-point $(-1, 5)$.

46. Express the equation $3X - Y = 0$ in terms of the xy-system, where the origin of the XY-system is at the xy-point $(4, 2)$.

In Problems 47 and 48, use translation of axes to discuss and sketch the graph of the given equation.

47. $x^2 - 6x - 2y + 8 = 0$

48. $x^2 - 8x + y^2 - 14y - 66 = 0$

In Problems 49 and 50, perform a suitable rotation of axes so that the resulting $x'y'$-equation has no $x'y'$-term. Sketch the graph.

49. $xy = -8$

50. $8x^2 - 4xy + 5y^2 = 36$

51. An automobile headlight is designed so that a cross section through its axis is a parabola and the light source is at the focus. Find the location of the light source if the headlamp is 6 in. across and 2 in. deep.

52. An ellipse is drawn by using a piece of string and two tacks, as shown in Figure 56. If the distance between the tacks is 10 cm and the string is 16 cm long, what are the lengths of the major and minor axes?

The eccentricity of an ellipse (or a hyperbola) can also be defined as the ratio of the distance between the foci to the length of the major axis (or the transverse axis). Stated in the notation of Section 7.4 (or 7.5), this is equivalent to $e = c/a$ (see Problem 28).

The magnitude of the eccentricity of an ellipse or a hyperbola determines its shape. The graph of an ellipse is elongated if e is near 1; but is nearly circular if e is near 0. See Figures 27(a) and (b), respectively. For the hyperbola, if e is near 1, its branches have a narrow opening; but if e is much greater than 1, its branches open widely. See Figures 27(c) and (d), respectively.

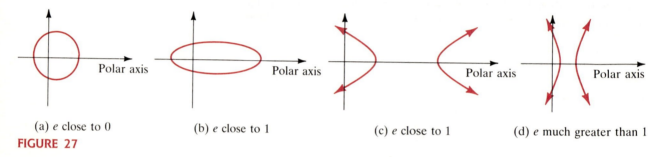

(a) e close to 0   (b) e close to 1   (c) e close to 1   (d) e much greater than 1

FIGURE 27

EXERCISE 8.3

In Problems 1–10, determine the eccentricity, identify the conic section, and sketch its graph.

1. $r = \dfrac{2}{1 - \sin \theta}$

2. $r = \dfrac{2}{2 - \cos \theta}$

3. $r = \dfrac{16}{4 + \cos \theta}$

4. $r = \dfrac{5}{2 + 2 \sin \theta}$

5. $r = \dfrac{4}{1 + 2 \sin \theta}$

6. $r = \dfrac{-4}{\cos \theta - 1}$

7. $r = \dfrac{18}{3 - 6 \cos \theta}$

8. $r = \dfrac{4 \csc \theta}{3 \csc \theta + 2}$

9. $r = \dfrac{6 \sec \theta}{\sec \theta - 1}$

10. $r = \sec \theta \, (\sec \theta + \tan \theta)$

In Problems 11–16, find a polar equation of the conic section with a focus at the pole and the given eccentricity and directrix.

11. $e = 1, \ r = \dfrac{3}{\cos \theta}$

12. $e = \dfrac{3}{2}, \ r = \dfrac{2}{\sin \theta}$

13. $e = \dfrac{2}{3}, \ r = -2 \csc \theta$

14. $e = \dfrac{1}{2}, \ r = 4 \sec \theta$

15. $e = 2, \ r = 6 \sec \theta$

16. $e = 1, \ r = -2 \csc \theta$

In Problems 17–20, convert the given polar form of a conic section to a Cartesian equation. Verify in each case that $e = c/a$.

17. $r = \dfrac{6}{1 + 2 \sin \theta}$

18. $r = \dfrac{10}{2 - 3 \cos \theta}$

19. $r = \dfrac{12}{3 - 2 \cos \theta}$

20. $r = \dfrac{2\sqrt{3}}{\sqrt{3} + \sin \theta}$

In Problems 21–26, find a polar equation of the parabola with focus at the pole and given vertex.

21. $\left(\dfrac{3}{2}, \dfrac{3\pi}{2} \right)$

22. $(2, \pi)$

23. $\left(\dfrac{1}{2}, \pi \right)$

24. $(2, 0)$

25. $\left(\dfrac{1}{4}, \dfrac{3\pi}{2} \right)$

26. $\left(\dfrac{3}{2}, \dfrac{\pi}{2} \right)$

27. By completing the square on x in equation (6), show that the graph of $r = ed/(1 - e \cos \theta)$ is an ellipse if $0 < e < 1$ and a hyperbola if $e > 1$.

28. Identify a, b, and c in the Cartesian equations of the conic sections found in Problem 27. Show that $e = c/a$.

29. If the orbit of a satellite around the earth is an ellipse with the earth at one focus, then an equation of the orbit is given by

$$r = \frac{ed}{1 - e \cos \theta}, \qquad 0 < e < 1.$$

Let r_p designate the length of r at its **perigee** (the point on the satellite's orbit that is closest to earth), and let r_a be the length of r at its **apogee** (the point on the satellite's orbit that is farthest from earth). Show that the eccentricity of the orbit

is given by

$$e = \frac{r_a - r_p}{r_a + r_p}$$

(see Figure 28).

30. A communications satellite is 15,000 km from the center of the earth at its perigee. If the eccentricity of its orbit is 0.25, use the result of Problem 29 to find its farthest distance from the center of the earth.

31. Determine a polar equation of the orbit of the satellite in Problem 30. [*Hint*: The apogee corresponds to $\theta = 0$.]

32. The orbits of the nine planets about the sun are, except for minor perturbations, ellipses with the sun at one focus. The orbits of seven of the nine planets have eccentricities less than 0.01 and therefore are nearly circular. The exceptions are Mercury and Pluto, which have eccentricities 0.206 and 0.249, respectively. Examine the shapes of these two orbits by sketching the graph of $r = ed/(1 - e \cos \theta)$ for $e = 0.2$ and for $e = 0.25$. Take $d = 10$.

33. Many asteroids have highly eccentric elliptical orbits. Sketch the shape of the graph of the orbit of the asteroid Hidalgo for which $e \approx 0.7$. Use $r = ed/(1 - e \cos \theta)$ with $d = 1$.

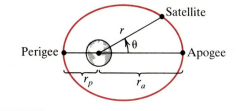

FIGURE 28

8.4

Plane Curves and Parametric Equations

Suppose that a ball is pushed off the edge of a table that is 4 ft high (see Figure 29(a)). If the ball is given an initial horizontal velocity of 2 ft/sec, then it can be shown using the laws of physics that the coordinates (x, y) of its position in a vertical plane at time t sec are given by the functions*

$$x = 2t \quad \text{and} \quad y = -16t^2 + 4, \qquad 0 \le t \le \tfrac{1}{2}. \tag{7}$$

The time $t = 0$ gives $(0, 4)$, which is the initial position of the ball on the table, and $t = \tfrac{1}{2}$ corresponds to $(1, 0)$, which is the ball's position when it hits the floor. (In other words, the ball strikes the floor 1 ft from the edge of the table.) By substituting any time t satisfying $0 < t < \tfrac{1}{2}$ into the equations in

---

*In this example we are ignoring the effects of air resistance.

irrespective of where the bead starts on the wire, the bead slides down to point *B* in the same time? The answer to both the **tautochrone** (Greek for "shortest time") problem and the **brachistochrone** ("equal time") problem was found to be an inverted arch of a **cycloid.**

■ EXAMPLE 7

A circular hoop of radius *a* rolls along the positive *x*-axis without slipping. A fixed point *P* on the hoop, initially at the origin, traces out a curve *C* called a **cycloid** (see Figure 37(a)). Find parametric equations for *C*.

Solution Suppose that the coordinates of the center O' of the circular hoop are (h, k) and that the coordinates of *P* are (x, y). Let *Q* denote the point of contact of the hoop with the *x*-axis. As the hoop rolls through a distance $d(O, Q)$ on the *x*-axis, the radius $O'Q$ rotates through an angle $QO'P$, shown in Figure 37(b). If *t* represents the positive radian measure of the angle $QO'P$, then we see from Figure 37(b) that

$$x = h - a \sin t, \qquad y = k - a \cos t.$$

But since the height of the center O' above the *x*-axis is always equal to the radius *a*, we have $k = a$. Also, the distance $d(O, Q) = h$ that the circle has rolled is the same as that portion of the circumference of the circular hoop corresponding to the central angle *t*, that is, $h = at$. Hence parametric equations for the cycloid are

$$x = at - a \sin t, \quad y = a - a \cos t, \qquad 0 \le t < \infty.$$

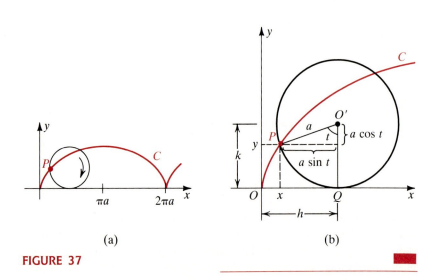

(a)   (b)

FIGURE 37

EXERCISE 8.4

In Problems 1–12, graph the curve C with the given parametric equations.

1. $x = t + 2$, $y = 4$, $-6 \le t \le 3$
2. $x = -3$, $y = t^2 + 1$, $0 \le t < \infty$
3. $x = t + 1$, $y = 2t + 2$, $0 \le t \le 5$
4. $x = 3t - 4$, $y = 2t + 1$, $-2 \le t \le 1$
5. $x = \frac{1}{2}t$, $y = -\frac{1}{2}t^2 + 1$, $-2 \le t \le 4$
6. $x = 2t + 1$, $y = t^2 + t$, $-3 \le t \le 3$
7. $x = 2 - t$, $y = -\sqrt{t}$, $0 \le t \le 9$
8. $x = t$, $y = \sqrt{25 - t^2}$, $-5 \le t \le 5$
9. $x = \frac{1}{2}t^3 - 6t$, $y = \frac{1}{2}t^2$, $-2 \le t \le 4$
10. $x = t^2$, $y = t^2 + t$, $-3 \le t \le 3$
11. $x = t$, $y = \dfrac{5}{1 + t^2}$, $-2 \le t \le 2$
12. $x = t^2$, $y = t^3$, $-1 \le t \le 2$

In Problems 13–22, eliminate the parameter from the given set of parametric equations.

13. $x = t + 1$, $y = 3t + 5$, $-\infty < t < \infty$
14. $x = 2t + 3$, $y = 4t + 5$, $0 \le t < \infty$
15. $x = t^2$, $y = t^4 - 3t^2 + 1$, $0 \le t < \infty$
16. $x = \sqrt{t}$, $y = 2t - 8$, $0 \le t < \infty$
17. $x = 6 \cos t$, $y = 6 \sin t$, $0 \le t \le 2\pi$
18. $x = 5 \cos t$, $y = 2 \sin t$, $0 \le t \le 2\pi$
19. $x = 1 + \cos t$, $y = -2 + \cos t$, $0 \le t \le \pi$
20. $x = 4 + 2 \sin t$, $y = 3 + \cos t$, $0 \le t \le \pi/2$
21. $x = \cos 2t$, $y = \sin t$, $\pi/2 \le t \le 3\pi/2$ [Hint: Use a trigonometric identity.]
22. $x = -\cos^2 t$, $y = 2 \sin t$, $\pi/2 \le t \le 3\pi/2$

In Problems 23–28, show graphically how the given curves differ.

23. C_1: $y = x$;
    C_2: $x = \cos t$, $y = \cos t$, $0 \le t \le \pi$
24. C_1: $y = 2x - 3$;
    C_2: $x = \frac{1}{2}t^2 + 2$, $y = t^2 + 1$, $0 \le t < \infty$
25. C_1: $y = \frac{1}{4}x^2 - 1$;
    C_2: $x = 2t$, $y = t^2 - 1$, $-1 \le t \le 1$
26. C_1: $x^2 + y^2 = 1$;
    C_2: $x = \sin t$, $y = \cos t$, $0 \le t \le \pi$
27. C_1: $y = x^2$;
    C_2: $x = e^t$, $y = e^{2t}$, $0 \le t < \infty$
28. C_1: $xy = 1$;
    C_2: $x = e^{-t}$, $y = e^t$, $-\infty < t < \infty$

In Problems 29 and 30, find parametric equations for the curve C defined by the given function.

29. $y = x^2 + 4x - 6$

30. $y = \dfrac{x}{x^2 + 10}$

31. Consider the rolling hoop of radius a in Figure 37 of Example 7. If the point $P(x, y)$ is fixed within the hoop such that $d(O', P) = b < a$, then the curve C traced by P is called a **curtate cycloid** (see Figure 38(a)). If the point $P(x, y)$ is fixed outside the hoop such that $d(O', P) = b > a$, then the curve C traced by P is called a **prolate cycloid** (see Figure 38(b)). Using the angle t shown in the figure, show that in either case parametric equations for C are

$$x = at - b \sin t, \quad y = a - b \cos t, \qquad 0 \le t < \infty.$$

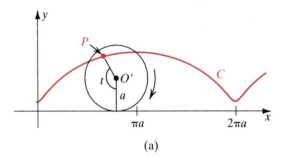

(a)

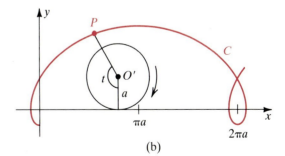

(b)

FIGURE 38

32. Show that a Cartesian equation for the cycloid in Example 7 is

$$x = a \arccos\left(\frac{a - y}{a}\right) \pm \sqrt{2ay - y^2}.$$

33. Suppose that a circle of radius a rolls on the inside of a circle of radius b, $b > a$. If a point $P(x, y)$ on the circumference of the inside circle starts from $(b, 0)$, the curve C traced out by P is called a **hypocycloid.** Using the angle t shown in Figure 39 as a parameter, show that parametric equations for C are

$$x = (b - a) \cos t + a \cos \frac{b - a}{a}t,$$

$$y = (b - a) \sin t - a \sin \frac{b - a}{a}t,$$

$0 \le t \le 2\pi.$

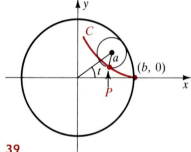

FIGURE 39

34. Suppose that a circle of radius a rolls on the outside of a circle of radius b. If a point $P(x, y)$ on the circumference of the outside circle starts at $(b, 0)$, then the curve C traced out by P is called an **epicycloid.** Using the angle t shown in Figure 40 as a parameter, show that parametric equations for C are

$$x = (a + b) \cos t - a \cos \frac{a + b}{a}t,$$

$$y = (a + b) \sin t - a \sin \frac{a + b}{a}t,$$

$0 \le t \le 2\pi.$

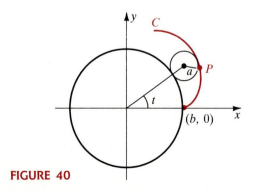

FIGURE 40

35. In Problem 33, when $b = 4a$, the curve is called a **hypocycloid of four cusps** (see Figure 41).
(a) Show that parametric equations for this curve are

$$x = b \cos^3 t, \quad y = b \sin^3 t, \qquad 0 \le t \le 2\pi.$$

(b) Obtain a Cartesian equation for the curve defined in part (a).

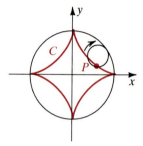

FIGURE 41

36. In Problem 34, when $b = 3a$, the curve is called an **epicycloid of three cusps** (see Figure 42). Show that parametric equations for this curve are

$$x = 4a \cos t - a \cos 4t, \quad y = 4a \sin t - a \sin 4t,$$

$0 \le t \le 2\pi.$

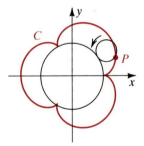

FIGURE 42

37. The **witch of Agnesi*** is the curve C shown in Figure 43. Suppose that a circle of radius a is tangent to the x-axis at the origin O, and suppose that the variable line OB intersects the circle at point A and the horizontal line $y = 2a$ at point B. Let $P(x, y)$ be the point of intersection of the vertical line

---

*This curve is named after the Italian mathematician Maria Gaetana Agnesi (1718–1799), who discussed it in an analytic geometry textbook that she wrote. Due to a translation error the curve became known as the "witch" of Agnesi.

through B and the horizontal line through A. The point P traces out the curve C as the line OA varies through the angle t shown in the figure. Using the angle t as a parameter, show that parametric equations for C are

$$x = 2a \cot t, \quad y = 2a \sin^2 t, \quad 0 < t < \pi.$$

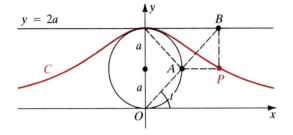

FIGURE 43

38. Eliminate the parameter in Problem 37 and show that a Cartesian equation for the witch of Agnesi is given by $y = 8a^3/(x^2 + 4a^2)$.

39. The parametric equations

$$x = \frac{1 - t^2}{1 + t^2}, \quad y = \frac{2t}{1 + t^2}, \quad -\infty < t < \infty$$

describe the circle $x^2 + y^2 = 1$ except for the point $(-1, 0)$. Eliminate the parameter from these equations.

40. As shown in Figure 44, a projectile is launched at an initial angle θ_0 with an initial velocity v_0. It can be shown that after t seconds the coordinates of its position are given by

$$x = (v_0 \cos \theta_0)t, \quad y = -16t^2 + (v_0 \sin \theta_0)t, \quad 0 \le t \le T.$$

Eliminate the parameter and identify the path of the projectile.

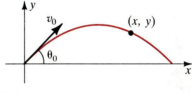

FIGURE 44

IMPORTANT CONCEPTS

Polar coordinate system	Limaçon	Plane curve
pole	Rose curve	Parametric equations
polar axis	Lemniscate	parameter
polar coordinates	Polar equations of conic sections	Graph of a plane curve
Polar graph	Eccentricity	Orientation of a plane curve
Tests for symmetry	Directrix	Eliminating the parameter
Cardioid	Focus	

CHAPTER 8 REVIEW EXERCISE

In Problems 1–10, fill in the blanks or answer true or false.

1. In polar coordinates, $(1, \pi)$ and $(-1, 0)$ represent the same point. ___

2. In polar coordinates, $(-1, -\pi/4)$ represents a point in the _____ quadrant.

3. The eccentricity of a parabola is 0. ___

4. The graph of an ellipse with eccentricity $e = 0.01$ is nearly circular. ___

5. The rectangular coordinates of the point $(2\sqrt{2}, 7\pi/4)$ in polar coordinates are _____.

6. The graph of the polar equation $r = 8 \sec \theta$ is a vertical line. ___

7. The graph of the polar equation $r = 10 \sin \theta$ is a circle with center at _____.

8. The graph of the polar equation $r(1 + \cos \theta) = 1$ is a parabola with a vertical axis. ___

9. The rose curve $r = 9 \sin 5\theta$ has _____ petals.

10. The y-intercept of the graph of the curve C:

$$x = \frac{2 - t}{t^2 + 4}, \quad y = \frac{t^4}{t^2 + 4}, \quad -\infty < t < \infty$$

is _____.

In Problems 11 and 12, find polar coordinates satisfying (a) $r > 0$, $-\pi < \theta \le \pi$, and (b) $r < 0$, $-\pi < \theta \le \pi$ for each point with the given rectangular coordinates.

11. $(3, -\sqrt{3})$

12. $(-\frac{1}{4}, \frac{1}{4})$

In Problems 13 and 14, convert the given polar equation to a Cartesian equation.

13. $r^2 = 25 \cos 2\theta$

14. $r = \tan \theta$

In Problems 15 and 16, convert the given Cartesian equation to a polar equation.

15. $x^2 - y^2 = 9$

16. $(x^2 + y^2 - 4y)^2 = 16(x^2 + y^2)$

In Problem 17–20, apply the tests for symmetry and graph the given polar equation.

17. $r \cos \theta = 5$

18. $\theta = \pi/6$

19. $r = 4 - 4 \cos \theta$

20. $r = 3 \sin 4\theta$

In Problems 21–24, determine the eccentricity, identify the conic section, and sketch its graph.

21. $r = \dfrac{16}{1 - \sin \theta}$

22. $r = \dfrac{8}{2 + 4 \sin \theta}$

23. $r = \dfrac{8}{4 + \cos \theta}$

24. $r = \dfrac{4 \sec \theta}{\sec \theta - 1}$

In Problems 25 and 26, convert the given conic from polar to Cartesian form. Verify that $e = c/a$ in each case.

25. $r = \dfrac{2}{1 + 4 \sin \theta}$

26. $r = \dfrac{2}{1 + \cos \theta}$

27. The center of the circle of radius a shown in Figure 45 has coordinates (b, α). Show that a polar equation of the circle

is

$$a^2 = r^2 + b^2 - 2br \cos (\theta - \alpha).$$

What does the equation become when $b = a$ and $\alpha = 0$? when $b = a$ and $\alpha = \pi/2$?

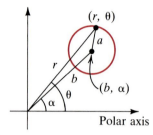

FIGURE 45

28. Graphs of $r = 1 + a \sin \theta$ are shown in Figure 46 for $a = 0$, $a = \frac{1}{2}$, and $a = 1$. In the figure, match the correct value of a with the indicated graph.

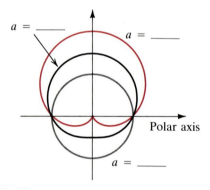

FIGURE 46

In Problems 29–34, eliminate the parameter from the given set of parametric equations.

29. $x = \sin t$, $y = \sin 2t$, $0 \le t \le 2\pi$

30. $x = \sec t$, $y = \tan t$, $-\pi/2 < t < \pi/2$

31. $x = 3 + 4 \sec t$, $y = -5 + 2 \tan t$, $-\pi/2 < t < \pi/2$

32. $x = (e^t + e^{-t})/2$, $y = (e^t - e^{-t})/2$, $0 \le t < \infty$ [*Hint:* Square each expression.]

33. $x = t^3$, $y = 3 \ln t$, $0 < t < \infty$

34. $x = \dfrac{2}{\sqrt{t + 2}}$, $y = \dfrac{t}{t + 2}$, $-2 < t < \infty$

35. Show that the line segment joining the points (x_1, y_1) and (x_2, y_2) has the parametric equations

$$x = x_1 + t(x_2 - x_1),$$

$$y = y_1 + t(y_2 - y_1),$$

$0 \le t \le 1$.

36. The **involute of a circle** of radius a is the curve C traced out by a point $P(x, y)$ at the end of a thread as the thread is held taut and unwound from a spool (the circle). Suppose the spool has its center at the origin and does not rotate as the thread is unwound starting from the point $(a, 0)$. Using the angle t shown in Figure 47, show that parametric equations for C are

$$x = a \cos t + at \sin t,$$

$$y = a \sin t - at \cos t,$$

$0 \le t \le 2\pi$.

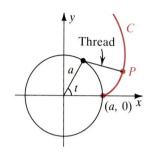

FIGURE 47

In Problems 37 and 38, show graphically how the given curves differ.

37. C_1: $x + y = 1$ and C_2: $x = \cos^2 t$, $y = \sin^2 t$, $0 \le t \le \pi/2$

38. C_1: $x = y^2$ and C_2: $x = \sin^2 t$, $y = \sin t$, $\pi/2 \le t \le 3\pi/2$

APPENDIX A
Linear and Angular Velocity

When an object travels a distance s along any path in time t, its **linear velocity** v or **linear speed** is defined to be

$$v = \frac{s}{t} \qquad (1)$$

■■■ EXAMPLE 1

The linear velocity of a runner who finishes a 10 km race in 1.25 hr is given by

$$v = \frac{10}{1.25} = 8 \text{ km/hr.} \qquad ■■■$$

You might recognize (1) in the familiar form $s = vt$, that is, distance = rate × time.

Now consider a point P on a circle of radius r shown in Figure 1. If P starts at point A and moves a distance s along the circumference of the circle, then its linear velocity is $v = s/t$. But as the point P changes position, the angle AOP also changes. The rate at which the angle AOP changes is called **angular velocity** and is denoted by the symbol ω (omega). More precisely, the angular velocity of the ray OP, the terminal side of the angle AOP, is defined by

$$\omega = \frac{\theta}{t}, \qquad (2)$$

where θ denotes the measure of the angle AOP in time t. Usually the measure of the angle AOP is expressed in radians so that the units for ω are radians per unit time.

■■■ EXAMPLE 2

What is the angular velocity ω of the second hand of a watch in **(a)** radians per minute, **(b)** radians per second?

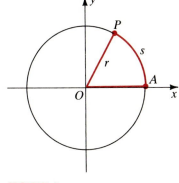

FIGURE 1

hand is 3 in.; (c) the tip of the second hand of a clock if the length of the hand is 2.8 in.

18. Find the angular velocity of the hour, minute, and second hands of a clock in (a) radians per hour; (b) degrees per minute.

19. Assume that the earth rotates once on its axis every 24 hours. (a) Find the angular velocity of the earth in radians per minute. (b) Find the linear velocity in feet per second of a point on the equator if the diameter of the earth is 8000 miles.

20. Assume that the orbit of the earth about the sun is a circle with the sun at the center. (a) Find the angular velocity of the earth in radians per hour using 365.25 days per year and 24 hours per day. (b) Find the linear velocity in miles per hour if the distance from the earth to the sun is 93 million miles.

21. Assume that the orbit of the moon about the earth is a circle with the earth at the center. (a) Find the linear velocity of the moon in miles per hour if one orbit takes 28 days and the distance from the earth to the moon is 2.4×10^5 miles. (b) Using the fact that the moon always keeps the same side facing the earth as it revolves around the earth, find the angular velocity of the moon in radians per hour.

22. A satellite orbits the earth at a constant altitude of 500 miles. Find its linear velocity in miles per hour if it makes 6 revolutions per day and the diameter of the earth is 8000 miles.

APPENDIX B
Some Challenging Problems in Trigonometry

Most of the problems we present in this section involve concepts from several different parts of the text. The problems are drawn from a variety of areas including distance and area measurement, space science, geometry, inequalities, and even tool-making.

The placement of a problem in a particular section or chapter of the text is a strong indicator of the intended method of solution. The location of this collection of problems will give you no such clue. Keep in mind, there is often more than one correct approach in solving a problem.

EXERCISE B

1. In Problem 32 of Exercise 2.3 we discussed a technique for measuring the height of a cloud that required knowing the distance from the observer to the point directly under the cloud. We now present a technique for measuring the height of an object that does not require knowing this distance. As shown in Figure 2 an observer stands at point P with point E at eye level. The object at point O whose altitude y is to be measured is directly above point G on the ground. A mirror is positioned horizontally on the ground at point M so that the reflection of O in the mirror is seen by the observer. If $\overline{EP} = x$, show that the height of the object is given by

$$y = x \frac{\tan \alpha + \tan \beta}{\tan \alpha - \tan \beta}$$

where α is the angle of depression from E to the mirror and β is the angle of elevation from E to the object.

2. Show that the formula given in Problem 1 for the height y is equivalent to

$$y = x \frac{\sin (\alpha + \beta)}{\sin (\alpha - \beta)}.$$

3. An observer sees the reflection of a balloon in a puddle and measures the angle of elevation to the balloon as 44° and the angle of depression to the reflection as 58.5°. Find the height of the balloon if the observer's eye level is 5 feet above the ground.

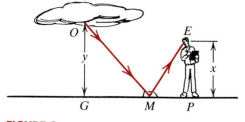

FIGURE 2

4. Tracking stations are to be positioned along the equator to observe a satellite in an equatorial orbit 500 miles above the earth. Each tracking station can scan 180° from horizon to horizon. What is the greatest separation allowed between two neighboring stations to avoid the situation shown in Figure 3 where the satellite cannot be observed by at least one station? Use 4000 mi as the radius of the earth.

5. Find the largest possible value of $f(x) = \sqrt{8} \sin 3x + \cos 3x$. At what values of x does this maximum occur?

6. Show that $\tan^{-1} 1 + \tan^{-1} 2 + \tan^{-1} 3 = \pi$.

7. Assume that the earth is a sphere of radius 4000 mi that rotates on its axis once every 24 hr. Find the linear velocity of the city of Minneapolis, Minnesota, if its latitude in 45°N.

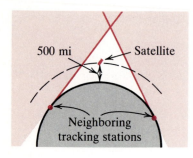

FIGURE 3

8. Show that if θ is measured in radians, then the area A of the shaded region in Figure 4 is given by

$$A = \frac{r^2}{2}(\theta - \sin \theta).$$

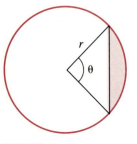

FIGURE 4

9. Show that the area A of the triangle shown in Figure 5 is given by

$$A = \frac{a^2 \sin \beta \sin \gamma}{2 \sin \alpha}.$$

10. Circles of radii 2 in., 4 in., and 5 in., shown in Figure 6, are tangent to each other. Find the area of the shaded region shown in the figure.

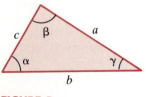

FIGURE 5

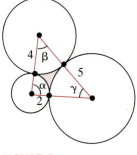

FIGURE 6

11. The triangle ABC shown in Figure 7 is circumscribed by a circle of radius r. Show that

$$r = \frac{a}{2 \sin \alpha} = \frac{b}{2 \sin \beta} = \frac{c}{2 \sin \gamma}.$$

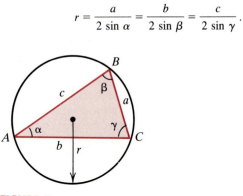

FIGURE 7

12. Find the lengths a and b indicated in Figure 8.

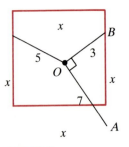

FIGURE 8

13. Find the length x of the side of the square shown in Figure 9 by showing that the angle AOB is a right angle.

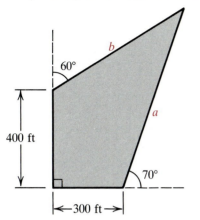

FIGURE 9

14. Show that if α and β are angles such that $0 \le \alpha \le 180°$, $0 \le \beta \le 180°$, then

$$\sin \alpha + \sin \beta \le 2 \sin \frac{\alpha + \beta}{2}.$$

15. If α, β, and γ are angles such that $0 \le \alpha \le 180°$, $0 \le \beta \le 180°, 0 \le \gamma \le 180°$, the result in Problem 14 generalizes to

$$\sin \alpha + \sin \beta + \sin \gamma \le 3 \sin \frac{\alpha + \beta + \gamma}{3}.$$

If α, β, and γ are angles in a triangle show that

$$\sin \alpha + \sin \beta + \sin \gamma \le \frac{3\sqrt{3}}{2}.$$

16. Show that if θ is any acute angle measured in radians then $\sin \theta < \theta$ and $\cos \theta > 1 - \theta$.

17. Find the angle θ indicated in Figure 10.

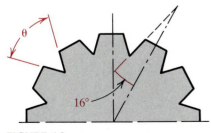

FIGURE 10

18. A notch is to be cut in a thick slab of metal. To verify dimensions a circular measuring cylinder is to be inserted in the notch. When the dimensions are correct the three sides of the notch are tangent to the circular cross section of the cylinder. See Figure 11. Show that the radius r of the circular cross section of the cylinder is given by

$$r = \frac{A \cos \theta - B \sin \theta}{1 - \sin \theta}.$$

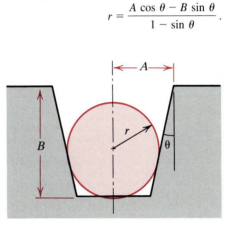

FIGURE 11

19. A circular driveway in front of a new house will curve from one corner of the lot to the next corner of the lot as shown in Figure 12. Find the length of the driveway.

20. Solve $\tan^2 3\theta \tan^2 7\theta = 1$.

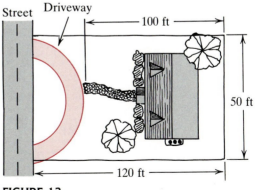

FIGURE 12

APPENDIX C
Using Logarithmic and Trigonometric Tables

Before the invention of the hand-held calculator, numerical calculations were often performed using a slide rule or tables. Like the dinosaurs, slide rules met with quick extinction, but tables of powers, roots, logarithms, and trigonometric functions are only slowly disappearing from the back pages of mathematics texts. Thus more for historical interest than practical need, we will, in this appendix, examine how to use Tables I and II (logarithms) and Table III (trigonometric functions).

COMMON LOGARITHMS

Logarithms to the base 10 are called **common logarithms.** In Section 1.3 we saw that every real number x can be written as a multiple of an integer power of ten; that is, in scientific notation,

$$x = a \cdot 10^c,$$

where $1 \le a < 10$ and c is an integer. Thus,

$$\log_{10} x = \log_{10} (a \cdot 10^c)$$
$$= \log_{10} a + \log_{10} 10^c$$
$$= \log_{10} a + c.$$

In other words, $\log_{10} x = m + c,$

where the numbers $m = \log_{10} a$ and c are called the **mantissa** and the **characteristic** of $\log_{10} x$, respectively. Note that the characteristic equals the integer exponent of 10 when x is written in scientific notation. Recall from Section 5.2 that $f(x) = \log_{10} x$ is an increasing function. Therefore, since $1 \le a < 10$, we have

$$\log_{10} 1 \le \log_{10} a < \log_{10} 10, \text{ or } 0 \le \log_{10} a < 1.$$

In other words, the mantissa is a nonnegative real number less than 1.

TABLE I

At this time it is appropriate to examine Table I (common logarithms). This table contains only the mantissas of logarithms of numbers $1 \le a < 10$ that have at most two decimal places. To locate the mantissa of the logarithm of a particular number, note that the integer portion and the first decimal place indicate the *row* in which the mantissa appears; the second decimal place determines the *column*. For example,

$$\text{if} \quad a = 1.18, \quad \text{then} \quad \log_{10} a \approx 0.0719$$

is obtained by locating the entry in the second row and ninth column.

▬ EXAMPLE 1

Approximate $\log_{10} 6840$.

Solution   We first write the number in scientific notation: $6840 = (6.84)10^3$. We then find its logarithm from Table I:

$$\begin{aligned}
\log_{10} 6840 &= \log_{10} 6.84 + \log_{10} 10^3 \\
&= \log_{10} 6.84 + 3 \log_{10} 10 \\
&\approx 0.8351 + 3 \\
&= 3.8351.
\end{aligned}$$
▬

▬ EXAMPLE 2

Approximate $\log_{10} (0.177)$.

Solution   Since $0.177 = (1.77)10^{-1}$, we have from Table I

$$\begin{aligned}
\log_{10} (0.177) &= \log_{10} 1.77 + \log_{10} 10^{-1} \\
&= \log_{10} 1.77 - \log_{10} 10 \\
&\approx 0.2480 - 1 \\
&= -0.7520.
\end{aligned}$$
▬

INTERPOLATION

The logarithm of a given real number may not appear in Table I. For example, the value of $\log_{10} 2.346$ is not given in the table. However, an approximation to the value of a logarithm can be obtained by **linear interpolation.** This technique is demonstrated in the next example.

▬ EXAMPLE 3

Approximate $\log_{10} 2.346$.

Solution   Since $f(x) = \log_{10} x$ is an increasing function, we know that $\log_{10} 2.340 < \log_{10} 2.346 < \log_{10} 2.350$. Using Figure 13, we reason that

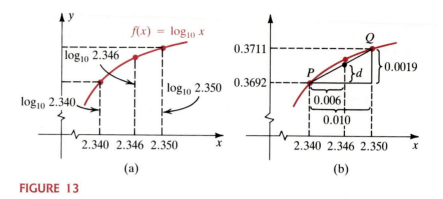

FIGURE 13

since 2.340 and 2.350 are very close on the x-axis, the line segment PQ joining $P(2.340, \log_{10} 2.340)$ and $Q(2.350, \log_{10} 2.350)$ should be close to the graph of $y = \log_{10} x$. Comparing Figure 13(a) with Figure 13(b), we see that $\log_{10} 2.346$ is approximated by the number $0.3692 + d$, which is the y-coordinate of the point on PQ corresponding to the x-coordinate 2.346. Referring again to the figure, it follows from similar triangles that

$$\frac{0.006}{0.010} = \frac{d}{0.0019},$$

and so
$$d = 0.6(0.0019)$$
$$= 0.00114 \approx 0.0011.$$

As is standard practice, we round to the same number of decimal places that appear in Table I. Thus we conclude that

$$\log_{10} 2.346 \approx 0.3692 + 0.0011$$
$$= 0.3703.$$

    Since it is not practical to draw a graph every time we use linear interpolation, it is helpful to arrange the work in a special manner. Using Example 3, we write

$$0.010 \left\{ \begin{array}{l} 0.006 \left\{ \begin{array}{l} \log_{10} 2.340 \approx 0.3692 \\ \log_{10} 2.346 \approx 0.3692 + d \end{array} \right\} d \\ \log_{10} 2.350 \approx 0.3711 \end{array} \right\} 0.0019$$

The numbers next to the left-hand braces are the differences shown next to the horizontal braces in Figure 13(b); the numbers next to the right-hand braces are the differences next to the vertical braces in Figure 13(b). We find d from the proportion 0.006 is to 0.010 as d is to 0.0019, that is, $0.006/0.010 = d/0.0019$.

In using Table I and linear interpolation, the number a in the scientific form $x = a \cdot 10^c$, $1 \le a < 10$, should be rounded to three decimal places. For example, if $x = 467{,}842 = 4.67842 \times 10^5$, then we round $a = 4.67842$ to 4.678. If $x = 0.0053468 = 5.3468 \times 10^{-3}$, then although $a = 5.3468$, we use 5.347.

THE ANTILOGARITHM

If we are given the logarithm of a real number x in the form

$$\log_{10} x = \log_{10} a + c, \quad 1 \le a < 10,$$

then we can write

$$\log_{10} x = \log_{10} a + \log_{10} 10^c$$
$$= \log_{10} (a \cdot 10^c).$$

Since the logarithm function is one-to-one, we have

$$x = a \cdot 10^c.$$

The following examples illustrate the use of Table I in finding a number when its logarithm is given. The number x is sometimes called the **antilogarithm** of $\log_{10} x$.

■ EXAMPLE 4

Find x if $\log_{10} x = 3.7118$.

Solution We first write the logarithm in the form of a mantissa plus a characteristic (that is, a nonnegative number less than 1 plus an integer):

$$\log_{10} x = 0.7118 + 3.$$

Now we can locate the mantissa in Table I:

$$\log_{10} 5.15 \approx 0.7118.$$

Thus,
$$\log_{10} x \approx \log_{10} 5.15 + 3$$
$$x \approx (5.15)10^3$$
$$= 5150. \qquad ■$$

■ EXAMPLE 5

Find x if $\log_{10} x = -2.3468$.

Solution To convert a negative logarithm into the correct form of a mantissa plus a characteristic, we should *add and subtract the smallest positive integer that will give a number in the form $m + c$, where $0 \le m < 1$. Thus in the* given problem we add 3 and subtract 3:

$$\log_{10} x = -2.3468$$
$$= -2.3468 + \underbrace{3 - 3}_{\textbf{zero}}$$
$$= (-2.3468 + 3) - 3$$
$$= \underbrace{0.6532}_{m} \underbrace{- 3}_{c}.$$

From Table I we find that $\log_{10} 4.50 \approx 0.6532$, so

$$x \approx (4.50)10^{-3}$$
$$= 0.0045.$$

Note of Caution: A common mistake when working with negative logarithms is to rewrite -2.3468 as $0.3468 - 2$. This is *incorrect* since $0.3468 - 2 = -1.6532 \neq -2.3468$.

■ EXAMPLE 6

Find x if $\log_{10} x = -0.1281$.

Solution We add 1 to, and subtract 1 from, the given logarithm to find the mantissa and the characteristic:

$$\log_{10} x = -0.1281 + 1 - 1$$
$$= 0.8719 - 1.$$

We then locate the mantissa 0.8719 between the entries 0.8716 and 0.8722 in Table I.

Interpolating,

$$0.01 \left\{ k \begin{cases} \log_{10} 7.44 & \approx 0.8716 \\ \log_{10} (7.44 + k) \approx 0.8719 \end{cases} 0.0003 \right\} 0.0006,$$
$$\log_{10} 7.45 \quad \approx 0.8722$$

gives

$$\frac{k}{0.01} = \frac{0.0003}{0.0006}, \quad \text{or} \quad k = 0.005.$$

Thus, $7.44 + k = 7.445$, so that

$$\log_{10} x \approx \log_{10} 7.445 - 1$$

and

$$x \approx (7.445)10^{-1}$$
$$= 0.7445.$$

TABLE II

Unlike using Table I, it is not necessary to write a number x in scientific notation in order to use Table II (natural logarithms). However, the procedure of linear interpolation is the same.

■ EXAMPLE 7

Approximate $\ln 3.64$.

Solution Inspection of Table II reveals that the natural logarithm of 3.64 is not given in the table. Linear interpolation,

$$0.1 \left\{ 0.04 \left\{ \begin{array}{l} \ln 3.6 \;\approx 1.2809 \\ \ln 3.64 \approx 1.2809 + d \end{array} \right\}^d \\ \ln 3.7 \;\approx 1.3083 \end{array} \right\} 0.0274,$$

yields $\dfrac{0.04}{0.1} = \dfrac{d}{0.0274}$,   or $d = 0.01096 \approx 0.0110$.

Consequently,

$$\ln 3.64 \approx 1.2809 + 0.0110$$
$$= 1.2919.$$

TABLE III

Table III contains four-decimal-place approximations to $\sin\theta$, $\cos\theta$, $\tan\theta$, and $\cot\theta$ for values of the angle θ in quadrant I. For values of θ between $0°$ and $45°$, we use the *left-hand* column together with the headings at the *top* of the table. For an angle θ between $45°$ and $90°$, we use the *right-hand* column with the headings at the *bottom* of the table. The table can be arranged in this fashion, since

$$\sin(90° - \theta) = \cos\theta \quad \text{and} \quad \cos(90° - \theta) = \sin\theta,$$

and, therefore,

$$\tan(90° - \theta) = \frac{\sin(90° - \theta)}{\cos(90° - \theta)} = \frac{\cos\theta}{\sin\theta} = \cot\theta.$$

■ EXAMPLE 8

Approximate **(a)** $\sin 37°50'$ and **(b)** $\cot 51°20'$.

Solution

(a) Since $37°50'$ is between $0°$ and $45°$, we locate the angle in the left-hand column and then read down the column labeled *sin* in the sample table shown here until we reach the row in which the angle appears.

read down

θ	sin	cos	tan	cot	
36°00′	0.5878	0.8090	0.7265	1.376	54°00′
10′	0.5901	0.8073	0.7310	1.368	53°50′
20′	0.5925	0.8056	0.7355	1.360	40′
30′	0.5948	0.8039	0.7400	1.351	30′
40′	0.5972	0.8021	0.7445	1.343	20′
36°50′	0.5995	0.8004	0.7490	1.335	10′
37°00′	0.6018	0.7986	0.7536	1.327	53°00′
10′	0.6041	0.7969	0.7581	1.319	52°50′
20′	0.6065	0.7951	0.7627	1.311	40′
30′	0.6088	0.7934	0.7673	1.303	30′
40′	0.6111	0.7916	0.7720	1.295	20′
37°50′	0.6134	0.7898	0.7766	1.288	10′
38°00′	0.6157	0.7880	0.7813	1.280	52°00′
10′	0.6180	0.7862	0.7860	1.272	51°50′
20′	0.6202	0.7844	0.7907	1.265	40′
30′	0.6225	0.7826	0.7954	1.257	30′
40′	0.6248	0.7808	0.8002	1.250	20′
38°50′	0.6271	0.7790	0.8050	1.242	10′
	cos	sin	cot	tan	θ

locate angle →

locate angle ←

read up

We see that

$$\sin 37°50′ \approx 0.6134.$$

(b) Since 51°20′ is between 45° and 90°, we locate the angle in the right-hand column and read up the column labeled *cot*. From the sample table we find that

$$\cot 51°20′ \approx 0.8002.$$

Table III lists angles from 0° to 90° in increments of 10′. To approximate the value of a trigonometric function of an angle not listed in the table, we proceed as in Examples 3 and 7 and use linear interpolation.

▬ EXAMPLE 9

Approximate sin 43°17′.

Solution   Since sin θ increases between 0° and 90°, it follows from Table III and interpolation,

$$10'\left\{7'\left\{\begin{array}{l} \sin 43°10' \approx 0.6841 \\ \sin 43°17' \approx 0.6841 + d \\ \sin 43°20' \approx 0.6862 \end{array}\right\}d\right\}0.0021,$$

that

$$\frac{7}{10} = \frac{d}{0.0021},$$

and so

$$d = 0.00147 \approx 0.0015.$$

Hence we obtain

$$\sin 43°17' \approx 0.6841 + 0.0015$$
$$= 0.6856.$$ ▬

▬ EXAMPLE 10

Approximate the degree measure of an angle θ between 0° and 90° such that tan $\theta = 0.5942$.

Solution   We first locate successive numbers in the *tan* columns of Table III such that 0.5942 is between these entries. We then interpolate:

$$10'\left\{k\left\{\begin{array}{l} \tan 30°40' \approx 0.5930 \\ \tan \theta \quad\;\; = 0.5942 \\ \tan 30°50' \approx 0.5969 \end{array}\right\}0.0012\right\}0.0039$$

$$\frac{k}{10} = \frac{0.0012}{0.0039}, \quad \text{or} \quad k \approx 3'.$$

Thus,
$$\theta \approx 30°40' + 3'$$
$$= 30°43'.$$ ▬

With the examples in this appendix, you should be able to use Table IV effectively and interpolate if necessary.

TABLE I

COMMON LOGARITHMS

x	0	1	2	3	4	5	6	7	8	9
1.0	0.0000	0.0043	0.0086	0.0128	0.0170	0.0212	0.0253	0.0294	0.0334	0.0374
1.1	0.0414	0.0453	0.0492	0.0531	0.0569	0.0607	0.0645	0.0682	0.0719	0.0755
1.2	0.0792	0.0828	0.0864	0.0899	0.0934	0.0969	0.1004	0.1038	0.1072	0.1106
1.3	0.1139	0.1173	0.1206	0.1239	0.1271	0.1303	0.1335	0.1367	0.1399	0.1430
1.4	0.1461	0.1492	0.1523	0.1553	0.1584	0.1614	0.1644	0.1673	0.1703	0.1732
1.5	0.1761	0.1790	0.1818	0.1847	0.1875	0.1903	0.1931	0.1959	0.1987	0.2014
1.6	0.2041	0.2068	0.2095	0.2122	0.2148	0.2175	0.2201	0.2227	0.2253	0.2279
1.7	0.2304	0.2330	0.2355	0.2380	0.2405	0.2430	0.2455	0.2480	0.2504	0.2529
1.8	0.2553	0.2577	0.2601	0.2625	0.2648	0.2672	0.2695	0.2718	0.2742	0.2765
1.9	0.2788	0.2810	0.2833	0.2856	0.2878	0.2900	0.2923	0.2945	0.2967	0.2989
2.0	0.3010	0.3032	0.3054	0.3075	0.3096	0.3118	0.3139	0.3160	0.3181	0.3201
2.1	0.3222	0.3243	0.3263	0.3284	0.3304	0.3324	0.3345	0.3365	0.3385	0.3404
2.2	0.3424	0.3444	0.3464	0.3483	0.3502	0.3522	0.3541	0.3560	0.3579	0.3598
2.3	0.3617	0.3636	0.3655	0.3674	0.3692	0.3711	0.3729	0.3747	0.3766	0.3784
2.4	0.3802	0.3820	0.3838	0.3856	0.3874	0.3892	0.3909	0.3927	0.3945	0.3962
2.5	0.3979	0.3997	0.4014	0.4031	0.4048	0.4065	0.4082	0.4099	0.4116	0.4133
2.6	0.4150	0.4166	0.4183	0.4200	0.4216	0.4232	0.4249	0.4265	0.4281	0.4298
2.7	0.4314	0.4330	0.4346	0.4362	0.4378	0.4393	0.4409	0.4425	0.4440	0.4456
2.8	0.4472	0.4487	0.4502	0.4518	0.4533	0.4548	0.4564	0.4579	0.4594	0.4609
2.9	0.4624	0.4639	0.4654	0.4669	0.4683	0.4698	0.4713	0.4728	0.4742	0.4757
3.0	0.4771	0.4786	0.4800	0.4814	0.4829	0.4843	0.4857	0.4871	0.4886	0.4900
3.1	0.4914	0.4928	0.4942	0.4955	0.4969	0.4983	0.4997	0.5011	0.5024	0.5038
3.2	0.5051	0.5065	0.5079	0.5092	0.5105	0.5119	0.5132	0.5145	0.5159	0.5172
3.3	0.5185	0.5198	0.5211	0.5224	0.5237	0.5250	0.5263	0.5276	0.5289	0.5302
3.4	0.5315	0.5328	0.5340	0.5353	0.5366	0.5378	0.5391	0.5403	0.5416	0.5428
3.5	0.5441	0.5453	0.5465	0.5478	0.5490	0.5502	0.5514	0.5527	0.5539	0.5551
3.6	0.5563	0.5575	0.5587	0.5599	0.5611	0.5623	0.5635	0.5647	0.5658	0.5670
3.7	0.5682	0.5694	0.5705	0.5717	0.5729	0.5740	0.5752	0.5763	0.5775	0.5786
3.8	0.5798	0.5809	0.5821	0.5832	0.5843	0.5855	0.5866	0.5877	0.5888	0.5899
3.9	0.5911	0.5922	0.5933	0.5944	0.5955	0.5966	0.5977	0.5988	0.5999	0.6010
4.0	0.6021	0.6031	0.6042	0.6053	0.6064	0.6075	0.6085	0.6096	0.6107	0.6117
4.1	0.6128	0.6138	0.6149	0.6160	0.6170	0.6180	0.6191	0.6201	0.6212	0.6222
4.2	0.6232	0.6243	0.6253	0.6263	0.6274	0.6284	0.6294	0.6304	0.6314	0.6325
4.3	0.6335	0.6345	0.6355	0.6365	0.6375	0.6385	0.6395	0.6405	0.6415	0.6425
4.4	0.6435	0.6444	0.6454	0.6464	0.6474	0.6484	0.6493	0.6503	0.6513	0.6522
4.5	0.6532	0.6542	0.6551	0.6561	0.6571	0.6580	0.6590	0.6599	0.6609	0.6618
4.6	0.6628	0.6637	0.6646	0.6656	0.6665	0.6675	0.6684	0.6693	0.6702	0.6712
4.7	0.6721	0.6730	0.6739	0.6749	0.6758	0.6767	0.6776	0.6785	0.6794	0.6803
4.8	0.6812	0.6821	0.6830	0.6839	0.6848	0.6857	0.6866	0.6875	0.6884	0.6893
4.9	0.6902	0.6911	0.6920	0.6928	0.6937	0.6946	0.6955	0.6964	0.6972	0.6981
5.0	0.6990	0.6998	0.7007	0.7016	0.7024	0.7033	0.7042	0.7050	0.7059	0.7067
5.1	0.7076	0.7084	0.7093	0.7101	0.7110	0.7118	0.7126	0.7135	0.7143	0.7152
5.2	0.7160	0.7168	0.7177	0.7185	0.7193	0.7202	0.7210	0.7218	0.7226	0.7235
5.3	0.7243	0.7251	0.7259	0.7267	0.7275	0.7284	0.7292	0.7300	0.7308	0.7316
5.4	0.7324	0.7332	0.7340	0.7348	0.7356	0.7364	0.7372	0.7380	0.7388	0.7396
x	0	1	2	3	4	5	6	7	8	9

TABLE I *(continued)*

x	0	1	2	3	4	5	6	7	8	9
5.5	0.7404	0.7412	0.7419	0.7427	0.7435	0.7443	0.7451	0.7459	0.7466	0.7474
5.6	0.7482	0.7490	0.7497	0.7505	0.7513	0.7520	0.7528	0.7536	0.7543	0.7551
5.7	0.7559	0.7566	0.7574	0.7582	0.7589	0.7597	0.7604	0.7612	0.7619	0.7627
5.8	0.7634	0.7642	0.7649	0.7657	0.7664	0.7672	0.7679	0.7686	0.7694	0.7701
5.9	0.7709	0.7716	0.7723	0.7731	0.7738	0.7745	0.7752	0.7760	0.7767	0.7774
6.0	0.7782	0.7789	0.7796	0.7803	0.7810	0.7818	0.7825	0.7832	0.7839	0.7846
6.1	0.7853	0.7860	0.7868	0.7875	0.7882	0.7889	0.7896	0.7903	0.7910	0.7917
6.2	0.7924	0.7931	0.7938	0.7945	0.7952	0.7959	0.7966	0.7973	0.7980	0.7987
6.3	0.7993	0.8000	0.8007	0.8014	0.8021	0.8028	0.8035	0.8041	0.8048	0.8055
6.4	0.8062	0.8069	0.8075	0.8082	0.8089	0.8096	0.8102	0.8109	0.8116	0.8122
6.5	0.8129	0.8136	0.8142	0.8149	0.8156	0.8162	0.8169	0.8176	0.8182	0.8189
6.6	0.8195	0.8202	0.8209	0.8215	0.8222	0.8228	0.8235	0.8241	0.8248	0.8254
6.7	0.8261	0.8267	0.8274	0.8280	0.8287	0.8293	0.8299	0.8306	0.8312	0.8319
6.8	0.8325	0.8331	0.8338	0.8344	0.8351	0.8357	0.8363	0.8370	0.8376	0.8382
6.9	0.8388	0.8395	0.8401	0.8407	0.8414	0.8420	0.8426	0.8432	0.8439	0.8445
7.0	0.8451	0.8457	0.8463	0.8470	0.8476	0.8482	0.8488	0.8494	0.8500	0.8506
7.1	0.8513	0.8519	0.8525	0.8531	0.8537	0.8543	0.8549	0.8555	0.8561	0.8567
7.2	0.8573	0.8579	0.8585	0.8591	0.8597	0.8603	0.8609	0.8615	0.8621	0.8627
7.3	0.8633	0.8639	0.8645	0.8651	0.8657	0.8663	0.8669	0.8675	0.8681	0.8686
7.4	0.8692	0.8698	0.8704	0.8710	0.8716	0.8722	0.8727	0.8733	0.8739	0.8745
7.5	0.8751	0.8756	0.8762	0.8768	0.8774	0.8779	0.8785	0.8791	0.8797	0.8802
7.6	0.8808	0.8814	0.8820	0.8825	0.8831	0.8837	0.8842	0.8848	0.8854	0.8859
7.7	0.8865	0.8871	0.8876	0.8882	0.8887	0.8893	0.8899	0.8904	0.8910	0.8915
7.8	0.8921	0.8927	0.8932	0.8938	0.8943	0.8949	0.8954	0.8960	0.8965	0.8971
7.9	0.8976	0.8982	0.8987	0.8993	0.8998	0.9004	0.9009	0.9015	0.9020	0.9025
8.0	0.9031	0.9036	0.9042	0.9047	0.9053	0.9058	0.9063	0.9069	0.9074	0.9079
8.1	0.9085	0.9090	0.9096	0.9101	0.9106	0.9112	0.9117	0.9122	0.9128	0.9133
8.2	0.9138	0.9143	0.9149	0.9154	0.9159	0.9165	0.9170	0.9175	0.9180	0.9186
8.3	0.9191	0.9196	0.9201	0.9206	0.9212	0.9217	0.9222	0.9227	0.9232	0.9238
8.4	0.9243	0.9248	0.9253	0.9258	0.9263	0.9269	0.9274	0.9279	0.9284	0.9289
8.5	0.9294	0.9299	0.9304	0.9309	0.9315	0.9320	0.9325	0.9330	0.9335	0.9340
8.6	0.9345	0.9350	0.9355	0.9360	0.9365	0.9370	0.9375	0.9380	0.9385	0.9390
8.7	0.9395	0.9400	0.9405	0.9410	0.9415	0.9420	0.9425	0.9430	0.9435	0.9440
8.8	0.9445	0.9450	0.9455	0.9460	0.9465	0.9469	0.9474	0.9479	0.9484	0.9489
8.9	0.9494	0.9499	0.9504	0.9509	0.9513	0.9518	0.9523	0.9528	0.9533	0.9538
9.0	0.9542	0.9547	0.9552	0.9557	0.9562	0.9566	0.9571	0.9576	0.9581	0.9586
9.1	0.9590	0.9595	0.9600	0.9605	0.9609	0.9614	0.9619	0.9624	0.9628	0.9633
9.2	0.9638	0.9643	0.9647	0.9652	0.9657	0.9661	0.9666	0.9671	0.9675	0.9680
9.3	0.9685	0.9689	0.9694	0.9699	0.9703	0.9708	0.9713	0.9717	0.9722	0.9727
9.4	0.9731	0.9736	0.9741	0.9745	0.9750	0.9754	0.9759	0.9763	0.9768	0.9773
9.5	0.9777	0.9782	0.9786	0.9791	0.9795	0.9800	0.9805	0.9809	0.9814	0.9818
9.6	0.9823	0.9827	0.9832	0.9836	0.9841	0.9845	0.9850	0.9854	0.9859	0.9863
9.7	0.9868	0.9872	0.9877	0.9881	0.9886	0.9890	0.9894	0.9899	0.9903	0.9908
9.8	0.9912	0.9917	0.9921	0.9926	0.9930	0.9934	0.9939	0.9943	0.9948	0.9952
9.9	0.9956	0.9961	0.9965	0.9969	0.9974	0.9978	0.9983	0.9987	0.9991	0.9996
x	0	1	2	3	4	5	6	7	8	9

TABLE II

NATURAL LOGARITHMS

x	$\ln x$	x	$\ln x$	x	$\ln x$
		4.5	1.5041	9.0	2.1972
0.1	−2.3026	4.6	1.5261	9.1	2.2083
0.2	−1.6094	4.7	1.5476	9.2	2.2192
0.3	−1.2040	4.8	1.5686	9.3	2.2300
0.4	−0.9163	4.9	1.5892	9.4	2.2407
0.5	−0.6931	5.0	1.6094	9.5	2.2513
0.6	−0.5108	5.1	1.6292	9.6	2.2618
0.7	−0.3567	5.2	1.6487	9.7	2.2721
0.8	−0.2231	5.3	1.6677	9.8	2.2824
0.9	−0.1054	5.4	1.6864	9.9	2.2925
1.0	0.0000	5.5	1.7047	10	2.3026
1.1	0.0953	5.6	1.7228	11	2.3979
1.2	0.1823	5.7	1.7405	12	2.4849
1.3	0.2624	5.8	1.7579	13	2.5649
1.4	0.3365	5.9	1.7750	14	2.6391
1.5	0.4055	6.0	1.7918	15	2.7081
1.6	0.4700	6.1	1.8083	16	2.7726
1.7	0.5306	6.2	1.8245	17	2.8332
1.8	0.5878	6.3	1.8405	18	2.8904
1.9	0.6419	6.4	1.8563	19	2.9444
2.0	0.6931	6.5	1.8718	20	2.9957
2.1	0.7419	6.6	1.8871	25	3.2189
2.2	0.7885	6.7	1.9021	30	3.4012
2.3	0.8329	6.8	1.9169	35	3.5553
2.4	0.8755	6.9	1.9315	40	3.6889
2.5	0.9163	7.0	1.9459	45	3.8067
2.6	0.9555	7.1	1.9601	50	3.9120
2.7	0.9933	7.2	1.9741	55	4.0073
2.8	1.0296	7.3	1.9879	60	4.0943
2.9	1.0647	7.4	2.0015	65	4.1744
3.0	1.0986	7.5	2.0149	70	4.2485
3.1	1.1314	7.6	2.0281	75	4.3175
3.2	1.1632	7.7	2.0412	80	4.3820
3.3	1.1939	7.8	2.0541	85	4.4427
3.4	1.2238	7.9	2.0669	90	4.4998
3.5	1.2528	8.0	2.0794	100	4.6052
3.6	1.2809	8.1	2.0919	110	4.7005
3.7	1.3083	8.2	2.1041	120	4.7875
3.8	1.3350	8.3	2.1163	130	4.8676
3.9	1.3610	8.4	2.1282	140	4.9416
4.0	1.3863	8.5	2.1401	150	5.0106
4.1	1.4110	8.6	2.1518	160	5.0752
4.2	1.4351	8.7	2.1633	170	5.1358
4.3	1.4586	8.8	2.1748	180	5.1930
4.4	1.4816	8.9	2.1861	190	5.2470

TABLE III

TRIGONOMETRIC FUNCTIONS (DEGREES)

θ	sin	cos	tan	cot		θ	sin	cos	tan	cot	
0°00′	0.0000	1.000	0.0000	—	90°00′	8°00′	0.1392	0.9903	0.1405	7.115	82°00′
10′	0.0029	1.000	0.0029	343.8	89°50′	10′	0.1421	0.9899	0.1435	6.968	81°50′
20′	0.0058	1.000	0.0058	171.9	40′	20′	0.1449	0.9894	0.1465	6.827	40′
30′	0.0087	1.000	0.0087	114.6	30′	30′	0.1478	0.9890	0.1495	6.691	30′
40′	0.0116	0.9999	0.0116	85.94	20′	40′	0.1507	0.9886	0.1524	6.561	20′
0°50′	0.0145	0.9999	0.0145	68.75	10′	8°50′	0.1536	0.9881	0.1554	6.435	10′
1°00′	0.0175	0.9998	0.0175	57.29	89°00′	9°00′	0.1564	0.9877	0.1584	6.314	81°00′
10′	0.0204	0.9998	0.0204	49.10	88°50′	10′	0.1593	0.9872	0.1614	6.197	80°50′
20′	0.0233	0.9997	0.0233	42.96	40′	20′	0.1622	0.9868	0.1644	6.084	40′
30′	0.0262	0.9997	0.0262	38.19	30′	30′	0.1650	0.9863	0.1673	5.976	30′
40′	0.0291	0.9996	0.0291	34.37	20′	40′	0.1679	0.9858	0.1703	5.871	20′
1°50′	0.0320	0.9995	0.0320	31.24	10′	9°50′	0.1708	0.9853	0.1733	5.769	10′
2°00′	0.0349	0.9994	0.0349	28.64	88°00′	10°00′	0.1736	0.9848	0.1763	5.671	80°00′
10′	0.0378	0.9993	0.0378	26.43	87°50′	10′	0.1765	0.9843	0.1793	5.576	79°50′
20′	0.0407	0.9992	0.0407	24.54	40′	20′	0.1794	0.9838	0.1823	5.485	40′
30′	0.0436	0.9990	0.0437	22.90	30′	30′	0.1822	0.9833	0.1853	5.396	30′
40′	0.0465	0.9989	0.0466	21.47	20′	40′	0.1851	0.9827	0.1883	5.309	20′
2°50′	0.0494	0.9988	0.0495	20.21	10′	10°50′	0.1880	0.9822	0.1914	5.226	10′
3°00′	0.0523	0.9986	0.0524	19.08	87°00′	11°00′	0.1908	0.9816	0.1944	5.145	79°00′
10′	0.0552	0.9985	0.0553	18.07	86°50′	10′	0.1937	0.9811	0.1974	5.066	78°50′
20′	0.0581	0.9983	0.0582	17.17	40′	20′	0.1965	0.9805	0.2004	4.989	40′
30′	0.0610	0.9981	0.0612	16.35	30′	30′	0.1994	0.9799	0.2035	4.915	30′
40′	0.0640	0.9980	0.0641	15.60	20′	40′	0.2022	0.9793	0.2065	4.843	20′
3°50′	0.0669	0.9978	0.0670	14.92	10′	11°50′	0.2051	0.9787	0.2095	4.773	10′
4°00′	0.0698	0.9976	0.0699	14.30	86°00′	12°00′	0.2079	0.9781	0.2126	4.705	78°00′
10′	0.0727	0.9974	0.0729	13.73	85°50′	10′	0.2108	0.9775	0.2156	4.638	77°50′
20′	0.0756	0.9971	0.0758	13.20	40′	20′	0.2136	0.9769	0.2186	4.574	40′
30′	0.0785	0.9969	0.0787	12.71	30′	30′	0.2164	0.9763	0.2217	4.511	30′
40′	0.0814	0.9967	0.0816	12.25	20′	40′	0.2193	0.9757	0.2247	4.449	20′
4°50′	0.0843	0.9964	0.0846	11.83	10′	12°50′	0.2221	0.9750	0.2278	4.390	10′
5°00′	0.0872	0.9962	0.0875	11.43	85°00′	13°00′	0.2250	0.9744	0.2309	4.331	77°00′
10′	0.0901	0.9959	0.0904	11.06	84°50′	10′	0.2278	0.9737	0.2339	4.275	76°50′
20′	0.0929	0.9957	0.0934	10.71	40′	20′	0.2306	0.9730	0.2370	4.219	40′
30′	0.0958	0.9954	0.0963	10.39	30′	30′	0.2334	0.9724	0.2401	4.165	30′
40′	0.0987	0.9951	0.0992	10.08	20′	40′	0.2363	0.9717	0.2432	4.113	20′
5°50′	0.1016	0.9948	0.1022	9.788	10′	13°50′	0.2391	0.9710	0.2462	4.061	10′
6°00′	0.1045	0.9945	0.1051	9.514	84°00′	14°00′	0.2419	0.9703	0.2493	4.011	76°00′
10′	0.1074	0.9942	0.1080	9.255	83°50′	10′	0.2447	0.9696	0.2524	3.962	75°50′
20′	0.1103	0.9939	0.1110	9.010	40′	20′	0.2476	0.9689	0.2555	3.914	40′
30′	0.1132	0.9936	0.1139	8.777	30′	30′	0.2504	0.9681	0.2586	3.867	30′
40′	0.1161	0.9932	0.1169	8.556	20′	40′	0.2532	0.9674	0.2617	3.821	20′
6°50′	0.1190	0.9929	0.1198	8.345	10′	14°50′	0.2560	0.9667	0.2648	3.776	10′
7°00′	0.1219	0.9925	0.1228	8.144	83°00′	15°00′	0.2588	0.9659	0.2679	3.732	75°00′
10′	0.1248	0.9922	0.1257	7.953	82′50′	10′	0.2616	0.9652	0.2711	3.689	74°50′
20′	0.1276	0.9918	0.1287	7.770	40′	20′	0.2644	0.9644	0.2742	3.647	40′
30′	0.1305	0.9914	0.1317	7.596	30′	30′	0.2672	0.9636	0.2773	3.606	30′
40′	0.1334	0.9911	0.1346	7.429	20′	40′	0.2700	0.9628	0.2805	3.566	20′
7°50′	0.1363	0.9907	0.1376	7.269	10′	15°50′	0.2728	0.9621	0.2836	3.526	10′
	cos	sin	cot	tan	θ		cos	sin	cot	tan	θ

TABLE III *(continued)*

θ	sin	cos	tan	cot		θ	sin	cos	tan	cot	
16°00′	0.2756	0.9613	0.2867	3.487	74°00′	24′00′	0.4067	0.9135	0.4452	2.246	66°00′
10′	0.2784	0.9605	0.2899	3.450	73°50′	10′	0.4094	0.9124	0.4487	2.229	65°50′
20′	0.2812	0.9596	0.2931	3.412	40′	20′	0.4120	0.9112	0.4522	2.211	40′
30′	0.2840	0.9588	0.2962	3.376	30′	30′	0.4147	0.9100	0.4557	2.194	30′
40′	0.2868	0.9580	0.2994	3.340	20′	40′	0.4173	0.9088	0.4592	2.177	20′
16°50′	0.2896	0.9572	0.3026	3.305	10′	24°50′	0.4200	0.9075	0.4628	2.161	10′
17°00′	0.2924	0.9563	0.3057	3.271	73°00′	25°00′	0.4226	0.9063	0.4663	2.145	65°00′
10′	0.2952	0.9555	0.3089	3.237	72°50′	10′	0.4253	0.9051	0.4699	2.128	64°50′
20′	0.2979	0.9546	0.3121	3.204	40′	20′	0.4279	0.9038	0.4734	2.112	40′
30′	0.3007	0.9537	0.3153	3.172	30′	30′	0.4305	0.9026	0.4770	2.097	30′
40′	0.3035	0.9528	0.3185	3.140	20′	40′	0.4331	0.9013	0.4806	2.081	20′
17°50′	0.3062	0.9520	0.3217	3.108	10′	25°50′	0.4358	0.9001	0.4841	2.066	10′
18°00′	0.3090	0.9511	0.3249	3.078	72°00′	26°00′	0.4384	0.8988	0.4877	2.050	64°00′
10′	0.3118	0.9502	0.3281	3.047	71°50′	10′	0.4410	0.8975	0.4913	2.035	63°50′
20′	0.3145	0.9492	0.3314	3.018	40′	20′	0.4436	0.8962	0.4950	2.020	40′
30′	0.3173	0.9483	0.3346	2.989	30′	30′	0.4462	0.8949	0.4986	2.006	30′
40′	0.3201	0.9474	0.3378	2.960	20′	40′	0.4488	0.8936	0.5022	1.991	20′
18°50′	0.3228	0.9465	0.3411	2.932	10′	26°50′	0.4514	0.8923	0.5059	1.977	10′
19°00′	0.3256	0.9455	0.3443	2.904	71°00′	27°00′	0.4540	0.8910	0.5095	1.963	63°00′
10′	0.3283	0.9446	0.3476	2.877	70°50′	10′	0.4566	0.8897	0.5132	1.949	62°50′
20′	0.3311	0.9436	0.3508	2.850	40′	20′	0.4592	0.8884	0.5169	1.935	40′
30′	0.3338	0.9426	0.3541	2.824	30′	30′	0.4617	0.8870	0.5206	1.921	30′
40′	0.3365	0.9417	0.3574	2.798	20′	40′	0.4643	0.8857	0.5243	1.907	20′
19°50′	0.3393	0.9407	0.3607	2.773	10′	27°50′	0.4669	0.8843	0.5280	1.894	10′
20°00′	0.3420	0.9397	0.3640	2.747	70°00′	28°00′	0.4695	0.8829	0.5317	1.881	62°00′
10′	0.3448	0.9387	0.3673	2.723	69°50′	10′	0.4720	0.8816	0.5354	1.868	61°50′
20′	0.3475	0.9377	0.3706	2.699	40′	20′	0.4746	0.8802	0.5392	1.855	40′
30′	0.3502	0.9367	0.3739	2.675	30′	30′	0.4772	0.8788	0.5430	1.842	30′
40′	0.3529	0.9356	0.3772	2.651	20′	40′	0.4797	0.8774	0.5467	1.829	20′
20°50′	0.3557	0.9346	0.3805	2.628	10′	28°50′	0.4823	0.8760	0.5505	1.816	10′
21°00′	0.3584	0.9336	0.3839	2.605	69°00′	29°00′	0.4848	0.8746	0.5543	1.804	61°00′
10′	0.3611	0.9325	0.3872	2.583	68°50′	10′	0.4874	0.8732	0.5581	1.792	60°50′
20′	0.3638	0.9315	0.3906	2.560	40′	20′	0.4899	0.8718	0.5619	1.780	40′
30′	0.3665	0.9304	0.3939	2.539	30′	30′	0.4924	0.8704	0.5658	1.767	30′
40′	0.3692	0.9293	0.3973	2.517	20′	40′	0.4950	0.8689	0.5696	1.756	20′
21°50′	0.3719	0.9283	0.4006	2.496	10′	29°50′	0.4975	0.8675	0.5735	1.744	10′
22°00′	0.3746	0.9272	0.4040	2.475	68°00′	30°00′	0.5000	0.8660	0.5774	1.732	60°00′
10′	0.3773	0.9261	0.4074	2.455	67°50′	10′	0.5025	0.8646	0.5812	1.720	59°50′
20′	0.3800	0.9250	0.4108	2.434	40′	20′	0.5050	0.8631	0.5851	1.709	40′
30′	0.3827	0.9239	0.4142	2.414	30′	30′	0.5075	0.8616	0.5890	1.698	30′
40′	0.3854	0.9228	0.4176	2.394	20′	40′	0.5100	0.8601	0.5930	1.686	20′
22°50′	0.3881	0.9216	0.4210	2.375	10′	30°50′	0.5125	0.8587	0.5969	1.675	10′
23°00′	0.3907	0.9205	0.4245	2.356	67°00′	31°00′	0.5150	0.8572	0.6009	1.664	59°00′
10′	0.3934	0.9194	0.4279	2.337	66°50′	10′	0.5175	0.8557	0.6048	1.653	58°50′
20′	0.3961	0.9182	0.4314	2.318	40′	20′	0.5200	0.8542	0.6088	1.643	40′
30′	0.3987	0.9171	0.4348	2.300	30′	30′	0.5225	0.8526	0.6128	1.632	30′
40′	0.4014	0.9159	0.4383	2.282	20′	40′	0.5250	0.8511	0.6168	1.621	20′
23°50′	0.4041	0.9147	0.4417	2.264	10′	31°50′	0.5275	0.8496	0.6208	1.611	10′
	cos	sin	cot	tan	θ		cos	sin	cot	tan	θ

TABLE III   *(continued)*

θ	sin	cos	tan	cot		θ	sin	cos	tan	cot	
32°00′	0.5299	0.8480	0.6249	1.600	58°00′	39°00′	0.6293	0.7771	0.8098	1.235	51°00′
10′	0.5324	0.8465	0.6289	1.590	57°50′	10′	0.6316	0.7753	0.8146	1.228	50°50′
20′	0.5348	0.8450	0.6330	1.580	40′	20′	0.6338	0.7735	0.8195	1.220	40′
30′	0.5373	0.8434	0.6371	1.570	30′	30′	0.6361	0.7716	0.8243	1.213	30′
40′	0.5398	0.8418	0.6412	1.560	20′	40′	0.6383	0.7698	0.8292	1.206	20′
32°50′	0.5422	0.8403	0.6453	1.550	10′	39°50′	0.6406	0.7679	0.8342	1.199	10′
33°00′	0.5446	0.8387	0.6494	1.540	57°00′	40°00′	0.6428	0.7660	0.8391	1.192	50°00′
10′	0.5471	0.8371	0.6536	1.530	56°50′	10′	0.6450	0.7642	0.8441	1.185	49°50′
20′	0.5495	0.8355	0.6577	1.520	40′	20′	0.6472	0.7623	0.8491	1.178	40′
30′	0.5519	0.8339	0.6619	1.511	30′	30′	0.6494	0.7604	0.8541	1.171	30′
40′	0.5544	0.8323	0.6661	1.501	20′	40′	0.6517	0.7585	0.8591	1.164	20′
33°50′	0.5568	0.8307	0.6703	1.492	10′	40°50′	0.6539	0.7566	0.8642	1.157	10′
34°00′	0.5592	0.8290	0.6745	1.483	56°00′	41°00′	0.6561	0.7547	0.8693	1.150	49°00′
10′	0.5616	0.8274	0.6787	1.473	55°50′	10′	0.6583	0.7528	0.8744	1.144	48°50′
20′	0.5640	0.8258	0.6830	1.464	40′	20′	0.6604	0.7509	0.8796	1.137	40′
30′	0.5664	0.8241	0.6873	1.455	30′	30′	0.6626	0.7490	0.8847	1.130	30′
40′	0.5688	0.8225	0.6916	1.446	20′	40′	0.6648	0.7470	0.8899	1.124	20′
34°50′	0.5712	0.8208	0.6959	1.437	10′	41°50′	0.6670	0.7451	0.8952	1.117	10′
35°00′	0.5736	0.8192	0.7002	1.428	55°00′	42°00′	0.6691	0.7431	0.9004	1.111	48°00′
10′	0.5760	0.8175	0.7046	1.419	54°50′	10′	0.6713	0.7412	0.9057	1.104	47°50′
20′	0.5783	0.8158	0.7089	1.411	40′	20′	0.6734	0.7392	0.9110	1.098	40′
30′	0.5807	0.8141	0.7133	1.402	30′	30′	0.6756	0.7373	0.9163	1.091	30′
40′	0.5831	0.8124	0.7177	1.393	20′	40′	0.6777	0.7353	0.9217	1.085	20′
35°50′	0.5854	0.8107	0.7221	1.385	10′	42°50′	0.6799	0.7333	0.9271	1.079	10′
36°00′	0.5878	0.8090	0.7265	1.376	54°00′	43°00′	0.6820	0.7314	0.9325	1.072	47°00′
10′	0.5901	0.8073	0.7310	1.368	53°50′	10′	0.6841	0.7294	0.9380	1.066	46°50′
20′	0.5925	0.8056	0.7355	1.360	40′	20′	0.6862	0.7274	0.9435	1.060	40′
30′	0.5948	0.8039	0.7400	1.351	30′	30′	0.6884	0.7254	0.9490	1.054	30′
40′	0.5972	0.8021	0.7445	1.343	20′	40′	0.6905	0.7234	0.9545	1.048	20′
36°50′	0.5995	0.8004	0.7490	1.335	10′	43°50′	0.6926	0.7214	0.9601	1.042	10′
37°00′	0.6018	0.7986	0.7536	1.327	53°00′	44°00′	0.6947	0.7193	0.9657	1.036	46°00′
10′	0.6041	0.7969	0.7581	1.319	52°50′	10′	0.6967	0.7173	0.9713	1.030	45°50′
20′	0.6065	0.7951	0.7627	1.311	40′	20′	0.6988	0.7153	0.9770	1.024	40′
30′	0.6088	0.7934	0.7673	1.303	30′	30′	0.7009	0.7133	0.9827	1.018	30′
40′	0.6111	0.7916	0.7720	1.295	20′	40′	0.7030	0.7112	0.9884	1.012	20′
37°50′	0.6134	0.7898	0.7766	1.288	10′	44°50′	0.7050	0.7092	0.9942	1.006	10′
38°00′	0.6157	0.7880	0.7813	1.280	52°00′	45°00′	0.7071	0.7071	1.000	1.000	45°00′
10′	0.6180	0.7862	0.7860	1.272	51°50′						
20′	0.6202	0.7844	0.7907	1.265	40′						
30′	0.6225	0.7826	0.7954	1.257	30′						
40′	0.6248	0.7808	0.8002	1.250	20′						
38°50′	0.6271	0.7790	0.8050	1.242	10′						
	cos	sin	cot	tan	θ		cos	sin	cot	tan	θ

TABLE IV

TRIGONOMETRIC FUNCTIONS (RADIANS)

t	sin	cos	tan	cot	t	sin	cos	tan	cot
0.00	0.0000	1.0000	0.0000	—	0.45	0.4350	0.9004	0.4831	2.070
0.01	0.0100	1.0000	0.0100	99.997	0.46	0.4439	0.8961	0.4954	2.018
0.02	0.0200	0.9998	0.0200	49.993	0.47	0.4529	0.8916	0.5080	1.969
0.03	0.0300	0.9996	0.0300	33.323	0.48	0.4618	0.8870	0.5206	1.921
0.04	0.0400	0.9992	0.0400	24.987	0.49	0.4706	0.8823	0.5334	1.875
0.05	0.0500	0.9988	0.0500	19.983	0.50	0.4794	0.8776	0.5463	1.830
0.06	0.0600	0.9982	0.0601	16.647	0.51	0.4882	0.8727	0.5594	1.788
0.07	0.0699	0.9976	0.0701	14.262	0.52	0.4969	0.8678	0.5726	1.747
0.08	0.0799	0.9968	0.0802	12.473	0.53	0.5055	0.8628	0.5859	1.707
0.09	0.0899	0.9960	0.0902	11.081	0.54	0.5141	0.8577	0.5994	1.668
0.10	0.0998	0.9950	0.1003	9.967	0.55	0.5227	0.8525	0.6131	1.631
0.11	0.1098	0.9940	0.1104	9.054	0.56	0.5312	0.8473	0.6269	1.595
0.12	0.1197	0.9928	0.1206	8.293	0.57	0.5396	0.8419	0.6410	1.560
0.13	0.1296	0.9916	0.1307	7.649	0.58	0.5480	0.8365	0.6552	1.526
0.14	0.1395	0.9902	0.1409	7.096	0.59	0.5564	0.8309	0.6696	1.494
0.15	0.1494	0.9888	0.1511	6.617	0.60	0.5646	0.8253	0.6841	1.462
0.16	0.1593	0.9872	0.1614	6.197	0.61	0.5729	0.8196	0.6989	1.431
0.17	0.1692	0.9856	0.1717	5.826	0.62	0.5810	0.8139	0.7139	1.401
0.18	0.1790	0.9838	0.1820	5.495	0.63	0.5891	0.8080	0.7291	1.372
0.19	0.1889	0.9820	0.1923	5.200	0.64	0.5972	0.8021	0.7445	1.343
0.20	0.1987	0.9801	0.2027	4.933	0.65	0.6052	0.7961	0.7602	1.315
0.21	0.2085	0.9870	0.2131	4.692	0.66	0.6131	0.7900	0.7761	1.288
0.22	0.2182	0.9759	0.2236	4.472	0.67	0.6210	0.7838	0.7923	1.262
0.23	0.2280	0.9737	0.2341	4.271	0.68	0.6288	0.7776	0.8087	1.237
0.24	0.2377	0.9713	0.2447	4.086	0.69	0.6365	0.7712	0.8253	1.212
0.25	0.2474	0.9689	0.2553	3.916	0.70	0.6442	0.7648	0.8423	1.187
0.26	0.2571	0.9664	0.2660	3.759	0.71	0.6518	0.7584	0.8595	1.163
0.27	0.2667	0.9638	0.2768	3.613	0.72	0.6594	0.7518	0.8771	1.140
0.28	0.2764	0.9611	0.2876	3.478	0.73	0.6669	0.7452	0.8949	1.117
0.29	0.2860	0.9582	0.2984	3.351	0.74	0.6743	0.7385	0.9131	1.095
0.30	0.2955	0.9553	0.3093	3.233	0.75	0.6816	0.7317	0.9316	1.073
0.31	0.3051	0.9523	0.3203	3.122	0.76	0.6889	0.7248	0.9505	1.052
0.32	0.3146	0.9492	0.3314	3.018	0.77	0.6961	0.7179	0.9697	1.031
0.33	0.3240	0.9460	0.3425	2.920	0.78	0.7033	0.7109	0.9893	1.011
0.34	0.3335	0.9428	0.3537	2.827	0.79	0.7104	0.7038	1.009	0.9908
0.35	0.3429	0.9394	0.3650	2.740	0.80	0.7174	0.6967	1.030	0.9712
0.36	0.3523	0.9359	0.3764	2.657	0.81	0.7243	0.6895	1.050	0.9520
0.37	0.3616	0.9323	0.3879	2.578	0.82	0.7311	0.6822	1.072	0.9331
0.38	0.3709	0.9287	0.3994	2.504	0.83	0.7379	0.6749	1.093	0.9146
0.39	0.3802	0.9249	0.4111	2.433	0.84	0.7446	0.6675	1.116	0.8964
0.40	0.3894	0.9211	0.4228	2.365	0.85	0.7513	0.6600	1.138	0.8785
0.41	0.3986	0.9171	0.4346	2.301	0.86	0.7578	0.6524	1.162	0.8609
0.42	0.4078	0.9131	0.4466	2.239	0.87	0.7643	0.6448	1.185	0.8437
0.43	0.4169	0.9090	0.4586	2.180	0.88	0.7707	0.6372	1.210	0.8267
0.44	0.4259	0.9048	0.4708	2.124	0.89	0.7771	0.6294	1.235	0.8100
t	sin	cos	tan	cot	t	sin	cos	tan	cot

TABLE IV *(continued)*

t	sin	cos	tan	cot	t	sin	cos	tan	cot
0.90	0.7833	0.6216	1.260	0.7936	1.25	0.9490	0.3153	3.010	0.3323
0.91	0.7895	0.6137	1.286	0.7774	1.26	0.9521	0.3058	3.113	0.3212
0.92	0.7956	0.6058	1.313	0.7615	1.27	0.9551	0.2963	3.224	0.3102
0.93	0.8016	0.5978	1.341	0.7458	1.28	0.9580	0.2867	3.341	0.2993
0.94	0.8076	0.5898	1.369	0.7303	1.29	0.9608	0.2771	3.467	0.2884
0.95	0.8134	0.5817	1.398	0.7151	1.30	0.9636	0.2675	3.602	0.2776
0.96	0.8192	0.5735	1.428	0.7001	1.31	0.9662	0.2579	3.747	0.2669
0.97	0.8249	0.5653	1.459	0.6853	1.32	0.9687	0.2482	3.903	0.2562
0.98	0.8305	0.5570	1.491	0.6707	1.33	0.9711	0.2385	4.072	0.2456
0.99	0.8360	0.5487	1.524	0.6563	1.34	0.9735	0.2288	4.256	0.2350
1.00	0.8415	0.5403	1.557	0.6421	1.35	0.9757	0.2190	4.455	0.2245
1.01	0.8468	0.5319	1.592	0.6281	1.36	0.9779	0.2092	4.673	0.2140
1.02	0.8521	0.5234	1.628	0.6142	1.37	0.9799	0.1994	4.913	0.2035
1.03	0.8573	0.5148	1.665	0.6005	1.38	0.9819	0.1896	5.177	0.1931
1.04	0.8624	0.5062	1.704	0.5870	1.39	0.9837	0.1798	5.471	0.1828
1.05	0.8674	0.4976	1.743	0.5736	1.40	0.9854	0.1700	5.798	0.1725
1.06	0.8724	0.4889	1.784	0.5604	1.41	0.9871	0.1601	6.165	0.1622
1.07	0.8772	0.4801	1.827	0.5473	1.42	0.9887	0.1502	6.581	0.1519
1.08	0.8820	0.4713	1.871	0.5344	1.43	0.9901	0.1403	7.055	0.1417
1.09	0.8866	0.4625	1.917	0.5216	1.44	0.9915	0.1304	7.602	0.1315
1.10	0.8912	0.4536	1.965	0.5090	1.45	0.9927	0.1205	8.238	0.1214
1.11	0.8957	0.4447	2.014	0.4964	1.46	0.9939	0.1106	8.989	0.1113
1.12	0.9001	0.4357	2.066	0.4840	1.47	0.9949	0.1006	9.887	0.1011
1.13	0.9044	0.4267	2.120	0.4718	1.48	0.9959	0.0907	10.983	0.0910
1.14	0.9086	0.4176	2.176	0.4596	1.49	0.9967	0.0807	12.350	0.0810
1.15	0.9128	0.4085	2.234	0.4475	1.50	0.9975	0.0707	14.101	0.0709
1.16	0.9168	0.3993	2.296	0.4356	1.51	0.9982	0.0608	16.428	0.0609
1.17	0.9208	0.3902	2.360	0.4237	1.52	0.9987	0.0508	19.670	0.0508
1.18	0.9246	0.3809	2.427	0.4120	1.53	0.9992	0.0408	24.498	0.0408
1.19	0.9284	0.3717	2.498	0.4003	1.54	0.9995	0.0308	32.461	0.0308
1.20	0.9320	0.3624	2.572	0.3888	1.55	0.9998	0.0208	48.078	0.0208
1.21	0.9356	0.3530	2.650	0.3773	1.56	0.9999	0.0108	92.620	0.0108
1.22	0.9391	0.3436	2.733	0.3659	1.57	1.0000	0.0008	1,255.8	0.0008
1.23	0.9425	0.3342	2.820	0.3546	1.58	1.0000	−0.0092	−108.65	−0.0092
1.24	0.9458	0.3248	2.912	0.3434	1.59	0.9998	−0.0192	−52.067	−0.0192
					1.60	0.9996	−0.0292	−34.233	−0.0292
t	sin	cos	tan	cot	t	sin	cos	tan	cot

Answers to Odd-Numbered Problems

CHAPTER 1

EXERCISE 1.1 (pages 11–12)

1. False **3.** True **5.** False **7.** True **9.** True
11. True **13.** True **15.**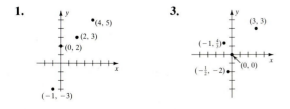

17. $x > 0$ **19.** $x + y \geq 0$ **21.** $b \geq 100$
23. $|t - 1| < 50$ **25.** $-3 < 15$ **27.** $1.33 < \frac{4}{3}$
29. $3.14 < \pi$ **31.** $-2 \geq -7$ **33.** $2.5 \geq \frac{5}{2}$ and $\frac{5}{2} \geq 2.5$
are both correct. **35.** $\frac{423}{157} \geq 2.6$
37. $(-\infty, 0)$; **39.** $[5, \infty)$;

41. $(8, 10]$; **43.** $[-2, 1)$;

45. 7 **47.** 22 **49.** $\frac{22}{7}$ **51.** $\sqrt{5}$ **53.** $4 - \pi$ **55.** 4
57. 4 **59.** $3 - \sqrt{5}$ **61.** $8 - \sqrt{7}$ **63.** $2.3 - \sqrt{5}$
65. $2\pi - 6.28$ **67.** $-h$ **69.** $2 - x$ **71.** $x - 2$
73. 0 **75.** 0 **77.** 1 **79.** $x \leq 0$ **83.** 4 **85.** 0.2
87. 3 **89.** 3 **91.** $-\frac{7}{3}$ **93.** $\frac{1}{3}$, 4
95. $2 - 2\sqrt{2}$, $2 + 2\sqrt{2}$ **97.** No real solutions
99. $0, -\dfrac{\sqrt{2}}{4}, \dfrac{\sqrt{2}}{4}$ **101.** $-\sqrt{15}, \sqrt{15}$ **103.** $0, -3, 3$

105. $0, -\dfrac{5}{2}, \dfrac{5}{2}$ **107. (a)** Larger;

(b) $[(\frac{16}{9})^2 - \pi]/\pi \approx 0.006 = 0.6\%$ **111.** Pythagoras lives.
113. No; $\sqrt{2}$ is irrational but $\sqrt{2}/\sqrt{2} = 1$ is a rational number.

EXERCISE 1.2 (pages 21–22)

1.

3.

5. II **7.** III **9.** II **11.** I **13.** III **15.** IV
17. **19.** **21.**

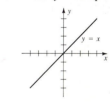

$xy > 0$

23. Symmetry with respect to the origin **25.** Symmetry with respect to the y-axis **27.** Symmetry with respect to the x- and y-axes and the origin **29.** Symmetry with respect to the origin
31. x-intercept: -3 **33.** x- and y-intercepts: 0

35. x-intercept: 1; y-intercept: -1 **37.** x- and y-intercepts: 0

39. x-intercepts: $-\sqrt{3}$, $\sqrt{3}$; y-intercept: -3 **41.** x-intercept: -1; y-intercept: 1

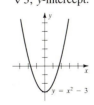
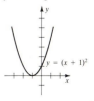

43. *x*-intercept: −9;
   y-intercepts: −3, 3

45. *x*- and *y*-intercepts: 0

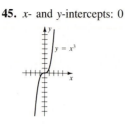

63. *x*- and *y*-intercepts: 0

65. *x*-intercepts: −4, 4;
   y-intercepts: −2, 2

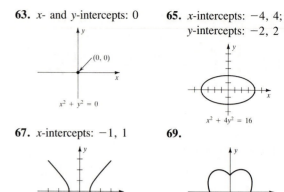

47. *x*- and *y*-intercepts: 0

49. *x*- and *y*-intercepts: 0

67. *x*-intercepts: −1, 1   **69.**

71.

51. *x*- and *y*-intercepts: 0

53. *x*- and *y*-intercepts: 0

55. *x*-intercepts: −2, 2;
   y-intercepts: −2, 2

57. *y*-intercepts: −3, 3

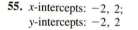

EXERCISE 1.3 (pages 26–27)

1. $2\sqrt{5}$ **3.** 10   **5.** 5   **7.** Not a right triangle   **9.** A right triangle   **11.** An isosceles triangle
13. (a) $2x + y - 5 = 0$. **(b)** The points (x, y) lie on the perpendicular bisector of the line segment joining A and B.
15. Through Des Moines   **17.** Center $(1, 3)$, $r = 7$
19. Center $(\frac{1}{2}, \frac{3}{2})$, $r = \sqrt{5}$ **21.** Center $(0, -4)$, $r = 4$
23. Center $(9, 3)$, $r = 10$ **25.** Center $(-1, -4)$, $r = \sqrt{22}$
27. Completing the square gives $x^2 + (y + 1)^2 = -8$. Since the sum of squares cannot be negative, this equation has no graph.   **29.** $x^2 + y^2 = 1$ **31.** $x^2 + (y - 2)^2 = 2$
33. $(x - 1)^2 + (y - 6)^2 = 8$ **35.** $x^2 + y^2 = 5$
37. $(x - 5)^2 + (y - 6)^2 = 36$
39.   **41.**   **43.**

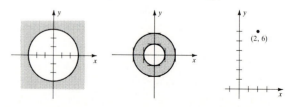

59. *x*-intercept: 1;
   y-intercept: −2

61. *x*-intercepts: 0, 1

EXERCISE 1.4 (pages 33–35)

1. Not a function   **3.** A function   **5.** A function
7. -1; 0; $x^2 + 2xh + h^2 - 1$; $4 + h$ **9.** 1; 2; 0; $\sqrt{6}$
11. 0; $\frac{3}{2}$; $\sqrt{2}$; $-\frac{3}{2}$

13. $3a^3 - a$; $3a^3 + 9a^2 + 8a + 2$; $3a^6 - a^2$; $\dfrac{3}{a^3} - \dfrac{1}{a}$; $-3a^3 + a$

15. $[-\frac{3}{2}, \infty)$ **17.** $\{x | x \neq \pm 2\}$ **19.** All real numbers
21. $\{x | x \geq -1 \text{ and } x \neq 0\}$ **23.** $[-1, 2)$ **25.** All real
numbers; all real numbers   **27.** All real numbers; all real
numbers   **29.** $[3, \infty)$; $[-1, \infty)$ **31.** $(-\infty, 16]$; $[0, \infty)$

33. $x = 20$ **35.** $-\frac{5}{2}, -1$ **37.** $P = 4\sqrt{A}$ **39.** $A = \dfrac{\sqrt{3}}{4}s^2$

41. $V = A^{3/2}$ **43.** $V = x(2 - 2x)(3 - 2x)$

45. $S = x^2 + \dfrac{64{,}000}{x}$ **47.** $F = 2x + \dfrac{9000}{x}$

49. $A = 10x - \left(\dfrac{1}{2} + \dfrac{\pi}{8}\right)x^2$

51. $d = \sqrt{2500t^2 + (4 - 30t)^2}$ **53.** $A = r - \pi r^2$

EXERCISE 1.5 (pages 44–45)

1. y-intercept: 3

3. x- and y-intercepts: 0

5. x-intercept: $-\frac{1}{2}$;
y-intercept: 1

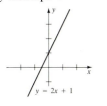

7. x- and y-intercepts: 0

9. x-intercept: 2

11. x- and y-intercepts: 0

13. x- and y-intercepts: 0

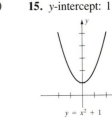

15. y-intercept: 1

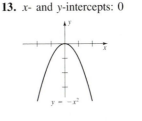

17. x-intercepts: 0, 2;
y-intercept: 0

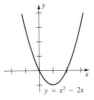

19. x-intercept: 2;
y-intercept: 4

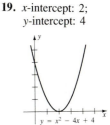

21. x- and y-intercepts: 0

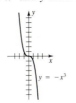

23. x-intercept: 1;
y-intercept: 1

25. x-intercepts: -3, 3;
y-intercept: -3

27. x-intercept: -2;
y-intercept: 2

29. No intercepts

31. No intercepts

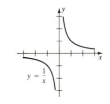

33. x- and y-intercepts: 0   **35.** x-intercepts: -1, 1

$y = x^4$

$y = \sqrt{x^2 - 1}$

37. x-intercepts: -9, 5; y-intercept: -45 **39.** x-intercepts: -1, 1; y-intercept: $1/\sqrt{2}$ **41.** x-intercept: $\frac{2}{3}$; no y-intercept
43. Not a function   **45.** Not a function   **47.** Not a function   **49.** Not a function   **51.** [1, 9]; [1, 6]
53. $(-\infty, \infty)$; $[-2, 2]$ **55.** -4, -3, -2, 0, 1, 2, 2, 3
57. Not one-to-one   **59.** Not one-to-one   **61.** One-to-one
63. (a)   **(b)**

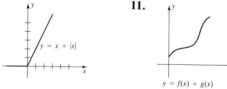

29. (a)   **(b)**

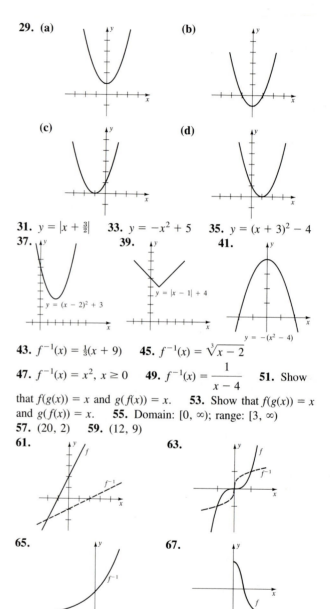

(c)   **(d)**

31. $y = |x + \frac{3}{2}|$ **33.** $y = -x^2 + 5$ **35.** $y = (x + 3)^2 - 4$
37.   **39.**   **41.**

$y = (x - 2)^2 + 3$ $y = |x - 1| + 4$

$y = -(x^2 - 4)$

43. $f^{-1}(x) = \frac{1}{3}(x + 9)$ **45.** $f^{-1}(x) = \sqrt[3]{x - 2}$
47. $f^{-1}(x) = x^2$, $x \geq 0$ **49.** $f^{-1}(x) = \dfrac{1}{x - 4}$ **51.** Show
that $f(g(x)) = x$ and $g(f(x)) = x$.   **53.** Show that $f(g(x)) = x$ and $g(f(x)) = x$.   **55.** Domain: [0, ∞); range: [3, ∞)
57. (20, 2)   **59.** (12, 9)
61.   **63.**

65.   **67.**

EXERCISE 1.6 (pages 53–55)

1. $(f + g)(x) = 3x^2 - x + 4$, domain: all real numbers; $(fg)(x) = 2x^4 - x^3 + 5x^2 - x + 3$, domain: all real numbers

3. $(f + g)(x) = 4x$, domain: (0, ∞); $(fg)(x) = 4x^2 - \dfrac{1}{x}$,

domain: (0, ∞)   **5.** $(fg)(x) = x^3 + 2x^2 - 4x - 8$, domain: all real numbers; $(f/g)(x) = x - 2$, domain: $\{x | x \neq -2\}$
7. $(f + g)(x) = \sqrt{1 - x} + \sqrt{x + 2}$, domain: $[-2, 1]$; $(fg)(x) = \sqrt{(1 - x)(x + 2)}$, domain: $[-2, 1]$
9.   **11.**

$y = x + |x|$

$y = f(x) + g(x)$

13. $(f \circ g)(x) = x$; $(g \circ f)(x) = |x|$ **15.** $(f \circ g)(x) = \dfrac{1}{2x^2 + 1}$;

$(g \circ f)(x) = \dfrac{4x^2 - 4x + 2}{4x^2 - 4x + 1}$ **17.** $(f \circ g)(x) = x$; $(g \circ f)(x) = x$

19. $(f \circ g)(x) = \dfrac{1}{x} + x^2$; $(g \circ f)(x) = \dfrac{x^2}{x^3 + 1}$

21. $(f \circ g)(x) = x + \sqrt{x - 1} + 1$; $(g \circ f)(x) = x + 1 + \sqrt{x}$
23. $[-1, \infty)$ **25.** $\{x | x \neq 0, x \neq 6\}$ **27.** $[0, \infty)$

CHAPTER 1
REVIEW EXERCISE (pages 56–57)

1. False   **3.** False   **5.** $|a - b|$ **7.** Third   **9.** (2, -3)
11. False   **13.** 2 and 4, respectively   **15.** True

17. True **19.** $(-7, 5)$ **21.** $-1.4 > -\sqrt{2}$
23. $\frac{2}{3} < 0.67$ **25.** $-\sqrt{8} + 3$ **27.** $x^2 + 5$
29. $-(t + 5)$ **31.** $\dfrac{3 + \sqrt{15}}{2}, \dfrac{3 - \sqrt{15}}{2}$
33. $0, -2\sqrt{2}, 2\sqrt{2}$ **35.** $-\sqrt{2}, \sqrt{2}, -\sqrt{6}, \sqrt{6}$
37. Right triangle **39.** $-5, 3$
41. **43.** **45.**

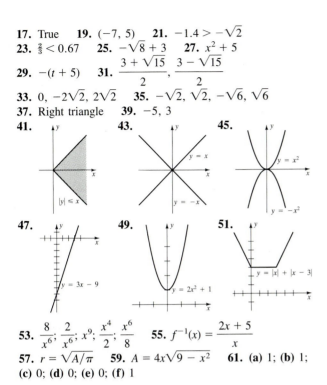

47. **49.** **51.**

53. $\dfrac{8}{x^6}; \dfrac{2}{x^6}; x^9; \dfrac{x^4}{2}; \dfrac{x^6}{8}$ **55.** $f^{-1}(x) = \dfrac{2x + 5}{x}$
57. $r = \sqrt{A/\pi}$ **59.** $A = 4x\sqrt{9 - x^2}$ **61. (a)** 1; **(b)** 1;
(c) 0; **(d)** 0; **(e)** 0; **(f)** 1

CHAPTER 2

EXERCISE 2.1 (pages 68–70)

1. **3.** **5.**

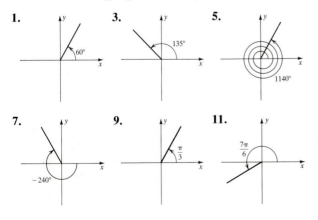

7. **9.** **11.**

13. **15.** **17.**

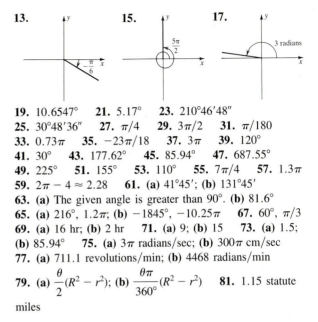

19. $10.6547°$ **21.** $5.17°$ **23.** $210°46'48''$
25. $30°48'36''$ **27.** $\pi/4$ **29.** $3\pi/2$ **31.** $\pi/180$
33. 0.73π **35.** $-23\pi/18$ **37.** 3π **39.** $120°$
41. $30°$ **43.** $177.62°$ **45.** $85.94°$ **47.** $687.55°$
49. $225°$ **51.** $155°$ **53.** $110°$ **55.** $7\pi/4$ **57.** 1.3π
59. $2\pi - 4 \approx 2.28$ **61. (a)** $41°45'$; **(b)** $131°45'$
63. (a) The given angle is greater than 90°. **(b)** $81.6°$
65. (a) $216°, 1.2\pi$; **(b)** $-1845°, -10.25\pi$ **67.** $60°, \pi/3$
69. (a) 16 hr; **(b)** 2 hr **71. (a)** 9; **(b)** 15 **73. (a)** 1.5;
(b) $85.94°$ **75. (a)** 3π radians/sec; **(b)** 300π cm/sec
77. (a) 711.1 revolutions/min; **(b)** 4468 radians/min
79. (a) $\dfrac{\theta}{2}(R^2 - r^2)$; **(b)** $\dfrac{\theta\pi}{360°}(R^2 - r^2)$ **81.** 1.15 statute
miles

EXERCISE 2.2 (pages 80–81)

1. $\sin\theta = \frac{4}{5}, \cos\theta = \frac{3}{5}, \tan\theta = \frac{4}{3}, \csc\theta = \frac{5}{4}, \sec\theta = \frac{5}{3},$
$\cot\theta = \frac{3}{4}$ **3.** $\sin\theta = \dfrac{3\sqrt{10}}{10}, \cos\theta = \dfrac{\sqrt{10}}{10}, \tan\theta = 3,$
$\csc\theta = \dfrac{\sqrt{10}}{3}, \sec\theta = \sqrt{10}, \cot\theta = \frac{1}{3}$ **5.** $\sin\theta = \frac{2}{5},$
$\cos\theta = \dfrac{\sqrt{21}}{5}, \tan\theta = \dfrac{2\sqrt{21}}{21}, \csc\theta = \frac{5}{2}, \sec\theta = \dfrac{5\sqrt{21}}{21},$
$\cot\theta = \dfrac{\sqrt{21}}{2}$ **7.** $\sin\theta = \frac{1}{3}, \cos\theta = \dfrac{2\sqrt{2}}{3}, \tan\theta = \dfrac{\sqrt{2}}{4},$
$\csc\theta = 3, \sec\theta = \dfrac{3\sqrt{2}}{4}, \cot\theta = 2\sqrt{2}$
9. $\sin\theta = \dfrac{y}{\sqrt{x^2 + y^2}}, \cos\theta = \dfrac{x}{\sqrt{x^2 + y^2}}, \tan\theta = \dfrac{y}{x},$
$\csc\theta = \dfrac{\sqrt{x^2 + y^2}}{y}, \sec\theta = \dfrac{\sqrt{x^2 + y^2}}{x}, \cot\theta = \dfrac{x}{y}$
11. $\tan\theta = \frac{2}{3}, \csc\theta = \dfrac{\sqrt{13}}{2}, \sec\theta = \dfrac{\sqrt{13}}{3}, \cot\theta = \frac{3}{2}$
13. $\tan\theta = \dfrac{2\sqrt{5}}{15}, \csc\theta = \frac{7}{2}, \sec\theta = \dfrac{7\sqrt{5}}{15}, \cot\theta = \dfrac{3\sqrt{5}}{2}$
15. $\cos\theta = \dfrac{8\sqrt{65}}{65}, \csc\theta = \sqrt{65}, \sec\theta = \dfrac{\sqrt{65}}{8}, \cot\theta = 8$
17. $\sin\theta = \frac{3}{5}, \cos\theta = \frac{4}{5}, \tan\theta = \frac{3}{4}, \cot\theta = \frac{4}{3}$

19. $\cos \theta = \frac{5}{13}$, $\tan \theta = \frac{12}{5}$, $\csc \theta = \frac{13}{12}$, $\sec \theta = \frac{13}{5}$, $\cot \theta = \frac{5}{12}$

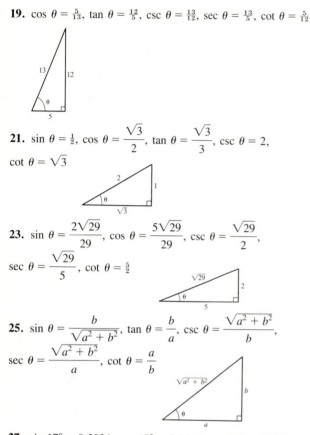

21. $\sin \theta = \frac{1}{2}$, $\cos \theta = \frac{\sqrt{3}}{2}$, $\tan \theta = \frac{\sqrt{3}}{3}$, $\csc \theta = 2$,

$\cot \theta = \sqrt{3}$

23. $\sin \theta = \frac{2\sqrt{29}}{29}$, $\cos \theta = \frac{5\sqrt{29}}{29}$, $\csc \theta = \frac{\sqrt{29}}{2}$,

$\sec \theta = \frac{\sqrt{29}}{5}$, $\cot \theta = \frac{5}{2}$

25. $\sin \theta = \frac{b}{\sqrt{a^2 + b^2}}$, $\tan \theta = \frac{b}{a}$, $\csc \theta = \frac{\sqrt{a^2 + b^2}}{b}$,

$\sec \theta = \frac{\sqrt{a^2 + b^2}}{a}$, $\cot \theta = \frac{a}{b}$

27. $\sin 17° = 0.2924$, $\cos 17° = 0.9563$, $\tan 17° = 0.3057$,
$\csc 17° = 3.4203$, $\sec 17° = 1.0457$, $\cot 17° = 3.2709$
29. $\sin 14.3° = 0.2470$, $\cos 14.3° = 0.9690$,
$\tan 14.3° = 0.2549$, $\csc 14.3° = 4.0486$, $\sec 14.3° = 1.0320$,

$\cot 14.3° = 3.9232$ **31.** $\sin \frac{\pi}{5} = 0.5878$, $\cos \frac{\pi}{5} = 0.8090$,

$\tan \frac{\pi}{5} = 0.7265$, $\csc \frac{\pi}{5} = 1.7013$, $\sec \frac{\pi}{5} = 1.2361$,

$\cot \frac{\pi}{5} = 1.3764$ **33.** $\sin 0.6725 = 0.6229$,

$\cos 0.6725 = 0.7823$, $\tan 0.6725 = 0.7963$,
$\csc 0.6725 = 1.6053$, $\sec 0.6725 = 1.2783$,
$\cot 0.6725 = 1.2558$ **35. (a)** 0.5539; **(b)** $31.74°$
37. (a) 1.1760; **(b)** $67.38°$ **39. (a)** 1.3694; **(b)** $78.46°$
41. (a) 0.3398; **(b)** $19.47°$ **43. (a)** 1.3052; **(b)** $74.78°$
45. True   **47.** True   **49.** True

51. $\dfrac{\cos \theta}{\sin \theta} = \dfrac{x/r}{y/r} = \dfrac{x}{y} = \cot \theta$

53. $\dfrac{1}{\cos \theta} = \dfrac{1}{x/r} = \dfrac{r}{x} = \sec \theta$

55. From Figure 33, we see that $\sin \theta = a/\sqrt{a^2 + b^2}$ and $\cos \theta = b/\sqrt{a^2 + b^2}$. Therefore,

$$(\sin \theta)^2 + (\cos \theta)^2 = \left(\frac{a}{\sqrt{a^2 + b^2}} \right)^2 + \left(\frac{b}{\sqrt{a^2 + b^2}} \right)^2$$

$$= \frac{a^2}{a^2 + b^2} + \frac{b^2}{a^2 + b^2}$$

$$= \frac{a^2 + b^2}{a^2 + b^2} = 1.$$

57. As shown in the figure below, let A be the point on the ground directly below the balloon. Let c_1 and c_2 be the distance from S_1 to A and from S_2 to A, respectively. Then $\cot \alpha = c_1/h$ and $\cot \beta = c_2/h$. It follows that

$$\cot \alpha + \cot \beta = \frac{c_1}{h} + \frac{c_2}{h} = \frac{c_1 + c_2}{h}.$$

Since $c_1 + c_2 = c$, we have

$$\cot \alpha + \cot \beta = \frac{c}{h} \text{ or } h = \frac{c}{\cot \alpha + \cot \beta}.$$

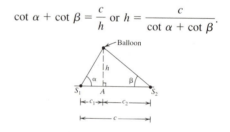

EXERCISE 2.3 (pages 89–92)

1. $b = 2.04$, $c = 4.49$ **3.** $a = 11.71$, $c = 14.18$
5. $b = 3.30$, $c = 6.85$ **7.** $\alpha = 60°$, $\beta = 30°$, $a = 2.60$
9. $\alpha = 21.8°$, $\beta = 68.2°$, $c = 10.77$
11. $\alpha = 48.58°$, $\beta = 41.42°$, $b = 7.94$
13. $a = 8.49$, $c = 21.73$ **15.** 52.1 m   **17.** 66.35 ft
19. 409.7 ft   **21.** Height: 15.54 ft; distance: 12.59 ft
23. $8.72°$ **25.** $34,158$ ft ≈ 6.47 mi   **27.** 13.47 ft
29. 3755.4 ft   **31.** Yes, since the altitude of the storm is 6.3 km.   **33.** Since R is the radius of the earth, we see from the figure below that $\cos \theta = r/R$ or $r = R \cos \theta$. From geometry, $C_e = 2\pi R$ and $C_\theta = 2\pi r$. Therefore,

$$\frac{C_\theta}{C_e} = \frac{2\pi r}{2\pi R}$$

or

$$C_\theta = \frac{C_e r}{R} = \frac{C_e(R \cos \theta)}{R} = C_e \cos \theta.$$

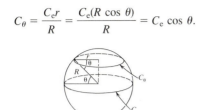

35. 20,911 km   **37.** 0.33 km

EXERCISE 2.4 (pages 102–105)

1. $\sin \theta = \frac{4}{5}$, $\cos \theta = \frac{3}{5}$, $\tan \theta = \frac{4}{3}$, $\csc \theta = \frac{5}{4}$, $\sec \theta = \frac{5}{3}$, $\cot \theta = \frac{3}{4}$ **3.** $\sin \theta = -\frac{12}{13}$, $\cos \theta = \frac{5}{13}$, $\tan \theta = -\frac{12}{5}$, $\csc \theta = -\frac{13}{12}$, $\sec \theta = \frac{13}{5}$, $\cot \theta = -\frac{5}{12}$ **5.** $\sin \theta = 1$, $\cos \theta = 0$, $\tan \theta$ is undefined, $\csc \theta = 1$, $\sec \theta$ is undefined, $\cot \theta = 0$ **7.** $\sin \theta = \dfrac{3\sqrt{13}}{13}$, $\cos \theta = -\dfrac{2\sqrt{13}}{13}$, $\tan \theta = -\frac{3}{2}$, $\csc \theta = \dfrac{\sqrt{13}}{3}$, $\sec \theta = -\dfrac{\sqrt{13}}{2}$, $\cot \theta = -\frac{2}{3}$

9. $\sin \theta = -\dfrac{\sqrt{3}}{3}$, $\cos \theta = -\dfrac{\sqrt{6}}{3}$, $\tan \theta = \dfrac{\sqrt{2}}{2}$, $\csc \theta = -\sqrt{3}$, $\sec \theta = -\dfrac{\sqrt{6}}{2}$, $\cot \theta = \sqrt{2}$ **11.** III

13. II   **15.** I   **17.** II   **19.** $\cos \theta = -\dfrac{\sqrt{15}}{4}$, $\tan \theta = -\dfrac{\sqrt{15}}{15}$, $\csc \theta = 4$, $\sec \theta = -\dfrac{4\sqrt{15}}{15}$, $\cot \theta = -\sqrt{15}$

21. $\sin \theta = -\dfrac{3\sqrt{10}}{10}$, $\cos \theta = -\dfrac{\sqrt{10}}{10}$, $\csc \theta = -\dfrac{\sqrt{10}}{3}$, $\sec \theta = -\sqrt{10}$, $\cot \theta = \frac{1}{3}$ **23.** $\sin \theta = -\frac{1}{10}$, $\cos \theta = \dfrac{3\sqrt{11}}{10}$, $\tan \theta = -\dfrac{\sqrt{11}}{33}$, $\sec \theta = \dfrac{10\sqrt{11}}{33}$, $\cot \theta = -3\sqrt{11}$ **25.** $\cos \theta = \dfrac{2\sqrt{6}}{5}$, $\tan \theta = -\dfrac{\sqrt{6}}{12}$, $\csc \theta = -5$, $\sec \theta = \dfrac{5\sqrt{6}}{12}$, $\cot \theta = -2\sqrt{6}$

27. $\sin \theta = \dfrac{8\sqrt{65}}{65}$, $\cos \theta = \dfrac{\sqrt{65}}{65}$, $\csc \theta = \dfrac{\sqrt{65}}{8}$, $\sec \theta = \sqrt{65}$, $\cot \theta = \frac{1}{8}$ **29.** $\pm\dfrac{\sqrt{91}}{10}$ **31.** $\sin \theta = \pm\dfrac{\sqrt{5}}{5}$, $\cos \theta = \pm\dfrac{2\sqrt{5}}{5}$ **33.** $\cos \theta = -\frac{1}{5}$, $\sin \theta = \pm\dfrac{2\sqrt{6}}{5}$

35.

θ (degrees)	θ (radians)	$\sin \theta$	$\cos \theta$	$\tan \theta$
0°	0	0	1	0
30°	$\dfrac{\pi}{6}$	$\dfrac{1}{2}$	$\dfrac{\sqrt{3}}{2}$	$\dfrac{\sqrt{3}}{3}$
45°	$\dfrac{\pi}{4}$	$\dfrac{\sqrt{2}}{2}$	$\dfrac{\sqrt{2}}{2}$	1
60°	$\dfrac{\pi}{3}$	$\dfrac{\sqrt{3}}{2}$	$\dfrac{1}{2}$	$\sqrt{3}$
90°	$\dfrac{\pi}{2}$	1	0	—
120°	$\dfrac{2\pi}{3}$	$\dfrac{\sqrt{3}}{2}$	$-\dfrac{1}{2}$	$-\sqrt{3}$
135°	$\dfrac{3\pi}{4}$	$\dfrac{\sqrt{2}}{2}$	$-\dfrac{\sqrt{2}}{2}$	-1
150°	$\dfrac{5\pi}{6}$	$\dfrac{1}{2}$	$-\dfrac{\sqrt{3}}{2}$	$-\dfrac{\sqrt{3}}{3}$
180°	π	0	-1	0
210°	$\dfrac{7\pi}{6}$	$-\dfrac{1}{2}$	$-\dfrac{\sqrt{3}}{2}$	$\dfrac{\sqrt{3}}{3}$
225°	$\dfrac{5\pi}{4}$	$-\dfrac{\sqrt{2}}{2}$	$-\dfrac{\sqrt{2}}{2}$	1
240°	$\dfrac{4\pi}{3}$	$-\dfrac{\sqrt{3}}{2}$	$-\dfrac{1}{2}$	$\sqrt{3}$
270°	$\dfrac{3\pi}{2}$	-1	0	—
300°	$\dfrac{5\pi}{3}$	$-\dfrac{\sqrt{3}}{2}$	$\dfrac{1}{2}$	$-\sqrt{3}$
315°	$\dfrac{7\pi}{4}$	$-\dfrac{\sqrt{2}}{2}$	$\dfrac{\sqrt{2}}{2}$	-1
330°	$\dfrac{11\pi}{6}$	$-\dfrac{1}{2}$	$\dfrac{\sqrt{3}}{2}$	$-\dfrac{\sqrt{3}}{3}$
360°	2π	0	1	0

37. -1 **39.** $\sqrt{3}$ **41.** $\sqrt{3}/2$ **43.** -2 **45.** -2 **47.** $\frac{1}{2}$ **49.** 1   **51.** Undefined   **53.** 60°, 240°   **55.** 135°, 225°   **57.** 270°   **59.** 0, π **61.** $3\pi/4$, $5\pi/4$ **63.** $5\pi/6$, $11\pi/6$ **71.** No, since $|\cos \theta| \leq 1$ for every angle θ in the domain of the cosine function.   **75.** 4.81 m

EXERCISE 2.5 (pages 112–114)

1. $\gamma = 80°$, $a = 20.16$, $c = 20.16$ **3.** $\alpha = 92°$, $b = 3.01$, $c = 3.89$ **5.** $\alpha = 79.6°$, $\gamma = 28.4°$, $a = 12.41$

7. No solution   **9.** $\alpha = 24.46°$, $\beta = 140.54°$, $b = 12.28$; $\alpha = 155.54°$, $\beta = 9.46°$, $b = 3.18$ **11.** No solution   **13.** $\alpha = 64°40'$, $a = 35.27$, $c = 38.17$ **15.** No solution   **17.** $\alpha = 45.58°$, $\gamma = 104.42°$, $c = 13.56$; $\alpha = 134.42°$, $\gamma = 15.58°$, $c = 3.76$ **19.** $\alpha = 70.5°$, $\beta = 29.5°$, $a = 7.66$ **21.** $\alpha = 26.3°$, $\gamma = 53.7°$, $c = 16.37$ **23.** No solution   **25.** $\alpha = 34.7°$, $\beta = 50.3°$, $b = 27.03$ **27.** 15.80 ft   **29.** 9.1 m   **31.** 8.81°   **33.** 11.82°   **35.** 290.3 m   **39.** If the lengths of the sides are a, b, and c, then a triangle will be determined if and only if $a + b > c$, $a + c > b$, and $b + c > a$.   **41. (a)** $b = \frac{5}{2}\sqrt{2}$; **(b)** $b < \frac{5}{2}\sqrt{2}$; **(c)** $\frac{5}{2}\sqrt{2} < b < 5$; **(d)** $b = \frac{5}{2}\sqrt{2}$ or $b \geq 5$ **43. (a)** $b \leq c$; **(b)** $b > c$

EXERCISE 2.6 (pages 120–122)

1. $\alpha = 37.59°$, $\beta = 77.41°$, $c = 7.43$ **3.** $\alpha = 52.62°$, $\beta = 83.33°$, $\gamma = 44.05°$ **5.** $\alpha = 36.87°$, $\beta = 53.13°$, $\gamma = 90°$ **7.** $\alpha = 76.45°$, $\beta = 57.1°$, $\gamma = 46.45°$ **9.** $\alpha = 27.66°$, $\beta = 40.54°$, $\gamma = 111.8°$ **11.** $\alpha = 25°$, $\beta = 57°40'$, $c = 7.04$ **13.** $\alpha = 26.38°$, $\beta = 36.34°$, $\gamma = 117.28°$ **15.** $\beta = 10.24°$, $\gamma = 147.76°$, $a = 6.32$ **17.** 35.94 nautical miles   **19. (a)** S 33.66°W; **(b)** S 2.82°E   **21.** $\alpha = 119.45°$, $\beta = 67.98°$ **23.** Front: 18.88°; back: 35.12°   **25.** 39.42 ft   **27.** 9.18 cm and 17.62 cm   **31.** 30   **33.** 18.94   **35.** 5188.67 ft²   **37.** From the Pythagorean theorem we have that $(2a)^2 = T^2 + (R + h)^2$. Solving for h, we obtain $h = \sqrt{4a^2 - T^2} - R^2$. By the law of cosines, $h^2 = 2a^2(1 - \cos\theta)$. Thus, $(\sqrt{4a^2 - T^2} - R^2)^2 = 2a^2(1 - \cos\theta)$. Squaring and solving for $\cos\theta$, we find $\cos\theta = (R/a)\sqrt{4 - (T/a)^2} + \frac{1}{2}[(T/a)^2 - (R/a)^2] - 1$.

CHAPTER 2
REVIEW EXERCISE (pages 123–127)

1. Clockwise   **3.** True   **5.** Tangent   **7.** Sines   **9.** Cosines

11.

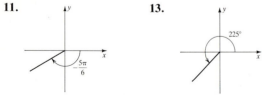

13.

15. $-2\pi/3$ **17.** 0.27π **19.** 20°   **21.** 131.78°   **23.** 70°30'   **25.** 177°37'1"   **27.** 445°, 805°, −275°,

−635° (*Note:* Other answers are possible.)   **29. (a)** 16 in.; **(b)** 1.6π in.   **31.** $\dfrac{790\pi}{3}$mi ≈ 827.29 mi   **33.** $\alpha = 36.87°$, $\beta = 53.13°$, $c = 50$ **35.** $\beta = 56°$, $a = 3.37$, $c = 6.03$ **37.** $\beta = 48°20'$, $a = 6.65$, $b = 7.47$ **39.** Angle at A: 33.69°; angle at B: 56.31°; angle at C: 90°   **41.** 42.61 m

43. 80.87°   **45.** $\sin\theta = \dfrac{2\sqrt{5}}{5}$, $\cos\theta = -\dfrac{\sqrt{5}}{5}$, $\tan\theta = -2$, $\csc\theta = \dfrac{\sqrt{5}}{2}$, $\sec\theta = -\sqrt{5}$, $\cot\theta = -\frac{1}{2}$

47. $\sin\theta = -\dfrac{3\sqrt{34}}{34}$, $\cos\theta = -\dfrac{5\sqrt{34}}{34}$, $\tan\theta = \frac{3}{5}$, $\csc\theta = -\dfrac{\sqrt{34}}{3}$, $\sec\theta = -\dfrac{\sqrt{34}}{5}$, $\cot\theta = \frac{5}{3}$

49. $\sin\theta = -\dfrac{4\sqrt{3}}{7}$, $\tan\theta = 4\sqrt{3}$, $\csc\theta = -\dfrac{7\sqrt{3}}{12}$, $\sec\theta = -7$, $\cot\theta = \dfrac{\sqrt{3}}{12}$ **51.** $\sin\theta = -\dfrac{\sqrt{26}}{26}$, $\cos\theta = \dfrac{5\sqrt{26}}{26}$, $\tan\theta = -\frac{1}{5}$, $\csc\theta = -\sqrt{26}$, $\sec\theta = \dfrac{\sqrt{26}}{5}$

53. $\sin\theta = -\frac{1}{7}$, $\cos\theta = -\dfrac{4\sqrt{3}}{7}$, $\tan\theta = \dfrac{\sqrt{3}}{12}$, $\csc\theta = -7$, $\sec\theta = -\dfrac{7\sqrt{3}}{12}$, $\cot\theta = 4\sqrt{3}$ **55.** $\sin\theta = \pm\dfrac{\sqrt{17}}{17}$, $\cos\theta = \mp\dfrac{4\sqrt{17}}{17}$, $\tan\theta = -\frac{1}{4}$, $\csc\theta = \pm\sqrt{17}$, $\sec\theta = \mp\dfrac{\sqrt{17}}{4}$ **57.** $\dfrac{\sqrt{2}}{2}$ **59.** −1   **61.** 45°, 135°

63. 120°, 240°   **65.** $\dfrac{\pi}{3}$, $\dfrac{2\pi}{3}$ **67.** $\dfrac{3\pi}{4}$, $\dfrac{7\pi}{4}$ **69.** $\gamma = 80°$, $a = 5.32$, $c = 10.48$ **71.** $\alpha = 81.09°$, $\gamma = 53.91°$, $a = 9.78$; $\alpha = 8.91°$, $\gamma = 126.09°$, $a = 1.53$ **73.** 796.28 ft   **75.** The vessel sailing at 5 knots arrives first in 0.22 hr; the vessel sailing at 10 knots arrives in 0.48 hr.   **77.** $a = 15.76$, $\beta = 99.44°$, $\gamma = 29.56°$ **79.** $\alpha = 36.34°$, $\beta = 117.28°$, $\gamma = 26.38°$ **81.** Angle at $(0, 0)$: 36.87°; angle at $(5, 0)$: 71.565°; angle at $(4, 3)$: 71.565°   **83.** 107.34 mi; 23.95°   **85. (a)** 42.35° **(b)** Let S represent the satellite and C the center of the earth, as shown in the figure below. Let PA be perpendicular to CS and let the distances d_1, d_2, and h be as labeled. From the right triangles SAP and PAC, we obtain $h = d_1 \tan\theta$ and

$h = R \sin \phi$. Thus, $d_1 \tan \theta = R \sin \phi$ or $\tan \theta = \dfrac{R \sin \phi}{d_1}$. It

remains to show that $d_1 = H + R(1 - \cos \phi)$. Since $d_1 + d_2 = H + R$, we have that $d_1 = H + R - d_2$. From right triangle PAC, $d_2 = R \cos \phi$. Thus, $d_1 = H + R - R \cos \phi = H + R(1 - \cos \phi)$ as desired.

CHAPTER 3

EXERCISE 3.1 (page 136)

1. (a)

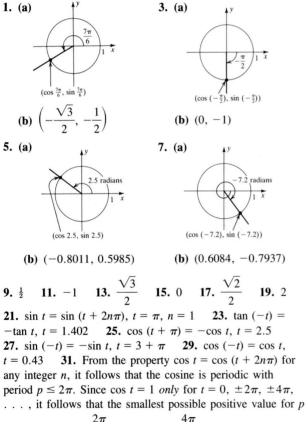

$\left(\cos \frac{7\pi}{6}, \sin \frac{7\pi}{6}\right)$

(b) $\left(-\dfrac{\sqrt{3}}{2}, -\dfrac{1}{2}\right)$

3. (a)

$\left(\cos \left(-\frac{\pi}{2}\right), \sin \left(-\frac{\pi}{2}\right)\right)$

(b) $(0, -1)$

5. (a)

2.5 radians

$(\cos 2.5, \sin 2.5)$

(b) $(-0.8011, 0.5985)$

7. (a)

-7.2 radians

$(\cos (-7.2), \sin (-7.2))$

(b) $(0.6084, -0.7937)$

9. $\frac{1}{2}$ **11.** -1 **13.** $\dfrac{\sqrt{3}}{2}$ **15.** 0 **17.** $\dfrac{\sqrt{2}}{2}$ **19.** 2

21. $\sin t = \sin (t + 2n\pi)$, $t = \pi$, $n = 1$ **23.** $\tan (-t) = -\tan t$, $t = 1.402$ **25.** $\cos (t + \pi) = -\cos t$, $t = 2.5$
27. $\sin (-t) = -\sin t$, $t = 3 + \pi$ **29.** $\cos (-t) = \cos t$, $t = 0.43$ **31.** From the property $\cos t = \cos (t + 2n\pi)$ for any integer n, it follows that the cosine is periodic with period $p \leq 2\pi$. Since $\cos t = 1$ *only* for $t = 0$, $\pm 2\pi$, $\pm 4\pi$, . . . , it follows that the smallest possible positive value for p is 2π. **39.** $t = \dfrac{2\pi}{3} + 2n\pi$ or $t = \dfrac{4\pi}{3} + 2n\pi$, $n = 0, \pm 1, \pm 2, \ldots$

EXERCISE 3.2 (pages 142–143)

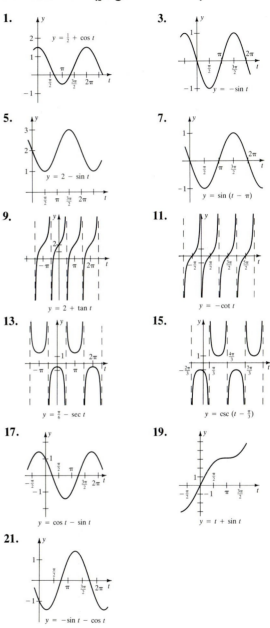

1. $y = \frac{1}{2} + \cos t$

3. $y = -\sin t$

5. $y = 2 - \sin t$

7. $y = \sin (t - \pi)$

9. $y = 2 + \tan t$

11. $y = -\cot t$

13. $y = \frac{\pi}{6} - \sec t$

15. $y = \csc (t - \frac{\pi}{3})$

17. $y = \cos t - \sin t$

19. $y = t + \sin t$

21. $y = -\sin t - \cos t$

23. $\tan 1.57 = 1255.77$, $\tan 1.58 = -108.65$. The great difference in these values is due to the fact that the tangent function has a vertical asymptote at $\pi/2$ and $1.57 < \pi/2 < 1.58$. **25.** No, because $|\csc t| \geq 1$ for all real numbers t.

27. (a) For t in the open intervals $(n\pi, m\pi)$, where $n = 2k$, $m = 2k + 1$, $k = 0, \pm 1, \pm 2, \ldots$, **(b)** For t in the open intervals $(n\pi, (n + \frac{1}{2})\pi)$ or $((n + \frac{1}{2})\pi, m\pi)$, where $n = 2k$, $m = 2k + 1$, $k = 0, \pm 1, \pm 2, \ldots$. **29.** Sine and cosine
33. (a) Odd; **(b)** even; **(c)** odd **39.** Domain: $\{t | t \neq \pi/2 + n\pi, n = 0, \pm 1, \pm 2, \ldots \}$; range: $\{y | |y| \geq 1\}$; asymptotes: $t = \pi/2 + n\pi, n = 0, \pm 1, \pm 2, \ldots$; period: 2π

EXERCISE 3.3 (pages 150–152)

1. (a) Amplitude 4, period 2π;
(b) amplitude $\frac{1}{4}$, period 2π

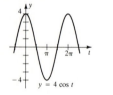

3. (a) Amplitude 1, period $\pi/2$;
(b) amplitude 1, period 8π

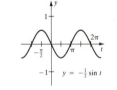

5. (a) Amplitude 1, period 1;
(b) amplitude 1, period π

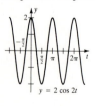

7. $y = 3 \sin 2t$ **9.** $y = \frac{1}{2} \cos \pi t$ **11.** $y = -\sin \pi t$
13. Amplitude: 4; period: 2π **15.** Amplitude: $\frac{1}{2}$; period: 2π

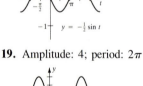

17. Amplitude: 2; period: π **19.** Amplitude: 4; period: 2π

21. Amplitude: 5; period: 1 **23.** Amplitude: 1; period: 3π

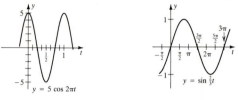

25. Amplitude: 3; period: π **27.**

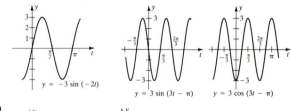

29.

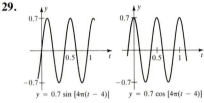

31. Amplitude: 1; period: 2π; phase shift: $\pi/6$

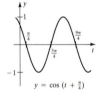

33. Amplitude: 1; period: 2π; phase shift: $-\pi/4$

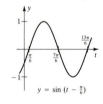

35. Amplitude: 4; period: π; phase shift: $3\pi/4$

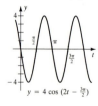

37. Amplitude: 3; period: 4π; phase shift: $2\pi/3$

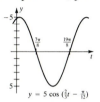

$y = 3 \sin\left(\frac{t}{2} - \frac{\pi}{3}\right)$

39. Amplitude: 5; period: 3π; phase shift: $\pi/8$

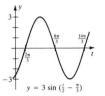

$y = 5 \cos\left(\frac{2}{3}t - \frac{\pi}{12}\right)$

41. Amplitude: 2; period: $\pi/2$; phase shift: $-\pi/3$

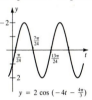

$y = 2 \cos\left(-4t - \frac{4\pi}{3}\right)$

43. Period: 1

45. Period: 4π

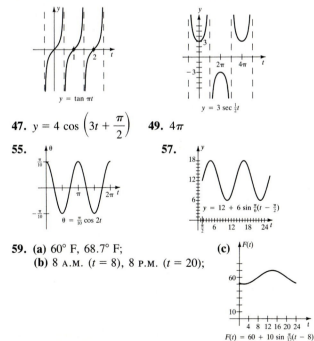

$y = \tan \pi t$

$y = 3 \sec \frac{1}{2}t$

47. $y = 4 \cos\left(3t + \dfrac{\pi}{2}\right)$ **49.** 4π

55.   **57.**

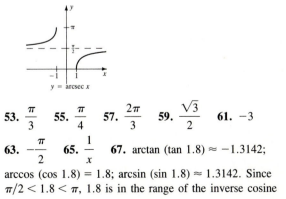

$\theta = \frac{\pi}{10} \cos 2t$

$y = 12 + 6 \sin \frac{\pi}{6}(t - \frac{\pi}{2})$

59. (a) 60° F, 68.7° F;
   (b) 8 A.M. ($t = 8$), 8 P.M. ($t = 20$);
   (c)

$F(t) = 60 + 10 \sin \frac{\pi}{12}(t - 8)$

(d) The maximum temperature of 70° F occurs at 2 P.M. ($t = 14$), and the minimum temperature of 50° F occurs at 2 a.m. ($t = 2$).

61.

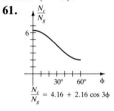

$\dfrac{N_c}{N_g} = 4.16 + 2.16 \cos 3\phi$

EXERCISE 3.4 (pages 161–163)

1. 0   **3.** π **5.** $\dfrac{\pi}{3}$ **7.** $-\dfrac{\pi}{3}$ **9.** $\dfrac{\pi}{4}$ **11.** $-\dfrac{\pi}{6}$

13. $-\dfrac{\pi}{4}$ **15.** $\frac{4}{5}$ **17.** $-\dfrac{\sqrt{5}}{2}$ **19.** $\dfrac{\sqrt{5}}{5}$ **21.** $\frac{5}{3}$

23. $\frac{1}{5}$ **25.** 1.2   **27.** $\dfrac{\pi}{16}$ **29.** 0   **31.** $\dfrac{\pi}{4}$

33. $\dfrac{x}{\sqrt{1+x^2}}$ **35.** $\dfrac{x}{\sqrt{1-x^2}}$ **37.** $\dfrac{\sqrt{1-x^2}}{x}$

39. $\dfrac{\sqrt{1+x^2}}{x}$ **41.** 0.7800   **43.** 1.400   **45.** 0.1307

47. 0.9001   **49.** Domain: $(-\infty, \infty)$; range: $(0, \pi)$

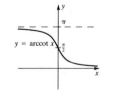

$y = \text{arccot } x$

51. Domain: $(-\infty, -1] \cup [1, \infty)$; range: $\left[0, \dfrac{\pi}{2}\right) \cup \left(\dfrac{\pi}{2}, \pi\right]$

$y = \text{arcsec } x$

53. $\dfrac{\pi}{3}$ **55.** $\dfrac{\pi}{4}$ **57.** $\dfrac{2\pi}{3}$ **59.** $\dfrac{\sqrt{3}}{2}$ **61.** -3

63. $-\dfrac{\pi}{2}$ **65.** $\dfrac{1}{x}$ **67.** arctan (tan 1.8) ≈ -1.3142;

arccos (cos 1.8) = 1.8; arcsin (sin 1.8) ≈ 1.3142. Since $\pi/2 < 1.8 < \pi$, 1.8 is in the range of the inverse cosine function, but 1.8 is not in the range of the inverse sine or the inverse tangent function.

69.   **71.**   **73.**

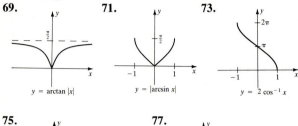

$y = \arctan |x|$ $y = |\arcsin x|$ $y = 2 \cos^{-1} x$

75.   **77.**

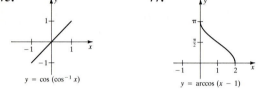

$y = \cos (\cos^{-1} x)$ $y = \arccos (x - 1)$

79. (a) All real numbers; **(b)** $0 < x < \pi$

81. Let $t = \operatorname{arccot} x$, where $0 < t < \pi$. Then

$x = \cot t = \tan \left(\dfrac{\pi}{2} - t \right)$, where $-\dfrac{\pi}{2} < \dfrac{\pi}{2} - t < \dfrac{\pi}{2}$. Hence

$\arctan x = \dfrac{\pi}{2} - t = \dfrac{\pi}{2} - \operatorname{arccot} t$. Therefore, $\operatorname{arccot} t =$

$\dfrac{\pi}{2} - \arctan t$.   **83.** Let $t = \arcsin \left(\dfrac{1}{x} \right)$, for $|x| \geq 1$. Then

$\sin t = \dfrac{1}{x}$ (where $-\dfrac{\pi}{2} \leq t \leq \dfrac{\pi}{2}$) and $\csc t = x$. It follows

that $\operatorname{arccsc} x = t = \arcsin \left(\dfrac{1}{x} \right)$.   **85.** 0.9273

87. -0.7297 **89.** 2.5559   **91.** 19.9°, 70.1°   **93.** 5.76°
95. $\phi \approx 0.5404$ radian $\approx 31°$

CHAPTER 3
REVIEW EXERCISE (pages 163–164)

1. True   **3.** False   **5.** True   **7.** True   **9.** True
11. Amplitude: 5; period: 2π **13.** Period: π

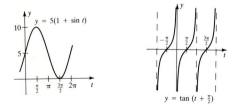

$y = 5(1 + \sin t)$ $y = \tan (t + \frac{\pi}{2})$

15. Period: 2π **17.**

$y = -\sec t$

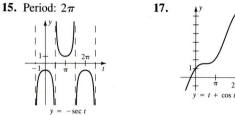

$y = t + \cos t$

19. Amplitude: 2; period: 8π

$y = -2 \cos \frac{1}{4} t$

21. Amplitude: 1; period: $\dfrac{2\pi}{3}$; phase shift: $-\dfrac{\pi}{3}$

$y = -\sin (3t + \pi)$

23. Period: $\dfrac{\pi}{2}$

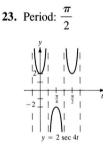

$y = 2 \sec 4t$

25. Amplitude: 10; period: $\dfrac{2\pi}{3}$; phase shift: $\dfrac{\pi}{6}$

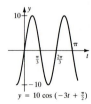

$y = 10 \cos (-3t + \frac{\pi}{2})$

27. $2\pi/3$ **29.** $\sqrt{7}/3$ **31.** 0   **33.** $\sqrt{1 - x^2}$

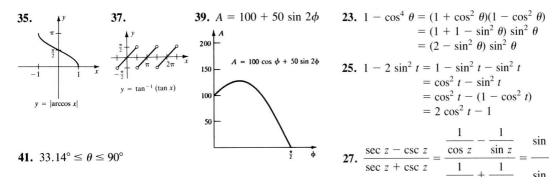

35. $y = |\arccos x|$

37. $y = \tan^{-1}(\tan x)$

39. $A = 100 + 50 \sin 2\phi$

$A = 100 \cos \phi + 50 \sin 2\phi$

41. $33.14° \leq \theta \leq 90°$

EXERCISE 4.1 (pages 171–172)

1. $\sec t \cos t = \dfrac{1}{\cos t} \cos t = 1$

3. $\dfrac{\sin \theta}{\csc \theta} + \dfrac{\cos \theta}{\sec \theta} = \dfrac{\sin \theta}{1/\sin \theta} + \dfrac{\cos \theta}{1/\cos \theta} = \sin^2 \theta + \cos^2 \theta = 1$

5. $\tan^2 t - \sec^2 t = \tan^2 t - (1 + \tan^2 t) = -1$

7. $\sin(-t) + \sin t = -\sin t + \sin t = 0$

9. $\sec(-x) \cos x = \sec x \cos x = \dfrac{1}{\cos x} \cdot \cos x = 1$

11. $\dfrac{\sin t + \sin t \cos t}{1 + \cos t} = \dfrac{\sin t (1 + \cos t)}{1 + \cos t} = \sin t$

13. $\dfrac{\sec^2 \alpha - 1}{\tan \alpha} = \dfrac{\tan^2 \alpha}{\tan \alpha} = \tan \alpha$

15. $\sin x + \cos x \cot x = \sin x + \cos x \left(\dfrac{\cos x}{\sin x}\right)$

$\qquad = \dfrac{\sin^2 x + \cos^2 x}{\sin x} = \dfrac{1}{\sin x} = \csc x$

17. $\dfrac{\sec^2 \alpha}{\cos \alpha + \cos \alpha \tan^2 \alpha} = \dfrac{\sec^2 \alpha}{\cos \alpha (1 + \tan^2 \alpha)}$

$\qquad = \dfrac{\sec^2 \alpha}{\cos \alpha (\sec^2 \alpha)} = \dfrac{1}{\cos \alpha} = \sec \alpha$

19. $\sin t \cos t \tan t \sec t \cot t$

$\qquad = \sin t \cos t \tan t \left(\dfrac{1}{\cos t}\right)\left(\dfrac{1}{\tan t}\right) = \sin t$

21. $\dfrac{\sin t}{\csc t} = \dfrac{\sin t}{\dfrac{1}{\sin t}} = \sin^2 t.$

$1 - \dfrac{\cos t}{\sec t} = 1 - \dfrac{\cos t}{\dfrac{1}{\cos t}} = 1 - \cos^2 t = \sin^2 t$

23. $1 - \cos^4 \theta = (1 + \cos^2 \theta)(1 - \cos^2 \theta)$

$\qquad = (1 + 1 - \sin^2 \theta) \sin^2 \theta$

$\qquad = (2 - \sin^2 \theta) \sin^2 \theta$

25. $1 - 2 \sin^2 t = 1 - \sin^2 t - \sin^2 t$

$\qquad = \cos^2 t - \sin^2 t$

$\qquad = \cos^2 t - (1 - \cos^2 t)$

$\qquad = 2 \cos^2 t - 1$

27. $\dfrac{\sec z - \csc z}{\sec z + \csc z} = \dfrac{\dfrac{1}{\cos z} - \dfrac{1}{\sin z}}{\dfrac{1}{\cos z} + \dfrac{1}{\sin z}} = \dfrac{\sin z \left(\dfrac{1}{\cos z} - \dfrac{1}{\sin z}\right)}{\sin z \left(\dfrac{1}{\cos z} + \dfrac{1}{\sin z}\right)}$

$\qquad = \dfrac{\tan z - 1}{\tan z + 1}$

29. $\dfrac{\sec^4 t - \tan^4 t}{1 + 2 \tan^2 t} = \dfrac{(\sec^2 t - \tan^2 t)(\sec^2 t + \tan^2 t)}{1 + \tan^2 t + \tan^2 t}$

$\qquad = \dfrac{1(\sec^2 t + \tan^2 t)}{\sec^2 t + \tan^2 t} = 1$

31. $\sin^2 x \cot^2 x + \cos^2 x \tan^2 x$

$\qquad = \sin^2 x \dfrac{\cos^2 x}{\sin^2 x} + \cos^2 x \dfrac{\sin^2 x}{\cos^2 x}$

$\qquad = \cos^2 x + \sin^2 x = 1$

33. $\sec t - \dfrac{\cos t}{1 + \sin t} = \dfrac{1}{\cos t} - \dfrac{\cos t}{1 + \sin t}$

$\qquad = \dfrac{1 + \sin t - \cos^2 t}{\cos t (1 + \sin t)}$

$\qquad = \dfrac{1 + \sin t - (1 - \sin^2 t)}{\cos t (1 + \sin t)}$

$\qquad = \dfrac{\sin t + \sin^2 t}{\cos t (1 + \sin t)}$

$\qquad = \dfrac{\sin t (1 + \sin t)}{\cos t (1 + \sin t)} = \tan t$

35. $\dfrac{\tan^2 \beta}{1 + \cos \beta} = \dfrac{\dfrac{\sin^2 \beta}{\cos^2 \beta}}{1 + \cos \beta} = \dfrac{\dfrac{1 - \cos^2 \beta}{\cos^2 \beta}}{1 + \cos \beta}$

$\qquad = \dfrac{(1 - \cos \beta)(1 + \cos \beta)}{\cos^2 \beta (1 + \cos \beta)} = \dfrac{1 - \cos \beta}{\cos^2 \beta},$

$\dfrac{\sec \beta - 1}{\cos \beta} = \dfrac{\dfrac{1}{\cos \beta} - 1}{\cos \beta}$

$\qquad = \dfrac{\dfrac{1 - \cos \beta}{\cos \beta}}{\cos \beta} = \dfrac{1 - \cos \beta}{\cos^2 \beta}$

37. $(\csc t - \cot t)^2 = \left(\dfrac{1}{\sin t} - \dfrac{\cos t}{\sin t}\right)^2 = \left(\dfrac{1 - \cos t}{\sin t}\right)^2$

$\qquad = \dfrac{(1 - \cos t)^2}{\sin^2 t} = \dfrac{(1 - \cos t)^2}{1 - \cos^2 t}$

$\qquad = \dfrac{(1 - \cos t)(1 - \cos t)}{(1 - \cos t)(1 + \cos t)} = \dfrac{1 - \cos t}{1 + \cos t}$

39. $\dfrac{\tan^2 x}{\sec x - 1} = \dfrac{\sec^2 x - 1}{\sec x - 1} = \dfrac{(\sec x - 1)(\sec x + 1)}{\sec x - 1}$

$\qquad = \sec x + 1 = 1 + \dfrac{1}{\cos x}$

41. $\dfrac{\cot t - \tan t}{\cot t + \tan t} = \dfrac{\dfrac{\cos t}{\sin t} - \dfrac{\sin t}{\cos t}}{\dfrac{\cos t}{\sin t} + \dfrac{\sin t}{\cos t}} = \dfrac{\cos^2 t - \sin^2 t}{\cos^2 t + \sin^2 t}$

$\qquad = 1 - \sin^2 t - \sin^2 t = 1 - 2 \sin^2 t$

43. $\cos(-t) \csc(-t) = \cos t \,(-\csc t)$

$\qquad\qquad = -\cos t \left(\dfrac{1}{\sin t}\right) = -\cot t$

45. $\sqrt{\dfrac{1 + \sin \theta}{1 - \sin \theta}} = \sqrt{\dfrac{1 + \sin \theta}{1 - \sin \theta} \cdot \dfrac{1 + \sin \theta}{1 + \sin \theta}}$

$\qquad = \sqrt{\dfrac{(1 + \sin \theta)^2}{1 - \sin^2 \theta}}$

$\qquad = \sqrt{\dfrac{(1 + \sin \theta)^2}{\cos^2 \theta}} = \dfrac{\sqrt{(1 + \sin \theta)^2}}{\sqrt{\cos^2 \theta}}$

$\qquad = \dfrac{1 + \sin \theta}{|\cos \theta|}$

47. $\sqrt{\dfrac{\sec t + \tan t}{\sec t - \tan t}} = \sqrt{\dfrac{\sec t + \tan t}{\sec t - \tan t} \cdot \dfrac{\sec t + \tan t}{\sec t + \tan t}}$

$\qquad = \sqrt{\dfrac{(\sec t + \tan t)^2}{\sec^2 t - \tan^2 t}} = \sqrt{\dfrac{(\sec t + \tan t)^2}{1}}$

$\qquad = |\sec t + \tan t| = \left|\dfrac{1}{\cos t} + \dfrac{\sin t}{\cos t}\right|$

$\qquad = \left|\dfrac{1 + \sin t}{\cos t}\right|$

49. $\sin^4 x + \cos^4 x = \sin^4 x + 2 \sin^2 x \cos^2 x + \cos^4 x -$
$2 \sin^2 x \cos^2 x = (\sin^2 x + \cos^2 x)^2 - 2 \sin^2 x \cos^2 x =$
$(1)^2 - 2 \sin^2 x \cos^2 x = 1 - 2 \sin^2 x \cos^2 x$

51. $\left(\dfrac{\sin^2 \theta}{\cot^4 \theta}\right)^4 \cdot \left(\dfrac{\csc \theta}{\tan^2 \theta}\right)^8 = \dfrac{\sin^8 \theta}{\cot^{16} \theta} \cdot \dfrac{\csc^8 \theta}{\tan^{16} \theta}$

$\qquad = \dfrac{\sin^8 \theta \,(1/\sin^8 \theta)}{(1/\tan^{16} \theta)(\tan^{16} \theta)} = 1$

53. $(\tan^2 t + 1)(\cos^2 t - 1) = (\sec^2 t)(-\sin^2 t)$

$\qquad = -\dfrac{\sin^2 t}{\cos^2 t} = -\tan^2 t = 1 - \sec^2 t$

55. $(1 - \tan \beta)^2(1 + \tan \beta)^2 + 4 \tan^2 \beta$

$\qquad = (1 - \tan^2 \beta)^2 + 4 \tan^2 \beta$

$\qquad = 1 - 2 \tan^2 \beta + \tan^4 \beta + 4 \tan^2 \beta$

$\qquad = 1 + 2 \tan^2 \beta + \tan^4 \beta$

$\qquad = (1 + \tan^2 \beta)^2 = (\sec^2 \beta)^2 = \sec^4 \beta$

57. $\dfrac{\sin \theta}{1 - \cot \theta} + \dfrac{\cos \theta}{1 - \tan \theta} = \dfrac{\sin \theta}{1 - \dfrac{\cos \theta}{\sin \theta}} + \dfrac{\cos \theta}{1 - \dfrac{\sin \theta}{\cos \theta}}$

$\qquad = \dfrac{\sin \theta}{\dfrac{\sin \theta - \cos \theta}{\sin \theta}} + \dfrac{\cos \theta}{\dfrac{\cos \theta - \sin \theta}{\cos \theta}}$

$\qquad = \dfrac{\sin^2 \theta}{\sin \theta - \cos \theta} + \dfrac{\cos^2 \theta}{\cos \theta - \sin \theta}$

$\qquad = \dfrac{\sin^2 \theta}{\sin \theta - \cos \theta} - \dfrac{\cos^2 \theta}{\sin \theta - \cos \theta}$

$\qquad = \dfrac{\sin^2 \theta - \cos^2 \theta}{\sin \theta - \cos \theta}$

$\qquad = \dfrac{(\sin \theta - \cos \theta)(\sin \theta + \cos \theta)}{\sin \theta - \cos \theta}$

$\qquad = \cos \theta + \sin \theta$

59. $\csc^4 t - \csc^2 t = \csc^2 t \,(\csc^2 t - 1) = \csc^2 t \cot^2 t$

$\qquad = (1 + \cot^2 t) \cot^2 t = \cot^2 t + \cot^4 t$

61. For $t = -\pi/6$, $\sin t = \sin(-\pi/6)$

$\qquad = -\frac{1}{2}$; but $\sqrt{1 - \cos^2 t} = \sqrt{1 - \cos^2(-\pi/6)}$

$\qquad = \sqrt{1 - (\sqrt{3}/2)^2} = \sqrt{1 - \frac{3}{4}} = \sqrt{\frac{1}{4}} = \frac{1}{2}$.

63. For $\theta = \pi/3$, $1 + \sec^2 \theta = 1 + \sec^2(\pi/3) = 1 + (2)^2 = 5$;
but $\tan^2 \theta = \tan^2(\pi/3) = (\sqrt{3})^2 = 3$. **65.** For $t = \pi/4$,
$\cot^2 t + \csc^2 t = \cot^2(\pi/4) + \csc^2(\pi/4) = (1)^2 + (\sqrt{2})^2 = 3 \neq 1$.

67. For $\theta = \pi/2$, $\sin(\csc \theta) = \sin\left(\csc \dfrac{\pi}{2}\right) = \sin 1 \approx$

$0.84 \neq 1$. **69.** For $x = \pi$, $\cos(-x) = \cos(-\pi) = -1$;
but $-\cos \pi = -(-1) = 1$.

EXERCISE 4.2 (pages 181–183)

1. $\dfrac{\sqrt{2}}{4}(1 + \sqrt{3})$ **3.** $\dfrac{\sqrt{2}}{4}(1 + \sqrt{3})$ **5.** $\dfrac{\sqrt{2}}{4}(1 + \sqrt{3})$

7. $2 + \sqrt{3}$ **9.** $\dfrac{\sqrt{2}}{4}(1 - \sqrt{3})$ **11.** $\dfrac{\sqrt{2}}{4}(\sqrt{3} - 1)$

13. $-\dfrac{\sqrt{2}}{4}(1 + \sqrt{3})$ **15.** $-2 + \sqrt{3}$ **17.** $\dfrac{\sqrt{2}}{4}(1 - \sqrt{3})$

19. $\dfrac{\sqrt{2}}{4}(\sqrt{3} + 1)$ **21.** $-\dfrac{\sqrt{2}}{4}(1 + \sqrt{3})$

23. $-\dfrac{\sqrt{2}}{4}(1 + \sqrt{3})$

25. $\cos(t + 2\pi) = \cos t \cos 2\pi - \sin t \sin 2\pi$
$$= (\cos t)(1) - (\sin t)(0) = \cos t$$

27. $\sin\left(t + \dfrac{\pi}{2}\right) = \sin t \cos \dfrac{\pi}{2} + \cos t \sin \dfrac{\pi}{2}$
$$= (\sin t)(0) + (\cos t)(1) = \cos t$$

29. $\tan\left(t + \dfrac{\pi}{2}\right) = \dfrac{\sin\left(t + \dfrac{\pi}{2}\right)}{\cos\left(t + \dfrac{\pi}{2}\right)} = \dfrac{\sin t \cos \dfrac{\pi}{2} + \cos t \sin \dfrac{\pi}{2}}{\cos t \cos \dfrac{\pi}{2} - \sin t \sin \dfrac{\pi}{2}}$

$$= \dfrac{\cos t}{-\sin t} = -\cot t$$

31. $\sin(t + \pi) = \sin t \cos \pi + \cos t \sin \pi$
$$= (\sin t)(-1) + (\cos t)(0) = -\sin t$$

33. $\tan(\pi - t) = \dfrac{\tan \pi - \tan t}{1 + \tan \pi \tan t} = \dfrac{0 - \tan t}{1 + (0)(\tan t)} = -\tan t$

35. $\sin\left(t + \dfrac{\pi}{4}\right) = \sin t \cos \dfrac{\pi}{4} + \cos t \sin \dfrac{\pi}{4}$

$$= (\sin t)\left(\dfrac{\sqrt{2}}{2}\right) + (\cos t)\left(\dfrac{\sqrt{2}}{2}\right)$$

$$= \dfrac{\sqrt{2}}{2}(\sin t + \cos t)$$

37. $\cot(t + \pi) = \dfrac{\cos(t + \pi)}{\sin(t + \pi)} = \dfrac{\cos t \cos \pi - \sin t \sin \pi}{\sin t \cos \pi + \cos t \sin \pi}$

$$= \dfrac{(\cos t)(-1) - (\sin t)(0)}{(\sin t)(-1) + (\cos t)(0)}$$

$$= \dfrac{\cos t}{\sin t} = \cot t$$

39. $\dfrac{\cos(x + h) - \cos x}{h} = \dfrac{\cos x \cos h - \sin x \sin h - \cos x}{h}$

$$= \dfrac{\cos x(\cos h - 1) - \sin x \sin h}{h}$$

$$= \cos x\left(\dfrac{\cos h - 1}{h}\right) - \sin x\left(\dfrac{\sin h}{h}\right)$$

41. Applying the addition formula yields $\dfrac{\sqrt{2}}{4}(\sqrt{3} - 1)$.

Using $\cos^2 t + \sin^2 t = 1$ with the results of Problem 5 yields $\frac{1}{4}\sqrt{8 - 4\sqrt{3}}$. To show that these are equal, we observe that

$$\left(\dfrac{1}{4}\sqrt{8 - 4\sqrt{3}}\right)\left(\dfrac{\sqrt{2}}{\sqrt{2}}\right) = \dfrac{\sqrt{2}}{4}\sqrt{\dfrac{8 - 4\sqrt{3}}{2}}$$

$$= \dfrac{\sqrt{2}}{4}\sqrt{4 - 2\sqrt{3}}$$

$$= \dfrac{\sqrt{2}}{4}\sqrt{3 - 2\sqrt{3} + 1}$$

$$= \dfrac{\sqrt{2}}{4}\sqrt{(\sqrt{3} - 1)^2}$$

$$= \dfrac{\sqrt{2}}{4}(\sqrt{3} - 1).$$

43. (a) $-2(\sqrt{10} + 1)/9$; (b) $(\sqrt{5} - 4\sqrt{2})/9$; (c) $2(1 - \sqrt{10})/9$; (d) $(\sqrt{5} + 4\sqrt{2})/9$

45. $y = \sqrt{2} \sin\left(\pi t + \dfrac{3\pi}{4}\right)$; amplitude: $\sqrt{2}$; period: 2; phase shift: $-\frac{3}{4}$

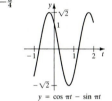

$y = \cos \pi t - \sin \pi t$

47. $y = 2 \sin\left(2t + \dfrac{\pi}{6}\right)$; amplitude: 2; period: π; phase shift: $-\dfrac{\pi}{12}$

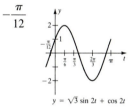

$y = \sqrt{3} \sin 2t + \cos 2t$

51. $a = \sqrt{17}/6$, $\phi \approx 1.8158$ radians

53.

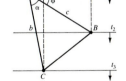

(a) Let BP be perpendicular to AP and let CQ be perpendicular to AQ, as shown in the figure. Let v be the velocity of the shock wave. Then the length of BP is $v(t_2 - t_1)$ and the length of CQ is $v(t_3 - t_1)$. From the right triangles APB and AQC, we obtain $\sin \phi = v(t_2 - t_1)/c$ and $\sin(\phi + \alpha) = v(t_3 - t_1)/b$. It follows that

$$R = \dfrac{t_3 - t_1}{t_2 - t_1} = \dfrac{\dfrac{b}{v}\sin(\phi + \alpha)}{\dfrac{c}{v}\sin \phi} = \dfrac{b \sin(\phi + \alpha)}{c \sin \phi}.$$

(b) Using the addition formula for the sine, we find

$$R = \frac{b \sin(\phi + \alpha)}{c \sin \phi} = \frac{b(\sin \phi \cos \alpha + \cos \phi \sin \alpha)}{c \sin \phi}$$

$$= \frac{b}{c}(\cos \alpha + \cot \phi \sin \alpha).$$

Thus, $\cot \phi = \left(\dfrac{c}{b}R - \cos \alpha\right)/\sin \alpha$. From the law of sines,

$\dfrac{\sin \beta}{b} = \dfrac{\sin \gamma}{c}$ or $\dfrac{c}{b} = \dfrac{\sin \gamma}{\sin \beta}$. It follows that

$$\cot \phi = \left(R\frac{\sin \gamma}{\sin \beta} - \cos \alpha\right)/\sin \alpha = \frac{R \sin \gamma}{\sin \alpha \sin \beta} - \cot \alpha.$$

55. You get an error message because the given expression equals $\tan(35° + 55°) = \tan 90°$, which is undefined.

EXERCISE 4.3 (pages 189–191)

1. $\sin 2\beta$ **3.** $\cos \dfrac{2\pi}{5}$ **5.** $\tfrac{1}{2}\tan 6t$ **7.** **(a)** $\tfrac{5}{9}$;

(b) $-\dfrac{2\sqrt{14}}{9}$; **(c)** $-\dfrac{2\sqrt{14}}{5}$ **9.** **(a)** $\tfrac{3}{5}$; **(b)** $\tfrac{4}{5}$; **(c)** $\tfrac{4}{3}$

11. **(a)** $-\tfrac{119}{169}$; **(b)** $-\tfrac{120}{169}$; **(c)** $\tfrac{120}{119}$ **13.** $\tfrac{1}{2}\sqrt{2 + \sqrt{3}}$

15. $\tfrac{1}{2}\sqrt{2 + \sqrt{2}}$ **17.** $\tfrac{1}{2}\sqrt{2 - \sqrt{2}}$ **19.** $-2 - \sqrt{3}$

21. $-2/\sqrt{2 - \sqrt{3}} = -2\sqrt{2 + \sqrt{3}}$ **23.** **(a)** $\dfrac{2\sqrt{13}}{13}$;

(b) $\dfrac{3\sqrt{13}}{13}$; **(c)** $\tfrac{3}{2}$ **25.** **(a)** $-\sqrt{(5 - \sqrt{5})/10}$;

(b) $\sqrt{(5 + \sqrt{5})/10}$; **(c)** $-(1 + \sqrt{5})/2$ **27.** **(a)** $\dfrac{\sqrt{30}}{6}$;

(b) $\dfrac{\sqrt{6}}{6}$; **(c)** $\dfrac{\sqrt{5}}{5}$

29. $\sin 4u = 2 \sin 2u \cos 2u = 2(2 \sin u \cos u)(1 - 2 \sin^2 u)$
$= 4 \sin u \cos u - 8 \cos u \sin^3 u$
$= 4 \cos u (\sin u - 2 \sin^3 u)$

31. $(\sin t + \cos t)^2 = \sin^2 t + 2 \sin t \cos t + \cos^2 t$
$= 1 + 2 \sin t \cos t = 1 + \sin 2t$

33. $\cot 2\theta = \dfrac{1}{\tan 2\theta} = \dfrac{1}{\dfrac{2 \tan \theta}{1 - \tan^2 \theta}} = \dfrac{1 - \tan^2 \theta}{2 \tan \theta}$

$= \dfrac{1}{2}\left(\dfrac{1}{\tan \theta} - \tan \theta\right) = \dfrac{1}{2}(\cot \theta - \tan \theta)$

35. $\tan \dfrac{\theta}{2} = \dfrac{1 - \cos \theta}{\sin \theta} = \dfrac{1}{\sin \theta} - \dfrac{\cos \theta}{\sin \theta} = \csc \theta - \cot \theta$

37. $\dfrac{2 \tan x}{1 + \tan^2 x} = \dfrac{2 \tan x}{\sec^2 x} = 2 \tan x \cos^2 x$

$= 2\left(\dfrac{\sin x}{\cos x}\right)(\cos^2 x)$

$= 2 \sin x \cos x = \sin 2x$

39. $\dfrac{\csc^2 z - 2}{2 \cot z} = \dfrac{\cot^2 z - 1}{2 \cot z} = \dfrac{\dfrac{1}{\tan^2 z} - 1}{2\dfrac{1}{\tan z}} = \dfrac{\dfrac{1 - \tan^2 z}{\tan^2 z}}{\dfrac{2}{\tan z}}$

$= \left(\dfrac{1 - \tan^2 z}{\tan^2 z}\right)\left(\dfrac{\tan z}{2}\right)$

$= \dfrac{1 - \tan^2 z}{2 \tan z} = \dfrac{1}{\tan 2z} = \cot 2z$

41. $\tan 4t = \dfrac{2 \tan 2t}{1 - \tan^2 2t} = \dfrac{2\left(\dfrac{2 \tan t}{1 - \tan^2 t}\right)}{1 - \left(\dfrac{2 \tan t}{1 - \tan^2 t}\right)^2}$

$= \dfrac{\dfrac{4 \tan t}{1 - \tan^2 t}}{\dfrac{(1 - \tan^2 t)^2 - 4 \tan^2 t}{(1 - \tan^2 t)^2}}$

$= \left(\dfrac{4 \tan t}{1 - \tan^2 t}\right)\left[\dfrac{(1 - \tan^2 t)^2}{(1 - \tan^2 t)^2 - 4 \tan^2 t}\right]$

$= \dfrac{4 \tan t (1 - \tan^2 t)}{(1 - \tan^2 t)^2 - 4 \tan^2 t}$

$= \dfrac{4 \tan t - 4 \tan^3 t}{1 - 6 \tan^2 t + \tan^4 t}$

45. Amplitude: 2; period: π **47.** Amplitude: 1; period; $\dfrac{\pi}{2}$

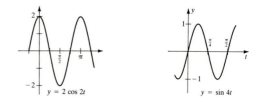

$y = 2 \cos 2t$ $y = \sin 4t$

49. **(a)** $d = 4 \cos 8t$; **(b)** amplitude: 4, period: $\pi/4$

51. $2\sqrt{2 + \sqrt{3}}$

53. From $\cos 2\phi = \dfrac{gh}{v_0^2 + gh}$ and the half-angle formulas, we find that

$$\sin^2 \phi = \frac{1}{2}(1 - \cos 2\phi) = \frac{1}{2}\left(1 - \frac{gh}{v_0^2 + gh}\right) = \frac{v_0^2}{2(v_0^2 + gh)}$$

and

$$\cos^2 \phi = \frac{1}{2}(1 + \cos 2\phi) = \frac{1}{2}\left(1 + \frac{gh}{v_0^2 + gh}\right) = \frac{v_0^2 + 2gh}{2(v_0^2 + gh)}.$$

Thus,

$$\sin \phi = \frac{v_0}{\sqrt{2(v_0^2 + gh)}} \text{ and } \cos \phi = \sqrt{\frac{v_0^2 + 2gh}{2(v_0^2 + gh)}}.$$

Substituting these values into the given expression for R and simplifying, we obtain

$$R = \frac{v_0^2 \cos \phi}{g}\left(\sin \phi + \sqrt{\sin^2 \phi + \frac{2gh}{v_0^2}}\right)$$

$$= \frac{v_0^2}{g}\sqrt{\frac{v_0^2 + 2gh}{2(v_0^2 + gh)}}\left(\frac{v_0}{\sqrt{2(v_0^2 + gh)}} + \sqrt{\frac{v_0^2}{2(v_0^2 + gh)} + \frac{2gh}{v_0^2}}\right)$$

$$= \frac{v_0^2}{g}\sqrt{\frac{v_0^2 + 2gh}{2(v_0^2 + gh)}}\left(\frac{v_0}{\sqrt{2(v_0^2 + gh)}} + \sqrt{\frac{v_0^4 + 4v_0^2gh + 4g^2h^2}{2(v_0^2 + gh)v_0^2}}\right)$$

$$= \frac{v_0^2}{g}\sqrt{\frac{v_0^2 + 2gh}{2(v_0^2 + gh)}}\left(\frac{v_0}{\sqrt{2(v_0^2 + gh)}} + \sqrt{\frac{(v_0^2 + 2gh)^2}{2(v_0^2 + gh)v_0^2}}\right)$$

$$= \frac{v_0^2}{g}\sqrt{\frac{v_0^2 + 2gh}{2(v_0^2 + gh)}}\left(\frac{v_0}{\sqrt{2(v_0^2 + gh)}} + \frac{v_0^2 + 2gh}{v_0\sqrt{2(v_0^2 + gh)}}\right)$$

$$= \frac{v_0^2}{g}\sqrt{\frac{v_0^2 + 2gh}{2(v_0^2 + gh)}}\left(\frac{2(v_0^2 + gh)}{v_0\sqrt{2(v_0^2 + gh)}}\right)$$

$$= \frac{v_0}{g}\sqrt{v_0^2 + 2gh}.$$

EXERCISE 4.4 (pages 195–196)

1. $\dfrac{2 - \sqrt{3}}{4}$ **3.** $\frac{1}{4}$ **5.** $\frac{1}{2}(\sin 6x - \sin 2x)$

7. $\frac{1}{2}(\sin 8\theta - \sin 2\theta)$ **9.** $\dfrac{1}{2}\left(\sin \dfrac{5x}{3} + \sin x\right)$

11.   **13.**   **15.**

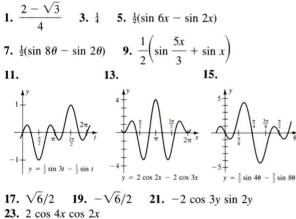

$y = \frac{1}{2}\sin 3t - \frac{1}{2}\sin t$ $y = 2 \cos 2x - 2 \cos 3x$ $y = \frac{5}{2}\sin 4\theta - \frac{5}{2}\sin 8\theta$

17. $\sqrt{6}/2$ **19.** $-\sqrt{6}/2$ **21.** $-2 \cos 3y \sin 2y$
23. $2 \cos 4x \cos 2x$
25. $2 \sin\left[\left(\dfrac{\omega_1 + \omega_2}{2}\right)t\right] \cos\left[\left(\dfrac{\omega_1 - \omega_2}{2}\right)t\right]$

27. $2 \sin\left(x + \dfrac{\pi}{4}\right) \sin\left(x - \dfrac{\pi}{4}\right) = 2 \cdot \dfrac{1}{2}\left\{\cos\left[\left(x + \dfrac{\pi}{4}\right) - \left(x - \dfrac{\pi}{4}\right)\right] - \cos\left[\left(x + \dfrac{\pi}{4}\right) + \left(x - \dfrac{\pi}{4}\right)\right]\right\}$

$= \cos \dfrac{\pi}{2} - \cos 2x = -\cos 2x$

29. $\dfrac{\sin 6\beta + \sin 2\beta}{\cos 2\beta - \cos 6\beta} = \dfrac{2 \sin 4\beta \cos 2\beta}{-2 \sin 4\beta \sin (-2\beta)}$

$= \dfrac{\cos 2\beta}{\sin 2\beta} = \cot 2\beta$

31. $2 \sin\left(t + \dfrac{\pi}{2}\right) \cos\left(t - \dfrac{\pi}{2}\right)$

$= 2 \cdot \dfrac{1}{2}\left\{\sin\left[\left(t + \dfrac{\pi}{2}\right) + \left(t - \dfrac{\pi}{2}\right)\right] + \sin\left[\left(t + \dfrac{\pi}{2}\right) - \left(t - \dfrac{\pi}{2}\right)\right]\right\}$

$= \sin 2t + \sin \pi = \sin 2t$

33. $\dfrac{\sin (x + h) - \sin x}{h}$

$= \dfrac{2 \cos\left(\dfrac{x + h + x}{2}\right) \sin\left(\dfrac{x + h - x}{2}\right)}{h}$

$= \dfrac{2}{h} \cos\left(x + \dfrac{h}{2}\right) \sin \dfrac{h}{2}$

35. $\dfrac{\sin 2\alpha + \sin 2\beta}{\sin 2\alpha - \sin 2\beta} = \dfrac{2 \sin\left(\dfrac{2\alpha + 2\beta}{2}\right) \cos\left(\dfrac{2\alpha - 2\beta}{2}\right)}{2 \cos\left(\dfrac{2\alpha + 2\beta}{2}\right) \sin\left(\dfrac{2\alpha - 2\beta}{2}\right)}$

$= \dfrac{2 \sin (\alpha + \beta) \cos (\alpha - \beta)}{2 \cos (\alpha + \beta) \sin (\alpha - \beta)}$

$= \tan (\alpha + \beta) \cot (\alpha - \beta)$

$= \dfrac{\tan (\alpha + \beta)}{\tan (\alpha - \beta)}$

39. $f(t) = 0.06 \sin 750\pi t \cos 250\pi t$
41. $\sin \omega t \sin (\omega t + \phi)$

$= \frac{1}{2}\{\cos [\omega t - (\omega t + \phi)] - \cos [\omega t + (\omega t + \phi)]\}$

$= \frac{1}{2}[\cos \phi - \cos (2\omega t + \phi)]$

EXERCISE 4.5 (pages 203–205)

1. $t = \dfrac{\pi}{3} + 2n\pi$ or $t = \dfrac{2\pi}{3} + 2n\pi$, $n = 0, \pm 1, \pm 2, \ldots$

3. $t = \dfrac{\pi}{4} + 2n\pi$ or $t = \dfrac{7\pi}{4} + 2n\pi$, $n = 0, \pm 1, \pm 2, \ldots$

5. $t = \dfrac{5\pi}{6} + n\pi$, $n = 0, \pm 1, \pm 2, \ldots$ **7.** $x = \pi + 2n\pi$,
$n = 0, \pm 1, \pm 2, \ldots$ **9.** $x = n\pi$, $n = 0, \pm 1, \pm 2, \ldots$

11. $x = \dfrac{3\pi}{2} + 2n\pi$, $n = 0, \pm 1, \pm 2, \ldots$

13. $\theta = 60° + 360°n$ or $\theta = 120° + 360°n$, $n = 0, \pm 1, \pm 2,$
$\ldots$ **15.** $\theta = 135° + 180°n$, $n = 0, \pm 1, \pm 2, \ldots$
17. $\theta = 120° + 360°n$ or $\theta = 240° + 360°n$, $n = 0, \pm 1, \pm 2,$
$\ldots$ **19.** $x = n\pi$, $n = 0, \pm 1, \pm 2, \ldots$ **21.** No
solutions   **23.** $\theta = 120° + 360°n$ or $\theta = 240° + 360°n$,
$n = 0, \pm 1, \pm 2, \ldots$ **25.** $\theta = 90° + 180°n$ or
$\theta = 135° + 180°n$, $n = 0, \pm 1, \pm 2, \ldots$

27. $x = \dfrac{\pi}{2} + n\pi$, $n = 0, \pm 1, \pm 2, \ldots$

29. $\theta = 10° + 120°n$ or $\theta = 50° + 120°n$, $n = 0, \pm 1, \pm 2,$

$\ldots$ **31.** $x = \dfrac{\pi}{2} + 2n\pi$, $n = 0, \pm 1, \pm 2, \ldots$

33. $x = n\pi$, $x = \dfrac{2\pi}{3} + 2n\pi$, or $x = \dfrac{4\pi}{3} + 2n\pi$, $n = 0, \pm 1,$

$\pm 2, \ldots$ **35.** $\theta = 30° + 360°n$, $\theta = 150° + 360°n$, or
$\theta = 270° + 360°n$, $n = 0, \pm 1, \pm 2, \ldots$

37. $x = \dfrac{\pi}{2} + n\pi$, $n = 0, \pm 1, \pm 2, \ldots$ **39.** $x = n\pi$,

$n = 0, \pm 1, \pm 2, \ldots$ **41.** $\theta = 90° + 180°n$, $n = 0, \pm 1,$

$\pm 2, \ldots$ **43.** $x = \dfrac{\pi}{6} + 2n\pi$ or $x = \dfrac{5\pi}{6} + 2n\pi$, $n = 0,$

$\pm 1, \pm 2, \ldots$ **45.** $\theta = 90° + 180°n$ or $\theta = 360°n$, $n = 0,$

$\pm 1, \pm 2, \ldots$ **47.** $t = \dfrac{\pi}{3}, \dfrac{2\pi}{3}, \pi$ **49.** $t = \frac{2}{3}, \frac{10}{3}, \frac{14}{3}$

51. $t = \pi, 2\pi, 3\pi$ **53.** $t = \dfrac{\pi}{3}, \pi, \dfrac{5\pi}{3}$ **55.** 1.37, 1.82

57. $-1.017, 0.55$ **59.** 0, 0.58, 1.81

61. The equation has infinitely many solutions.

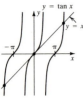

63. The equation has infinitely many solutions.

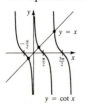

65. 30° or 150°   **67.** 60°   **69.** $t = \frac{1}{120}(\frac{1}{6} + n)$, $n = 0, \pm 1,$
$\pm 2, \ldots$ **71.** (a) 36.93 million square kilometers;
(b) $w = 31$ weeks; (c) August

CHAPTER 4
REVIEW EXERCISE (pages 205–206)

1. True   **3.** False   **5.** True   **7.** True   **9.** False
11. $(\sin \alpha + \cos \alpha)^2 + (\sin \alpha - \cos \alpha)^2$
$= \sin^2 \alpha + 2 \sin \alpha \cos \alpha + \cos^2 \alpha + \sin^2 \alpha - 2 \sin \alpha \cos \alpha + \cos^2 \alpha$
$= 2 \sin^2 \alpha + 2 \cos^2 \alpha$
$= 2(\sin^2 \alpha + \cos^2 \alpha) = 2$

13. $\dfrac{\tan x}{\sec x + 1} = \dfrac{\tan x}{\sec x + 1} \cdot \dfrac{\sec x - 1}{\sec x - 1} = \dfrac{\tan x (\sec x - 1)}{\sec^2 x - 1}$

$\qquad = \dfrac{\tan x (\sec x - 1)}{\tan^2 x} = \dfrac{\sec x - 1}{\tan x}$

15. $\sin^2 \theta \sec \theta = (1 - \cos^2 \theta) \sec \theta = \sec \theta - \cos \theta$

17. $\dfrac{\tan^2 \beta}{\sec \beta - 1} = \dfrac{\sec^2 \beta - 1}{\sec \beta - 1} = \dfrac{(\sec \beta + 1)(\sec \beta - 1)}{\sec \beta - 1} = 1 + \sec \beta$

19. For $\theta = \pi$, $\sin \theta + \cos \theta = \sin \pi + \cos \pi = 0 - 1$
$\qquad\qquad\qquad\qquad\qquad\qquad\qquad = -1 \neq 1.$

21. $\frac{1}{2}\sqrt{2 - \sqrt{2}}$ **23.** $\dfrac{\sqrt{2}}{4}(1 - \sqrt{6})$ **25.** $1 - \sqrt{2}$

27. $(\sqrt{3} - 1)/4$ **29.** $\sqrt{6}/2$
31. $\sin (t/2) = -\sqrt{(3 - \sqrt{2})/6}$; $\cos (t/2) = \sqrt{(3 + \sqrt{2})/6}$;
$\tan (t/2) = \sqrt{7}(\sqrt{2} - 3)/7$; $\sin 2t = -2\sqrt{14}/9$;
$\cos 2t = -5/9$; $\tan 2t = 2\sqrt{14}/5$ **33.** $\theta = 180°n$,
$\theta = 30° + 360°n$, or $\theta = 150° + 360°n$, $n = 0, \pm 1, \pm 2, \ldots$
35. $x = \pi + 2n\pi$, $n = 0, \pm 1, \pm 2, \ldots$ **37.** $-0.68, 0.92$
39. Using the addition formula for the sine function, we find
that

$$\sin \left(\arcsin \frac{3}{5} + \arcsin \frac{5}{13}\right) = \sin \left(\arcsin \frac{3}{5}\right) \cos \left(\arcsin \frac{5}{13}\right)$$
$$+ \sin \left(\arcsin \frac{5}{13}\right) \cos \left(\arcsin \frac{3}{5}\right)$$
$$= \frac{3}{5} \cdot \frac{12}{13} + \frac{5}{13} \cdot \frac{4}{5}$$
$$= \frac{36 + 20}{65} = \frac{56}{65}.$$

Thus,

$$\arcsin \frac{56}{65} = \arcsin \frac{3}{5} + \arcsin \frac{5}{13}$$

provided that

$$-\frac{\pi}{2} \le \arcsin\frac{3}{5} + \arcsin\frac{5}{13} \le \frac{\pi}{2}.$$

To see that this inequality holds, we observe that

$0 < \arcsin\dfrac{3}{5} < \dfrac{\pi}{4}$ and $0 < \arcsin\dfrac{5}{13} < \dfrac{\pi}{4}$, since

$0 < \dfrac{3}{5} < \dfrac{\sqrt{2}}{2}$ and $0 < \dfrac{5}{13} < \dfrac{\sqrt{2}}{2}$. **41.** $A = 2$, $b = \sqrt{g/L}$, $\phi = \pi/3$

■ CHAPTER 5

EXERCISE 5.1 (pages 215–216)

1. $|\mathbf{v}| = 2$, $\theta = \dfrac{11\pi}{6}$ **3.** $|\mathbf{v}| = 5$, $\theta = 2\pi$

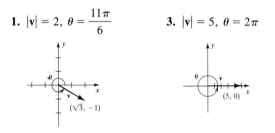

5. $|\mathbf{v}| = 8$, $\theta = \dfrac{2\pi}{3}$ **7.** $|\mathbf{v}| = 10\sqrt{2}$, $\theta = \dfrac{3\pi}{4}$

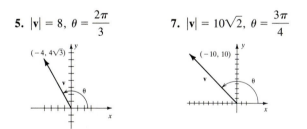

9. $\langle 3, 2\rangle$; $\langle 1, 4\rangle$; $\langle -6, -9\rangle$; $\langle 2, 13\rangle$ **11.** $\langle 0, 3\rangle$; $\langle -8, 1\rangle$; $\langle 12, -6\rangle$; $\langle -28, 2\rangle$ **13.** $\langle -\frac{9}{2}, -\frac{29}{4}\rangle$; $\langle -\frac{11}{2}, -\frac{27}{4}\rangle$; $\langle 15, 21\rangle$; $\langle -17, -20\rangle$ **15.** $-31\mathbf{i} - 14\mathbf{j}$; $42\mathbf{i} + 11\mathbf{j}$ **17.** $-\frac{15}{2}\mathbf{i} - \frac{3}{2}\mathbf{j}$; $11\mathbf{i} - 3\mathbf{j}$ **19.** $5.8\mathbf{i} + 8.5\mathbf{j}$; $-6.6\mathbf{i} - 10.3\mathbf{j}$ **21.**   **23.**

25.   **27.**

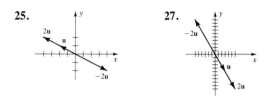

29. Horizontal component: 4; vertical component: -6
31. Horizontal component: -10; vertical component: 8
33. (a) $2\left(\cos\dfrac{3\pi}{4}\mathbf{i} + \sin\dfrac{3\pi}{4}\mathbf{j}\right)$; (b) $-\sqrt{2}\mathbf{i} + \sqrt{2}\mathbf{j}$

35. (a) $6\left(\cos\dfrac{5\pi}{6}\mathbf{i} + \sin\dfrac{5\pi}{6}\mathbf{j}\right)$; (b) $-3\sqrt{3}\mathbf{i} + 3\mathbf{j}$

37. 10.16, 30.26° if $\mathbf{F_1} = 4\mathbf{i}$.   **39.** $2\mathbf{i} - 10\mathbf{j}$ **41.** The swimmer's actual course makes an angle of 71.6° with the original heading.   **43.** 52 mph

EXERCISE 5.2 (pages 223–224)

1. 10   **3.** 42   **5.** 1   **7.** 0   **9.** 102.52°   **11.** 63.43°   **13.** 45°   **19.** $k = 15$ **21.** -8 **23.** 18   **25.** 180   **27.** 23   **29.** 8   **31.** $-13\mathbf{i} - 26\mathbf{j}$ **33.** $\dfrac{-8\sqrt{13}}{13}$

35. $\frac{18}{5}$ **37.** 55 ft-lb   **39.** 48 ft-lb   **41.** Approximately 466.7 joules

EXERCISE 5.3 (page 230)

1. $10i$ **3.** $-3 - \sqrt{3}i$ **5.** $-1 + 4i$ **7.** $2 - 13i$ **9.** $-9 - 15i$ **11.** $-11 + 7i$ **13.** $1 - 5i$ **15.** $35i$ **17.** $1 + 4i$ **19.** $-1 + 12i$ **21.** -10 **23.** $-18 - 16i$ **25.** 1   **27.** -1 **29.** $\frac{4}{25} + \frac{3}{25}i$ **31.** $\frac{20}{41} - \frac{16}{41}i$ **33.** $\frac{1}{2} + \frac{1}{2}i$ **35.** $6 - 4i$ **37.** i **39.** $-\frac{6}{53} + \frac{32}{53}i$ **41.** $\frac{11}{2} + \frac{9}{2}i$ **43.** $-\frac{6}{13} - \frac{4}{13}i$ **45.** $9 + i$ **47.** $x = 2$, $y = \frac{3}{2}$ **49.** $x = -9$, $y = -20$ **51.** $x = -4$, $y = -5$ **53.** $\pm 3i$ **55.** $\pm\dfrac{\sqrt{10}}{2}i$ **57.** $\dfrac{1}{4} - \dfrac{\sqrt{7}}{4}i$, $\dfrac{1}{4} + \dfrac{\sqrt{7}}{4}i$ **59.** $-4 - 6i$, $-4 + 6i$ **61.** $\dfrac{1}{8} - \dfrac{\sqrt{31}}{8}i$, $\dfrac{1}{8} + \dfrac{\sqrt{31}}{8}i$ **63.** $\pm i$, $\pm\sqrt{2}i$ **65.** $\dfrac{\sqrt{2}}{2} + \dfrac{\sqrt{2}}{2}i$ **71.** $\frac{65}{68} - \frac{5}{68}i$

EXERCISE 5.4 (pages 235–236)

1. $\bar{z}_1 = 2 - 5i$ **3.** $z_1 + z_2 = 3 - i$ **5.** $\bar{z}_1 + z_2 = 6 + 2i$

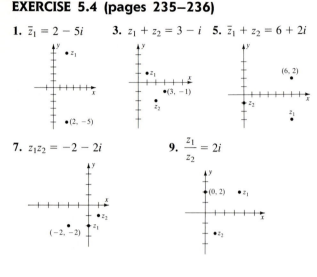

7. $z_1z_2 = -2 - 2i$ **9.** $\dfrac{z_1}{z_2} = 2i$

11. $r = 1,\ \theta = -\dfrac{\pi}{3}$ **13.** $r = 3\sqrt{2},\ \theta = -70.53°$

15. $r = \dfrac{\sqrt{10}}{4},\ \theta = -18.43°$ **17.** $r = 3\sqrt{2},\ \theta = \dfrac{\pi}{4}$

19. $r = 2,\ \theta = \dfrac{\pi}{6}$ **21.** $r = \sqrt{5},\ \theta = -26.57°$

23. $4\left[\cos\left(-\dfrac{\pi}{2}\right) + i\sin\left(-\dfrac{\pi}{2}\right)\right]$

25. $10\left[\cos\dfrac{\pi}{6} + i\sin\dfrac{\pi}{6}\right]$

27. $\sqrt{29}[\cos 111.8° + i\sin 111.8°]$

29. $\sqrt{34}[\cos(-59.04°) + i\sin(-59.04°)]$

31. $2\sqrt{2}\left[\cos\left(-\dfrac{3\pi}{4}\right) + i\sin\left(-\dfrac{3\pi}{4}\right)\right]$ **33.** $1 + i$

35. $-5\sqrt{3} - 5i$ **37.** $\sqrt{3} + i$ **39.** $\dfrac{1}{2} + \dfrac{\sqrt{3}}{2}i$

41. $\dfrac{4\sqrt{5}}{5} + \dfrac{8\sqrt{5}}{5}i$

43. $z_1z_2 = 18\sqrt{2}\left[\cos\dfrac{3\pi}{4} + i\sin\dfrac{3\pi}{4}\right]$;

$\dfrac{z_1}{z_2} = \dfrac{\sqrt{2}}{4}\left[\cos\dfrac{\pi}{4} + i\sin\dfrac{\pi}{4}\right]$

45. $z_1z_2 = 8\left[\cos\dfrac{\pi}{2} + i\sin\dfrac{\pi}{2}\right]$;

$\dfrac{z_1}{z_2} = \dfrac{1}{2}\left[\cos\dfrac{\pi}{6} + i\sin\dfrac{\pi}{6}\right]$

47. $z_1z_2 = 10\sqrt{2}\left[\cos\left(-\dfrac{\pi}{12}\right) + i\sin\left(-\dfrac{\pi}{12}\right)\right]$;

$\dfrac{z_1}{z_2} = \dfrac{\sqrt{2}}{5}\left[\cos\dfrac{5\pi}{12} + i\sin\dfrac{5\pi}{12}\right]$

49. $z_1z_2 = 2\sqrt{3}\left[\cos\dfrac{\pi}{12} + i\sin\dfrac{\pi}{12}\right]$;

$\dfrac{z_1}{z_2} = \sqrt{3}\left[\cos\dfrac{7\pi}{12} + i\sin\dfrac{7\pi}{12}\right]$

51. $z_1z_2 = 12\left[\cos\dfrac{\pi}{8} + i\sin\dfrac{\pi}{8}\right]$;

$\dfrac{z_1}{z_2} = \dfrac{3}{4}\left[\cos\dfrac{3\pi}{8} + i\sin\dfrac{3\pi}{8}\right]$

53. Let $z = a + bi$. Then

$$|z| = \sqrt{a^2 + b^2} \text{ and } \sqrt{z\bar{z}} = \sqrt{(a + bi)(a - bi)}$$
$$= \sqrt{a^2 + b^2}.$$

Therefore, $|z| = \sqrt{z\bar{z}}$.

EXERCISE 5.5 (pages 240–241)

1. -1 **3.** $-8i$ **5.** -64 **7.** $-16\sqrt{3} + 16i$ **9.** $-2^{9/2}$

11. $-7 - 24i$ **13.** -1 **15.** $1 + \sqrt{3}i,\ -2,\ 1 - \sqrt{3}i$

17. $0.9239 + 0.3827i,\ -0.3827 + 0.9239i,$

$-0.9239 - 0.3827i,\ 0.3827 - 0.9239i$

19. $\sqrt[4]{2}\left[\dfrac{1}{2} + \dfrac{\sqrt{3}}{2}i\right],\ \sqrt[4]{2}\left[-\dfrac{\sqrt{3}}{2} + \dfrac{1}{2}i\right],$

$\sqrt[4]{2}\left[-\dfrac{1}{2} - \dfrac{\sqrt{3}}{2}i\right],\ \sqrt[4]{2}\left[\dfrac{\sqrt{3}}{2} - \dfrac{1}{2}i\right]$

21. $2^{1/4}[0.9239 + 0.3827i],\ 2^{1/4}[-0.9239 - 0.3827i]$

23. $1.9754 + 0.3128i,\ 0.7167 + 1.8672i,$

$-1.2586 + 1.5542i,\ -1.9754 - 0.3128i,$

$-0.7167 - 1.8672i,\ 1.2586 - 1.5542i$

25. $(x + 2 - 2\sqrt{3}i)(x - 2 + 2\sqrt{3}i)$ **27.** $n = 8k,\ k = 1,\ 2,$

$3, \ldots;\ n = 2 + 8k,\ k = 0,\ 1,\ 2, \ldots;\ n = 5 + 8k,\ k = 0,$

$1,\ 2, \ldots;\ n = 1 + 8k,\ k = 0,\ 1,\ 2, \ldots$ **29.** $-16i$

31. $\cos 2\theta = \cos^2\theta - \sin^2\theta,\ \sin 2\theta = 2\sin\theta\cos\theta$;

$\cos 3\theta = \cos^3\theta - 3\cos\theta\sin^2\theta,$

$\sin 3\theta = -\sin^3\theta + 3\sin\theta\cos^2\theta$

CHAPTER 5
REVIEW EXERCISE (pages 241–242)

1. True   **3.** $k = -8$ **5.** True   **7.** $4 + 9i$ **9.** 15

11. $\langle -2, 5\rangle$ **13.** $\langle -6, 8\rangle$ **15.** -138 **17.** -86

19. $-\dfrac{23}{29}$ **21.** $2\sqrt{3};\ 5\pi/3$ **23.** $2\sqrt{2};\ 7\pi/4$

25. Horizontal component: 6; vertical component: -5

27. Horizontal component: -17; vertical component: 51

29. $4\mathbf{i} - 2\mathbf{j}$ **31.** $5\mathbf{i} + 20\mathbf{j}$ **33.** $2\sqrt{5}\left(\cos\dfrac{\pi}{6}\mathbf{i} + \sin\dfrac{\pi}{6}\mathbf{j}\right)$

35. $|\mathbf{F}_2| = 5\sqrt{2}$; $135°$ **37.** $146.31°$ **39.** -24

41. $\overline{z_1} = 2 + 3i$, $z_1 + z_2 = 2 + i$, $z_1 z_2 = 12 + 8i$

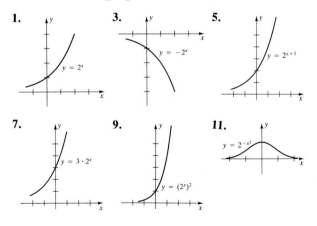

43. $r = 3\sqrt{2}$, $\theta = -\pi/4$, $3\sqrt{2}\left[\cos\left(-\dfrac{\pi}{4}\right) + i\sin\left(-\dfrac{\pi}{4}\right)\right]$

45. $r = \sqrt{13}$, $\theta = -56.31°$,

$\sqrt{13}[\cos(-56.31°) + i\sin(-56.31°)]$ **47.** $-2 - 2\sqrt{3}i$

49. $z_1 z_2 = 2\sqrt{2}\left[\cos\dfrac{19\pi}{12} + i\sin\dfrac{19\pi}{12}\right]$;

$\dfrac{z_1}{z_2} = \dfrac{\sqrt{2}}{2}\left[\cos\dfrac{11\pi}{12} + i\sin\dfrac{11\pi}{12}\right]$

51. $(\sqrt{2})^7\left[\cos\dfrac{35\pi}{4} + i\sin\dfrac{35\pi}{4}\right] = -8 + 8i$ **53.** 2,

$0.6180 + 1.9021i$, $-1.6180 + 1.1756i$, $-1.6180 - 1.1756i$,

$0.6180 - 1.9021i$

■ **CHAPTER 6**

EXERCISE 6.1 (pages 250–251)

1.   **3.**   **5.**

7.   **9.**   **11.**

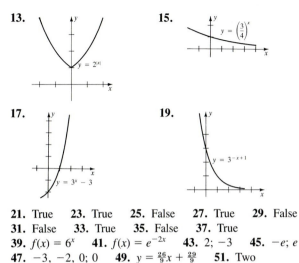

13.   **15.**

17.   **19.**

21. True   **23.** True   **25.** False   **27.** True   **29.** False
31. False   **33.** True   **35.** False   **37.** True
39. $f(x) = 6^x$ **41.** $f(x) = e^{-2x}$ **43.** 2; -3 **45.** $-e$; e
47. -3, -2, 0; 0   **49.** $y = \frac{26}{9}x + \frac{29}{9}$ **51.** Two
53. None   **55.** $(5, \infty)$ **57.**

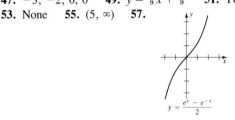

59. (a) 10.5561; **(b)** 11.0043; **(c)** 11.0349; **(d)** 11.0364;
(e) 11.0357   **61.** 245.1046   **63.** 2.6900   **65.** 0.9744

67.
2.7048138
2.7169239
2.7181459
2.7182682
2.7182805
2.7182817

EXERCISE 6.2 (pages 260–261)

1. $\log_4 \frac{1}{2} = -\frac{1}{2}$ **3.** $\log_9 1 = 0$ **5.** $\log_{10} x = y$
7. $\log_{1/64} 8 = -\frac{1}{2}$ **9.** $\log_{36}\frac{1}{216} = -\frac{3}{2}$ **11.** $\log_8 4 = \frac{2}{3}$
13. $81 = 3^4$ **15.** $10 = 10^1$ **17.** $\frac{1}{25} = 5^{-2}$
19. $2 = 16^{1/4}$ **21.** $b^2 = b^2$ **23.** $1 = e^0$ **25.** -6
27. 3   **29.** $-\frac{5}{6}$ **31.** -4 **33.** -3 **35.** e **37.** $b = 5$
39. $x = 3$ **41.** $N = \frac{1}{32}$ **43.** $N = 3$ **45.** $x = -3$
47. $N = e^9$ **49.** $6^2 = 36$ **51.** 7^{-1} **53.** 0.3011
55. 1.8063   **57.** 0.3495   **59.** 0.2007   **61.** -0.0969
63. 1.6609   **65.** $\log_{10} 10 = 1$ **67.** $\log_{10}(x^2 - 2)$
69. $\log_2 1 = 0$

71. Domain: $(0, \infty)$

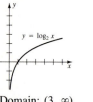

73. Domain: $(-\infty, 0) \cup (0, \infty)$

75. Domain: $(3, \infty)$

77. Domain: $(-1, \infty)$

79. Domain: $(0, \infty)$

81. Domain: $(0, \infty)$

83. Domain: $(0, \infty)$

85. 1.6113   **87.** 3.5835   **89.** 1.7227   **91.** 2.4232

93. $-10^7 \ln\left(\dfrac{x}{10^7}\right)$ **95.** (c)

EXERCISE 6.3 (pages 267–268)

1. 2   **3.** 2   **5.** 0.3495   **7.** 1.0802   **9.** 0   **11.** 3
13. -1 **15.** ± 4 **17.** ± 3 **19.** -0.8782 **21.** 0, 2
23. 32   **25.** 45   **27.** $\pm\frac{1}{10}$ **29.** 81   **31.** $\frac{3}{2}$ **33.** 100
35. 2, 8   **37.** 1   **39.** 4   **41.** 1, 16   **43.** e^{-3}, e^3
45. -0.6,   **47.** -0.5, 1.5,   **49.** 3,

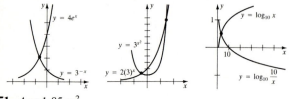

51. $A \approx 1.85 \text{ m}^2$

EXERCISE 6.4 (pages 276–278)

1. 200; 800; 3200; 204,800; 2.3×10^{20} **3.** 2344
5. (a) $f(t) = 100(\frac{1}{2})^t$; (b) 0.7813   **7.** 24,151 yr
9. $(0.6730)A_0$; $(0.2264)A_0$ **11.** Between 18,600 yr and 24,200 yr

13.

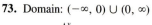

| 1100.00 |
| 1102.50 |
| 1103.81 |
| 1104.71 |
| 1105.06 |
| 1105.16 |
| 1105.17 |
| 1105.17 |

15. \$1885.64   **17.** 7   **19.** 7.4   **21.** 2.1
23. 1.3×10^{-3} **25.** 5.0×10^{-11} **27.** 1.6×10^{-9}
29. About 50 times more acidic   **31.** About 10^{14} times more acidic   **33.** Approximately 2 times as intense; approximately 5 times as intense; approximately 79 times as intense

35.

| 20 |
| 50 |
| 60 |
| 70 |
| 90 |
| 120 |

37. 65 db   **39.** 84; 2000
41. 211° F; 70° F   **43.** -7.75 **45.**

 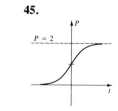

CHAPTER 6
REVIEW EXERCISE (pages 279–281)

1. $b > 1$ **3.** 10   **5.** $\frac{4}{3}$ **7.** $10^{0.6990}$ **9.** $\log_8 2 = \frac{1}{3}$
11. 4   **13.** 0.9542   **15.** $\ln 5/\ln 4$ **17.** 4.89   **19.** True
21. $\frac{1}{27}$ **23.** 16   **25.** 16   **27.** $\log_5 0.2 = -1$
29. $27 = 9^{1.5}$ **31.** $\log_3 \frac{8}{9}$ **33.** $(-\infty, 0) \cup (0, \infty)$
35. $(-\infty, 5) \cup (5, \infty)$

37. 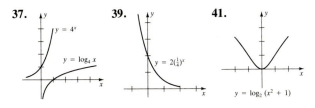 **39.** **41.**

$y = 4^x$

$y = \log_4 x$

$y = 2(\frac{1}{4})^x$

$y = \log_2 (x^2 + 1)$

43. $\frac{2}{3}$ **45.** -2.1729 **47.** $\frac{1}{25}$ **49.** $\frac{3}{2}$ **51.** $\frac{5}{2}$
53. -0.6770 **55.** C, D, A, B **57.** $x > -2$ **59.** 5.5
61. $\frac{1}{16}$ of the initial quantity **63.** (a) 6 days; (b) 19.84%

65. $x = -\dfrac{1}{c} \ln \left(-\dfrac{1}{b} \ln \dfrac{y}{a} \right)$

▪ CHAPTER 7

EXERCISE 7.1 (pages 295–297)

1. $-\frac{7}{2}$ **3.** 5 **5.** -1
7. **9.** **11.**

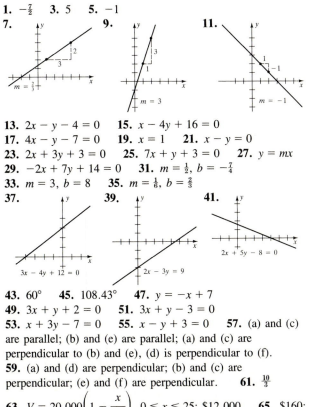

$m = \frac{2}{3}$ $m = 3$ $m = -1$

13. $2x - y - 4 = 0$ **15.** $x - 4y + 16 = 0$
17. $4x - y - 7 = 0$ **19.** $x = 1$ **21.** $x - y = 0$
23. $2x + 3y + 3 = 0$ **25.** $7x + y + 3 = 0$ **27.** $y = mx$
29. $-2x + 7y + 14 = 0$ **31.** $m = \frac{1}{2}, b = -\frac{7}{4}$
33. $m = 3, b = 8$ **35.** $m = \frac{1}{6}, b = \frac{2}{3}$
37. **39.** **41.**

$3x - 4y + 12 = 0$ $2x - 3y = 9$ $2x + 5y - 8 = 0$

43. 60° **45.** 108.43° **47.** $y = -x + 7$
49. $3x + y + 2 = 0$ **51.** $3x + y - 3 = 0$
53. $x + 3y - 7 = 0$ **55.** $x - y + 3 = 0$ **57.** (a) and (c)
are parallel; (b) and (e) are parallel; (a) and (c) are
perpendicular to (b) and (e), (d) is perpendicular to (f).
59. (a) and (d) are perpendicular; (b) and (c) are
perpendicular; (e) and (f) are perpendicular. **61.** $\frac{10}{3}$

63. $V = 20,000\left(1 - \dfrac{x}{25}\right), 0 \le x \le 25; \$12,000$ **65.** $\$160$;

after 30 years **67.** Let f represent distance measured in feet
and let m represent distance measured in meters. Then

$f = 3.28m$ converts meters into feet and $m = \dfrac{1}{3.28}f$ converts

feet into meters. **69.** $T_K = T_C + 273; -273°$ C; $-459.4°$ F

EXERCISE 7.3 (pages 307–308)

1. Vertex: $(0, 0)$;
Focus: $(1, 0)$;
Directrix: $x = -1$;
Axis: $y = 0$

3. Vertex: $(0, 0)$;
Focus: $(-\frac{1}{3}, 0)$;
Directrix: $x = \frac{1}{3}$;
Axis: $y = 0$

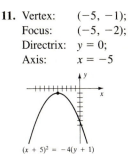

$y^2 = 4x$

$y^2 = -\frac{4}{3}x$

5. Vertex: $(0, 0)$;
Focus: $(0, -4)$;
Directrix: $y = 4$;
Axis: $x = 0$

7. Vertex: $(0, 0)$;
Focus: $(0, 7)$;
Directrix: $y = -7$;
Axis: $x = 0$

$x^2 = -16y$

$x^2 = 28y$

9. Vertex: $(0, 1)$;
Focus: $(4, 1)$;
Directrix: $x = -4$;
Axis: $y = 1$

11. Vertex: $(-5, -1)$;
Focus: $(-5, -2)$;
Directrix: $y = 0$;
Axis: $x = -5$

$(y - 1)^2 = 16x$

$(x + 5)^2 = -4(y + 1)$

13. Vertex: $(-5, -6)$;
Focus: $(-4, -6)$;
Directrix: $x = -6$;
Axis: $y = -6$

15. Vertex: $(-\frac{5}{2}, -1)$;
Focus: $(-\frac{5}{2}, -\frac{15}{16})$;
Directrix: $y = -\frac{17}{16}$;
Axis: $x = -\frac{5}{2}$

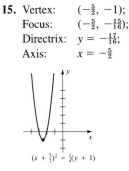

$(y + 6)^2 = 4(x + 5)$

$(x + \frac{5}{2})^2 = \frac{1}{4}(y + 1)$

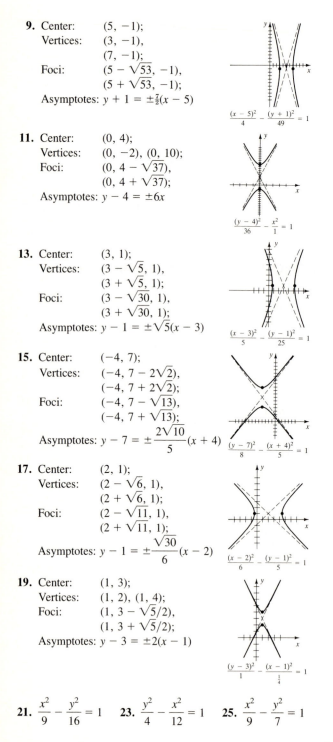

9. Center: $(5, -1)$;
Vertices: $(3, -1)$,
  $(7, -1)$;
Foci: $(5 - \sqrt{53}, -1)$,
  $(5 + \sqrt{53}, -1)$;
Asymptotes: $y + 1 = \pm\frac{7}{2}(x - 5)$

$\frac{(x-5)^2}{4} - \frac{(y+1)^2}{49} = 1$

11. Center: $(0, 4)$;
Vertices: $(0, -2)$, $(0, 10)$;
Foci: $(0, 4 - \sqrt{37})$,
  $(0, 4 + \sqrt{37})$;
Asymptotes: $y - 4 = \pm 6x$

$\frac{(y-4)^2}{36} - \frac{x^2}{1} = 1$

13. Center: $(3, 1)$;
Vertices: $(3 - \sqrt{5}, 1)$,
  $(3 + \sqrt{5}, 1)$;
Foci: $(3 - \sqrt{30}, 1)$,
  $(3 + \sqrt{30}, 1)$;
Asymptotes: $y - 1 = \pm\sqrt{5}(x - 3)$

$\frac{(x-3)^2}{5} - \frac{(y-1)^2}{25} = 1$

15. Center: $(-4, 7)$;
Vertices: $(-4, 7 - 2\sqrt{2})$,
  $(-4, 7 + 2\sqrt{2})$;
Foci: $(-4, 7 - \sqrt{13})$,
  $(-4, 7 + \sqrt{13})$;
Asymptotes: $y - 7 = \pm\dfrac{2\sqrt{10}}{5}(x + 4)$

$\frac{(y-7)^2}{8} - \frac{(x+4)^2}{5} = 1$

17. Center: $(2, 1)$;
Vertices: $(2 - \sqrt{6}, 1)$,
  $(2 + \sqrt{6}, 1)$;
Foci: $(2 - \sqrt{11}, 1)$,
  $(2 + \sqrt{11}, 1)$;
Asymptotes: $y - 1 = \pm\dfrac{\sqrt{30}}{6}(x - 2)$

$\frac{(x-2)^2}{6} - \frac{(y-1)^2}{5} = 1$

19. Center: $(1, 3)$;
Vertices: $(1, 2)$, $(1, 4)$;
Foci: $(1, 3 - \sqrt{5}/2)$,
  $(1, 3 + \sqrt{5}/2)$;
Asymptotes: $y - 3 = \pm 2(x - 1)$

$\frac{(y-3)^2}{1} - \frac{(x-1)^2}{\frac{1}{4}} = 1$

21. $\dfrac{x^2}{9} - \dfrac{y^2}{16} = 1$ **23.** $\dfrac{y^2}{4} - \dfrac{x^2}{12} = 1$ **25.** $\dfrac{x^2}{9} - \dfrac{y^2}{7} = 1$

27. $\dfrac{y^2}{25/4} - \dfrac{x^2}{11/4} = 1$ **29.** $\dfrac{x^2}{4} - \dfrac{y^2}{5} = 1$

31. $\dfrac{y^2}{64} - \dfrac{x^2}{16} = 1$ **33.** $\dfrac{x^2}{4} - \dfrac{y^2}{64/9} = 1$

35. $\dfrac{(y+3)^2}{4} - \dfrac{(x-1)^2}{5} = 1$ **37.** $\dfrac{(x+1)^2}{4} - \dfrac{(y-2)^2}{5} = 1$

39. $\dfrac{x^2}{4} - \dfrac{y^2}{8} = 1$ **41.** $\dfrac{(y-3)^2}{1} - \dfrac{(x+1)^2}{4} = 1$

43. $(-7, 12)$

47. (a) 9 **(b)** The x-coordinate of a point on a focal chord is $\pm c$. We substitute $x = c$ into the equation of the hyperbola and solve for y:

$$\frac{c^2}{a^2} - \frac{y^2}{b^2} = 1$$

$$y^2 = b^2\left(\frac{c^2}{a^2} - 1\right)$$

$$= \frac{b^2}{a^2}(c^2 - a^2) = \frac{b^4}{a^2}$$

$$y = \pm\frac{b^2}{a}.$$

Therefore, the endpoints of the right-hand focal chord are $(c, b^2/a)$ and $(c, -b^2/a)$. It follows that the focal width of the hyperbola is $2b^2/a$.

49. One asymptote has slope 1; the other asymptote has slope -1.   **51.** One branch of a hyperbola   **53.** In the definition of an ellipse (Definition 3), the *sum* of the distances from any point on the ellipse to the foci is constant, whereas in the definition of a hyperbola (Definition 4), the *difference* of the distances (in absolute value) from any point on the hyperbola to the foci is constant.

EXERCISE 7.6 (pages 338–339)

1. $(2, 4)$ **3.** $(-6, -3)$ **5.** $(0, -3)$ **7.** $(6, 2)$
9. $(-4, -11)$ **11.** $(4, -3)$ **13.** $X^2 + Y^2 = 9$
15. $X^2 + Y^2 = 5$ **17.** $X^2 + Y^2 = 8$ **19.** $3X + 2Y = 0$
21. $(x + 3)^2 + (y + 1)^2 = 25$ **23.** $(x - 4)^2 + (y + 3)^2 = 1$
25. $2x - 3y - 10 = 0$ **27.** $3x + 5y = 0$
29. Parabola   **31.** Circle   **33.** A point

$(y + 1)^2 = 6(x + \frac{1}{6})$

$(x - 2)^2 + (y + 1)^2 = \frac{5}{2}$

$(1, -1)$

35. Ellipse   **37.** Ellipse   **39.** No graph

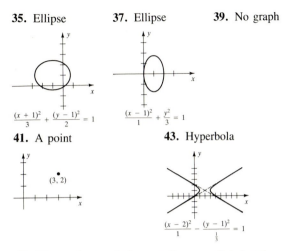

$\dfrac{(x+1)^2}{3} + \dfrac{(y-1)^2}{2} = 1$ $\dfrac{(x-1)^2}{1} + \dfrac{y^2}{3} = 1$

41. A point   **43.** Hyperbola

$\dfrac{(x-2)^2}{1} - \dfrac{(y-1)^2}{\frac{1}{3}} = 1$

45. We use the translation equations $x = X + h$ and $y = Y + k$ to write the given equation with respect to the XY-system with origin at the xy-point (h, k):

$$(X + h)(Y + k) + a(X + h) + b(Y + k) + c = 0.$$

This simplifies to

$$XY + (h + b)Y + (k + a)X + hk + ah + bk + c = 0.$$

If we choose $h = -b$, $k = -a$, the coefficients of X and Y will be zero and the resulting equation will have the desired form:

$$XY + (-b + b)Y + (-a + a)X + (-b)(-a) + a(-b) \\ + b(-a) + c = 0,$$

or $XY = ab - c$. Therefore, if the origin of the XY-system is at the xy-point $(-b, -a)$, then in the XY-system the given equation has the form $XY = d$, where $d = ab - c$.

47. Suppose that $A > 0$ and $C < 0$ (the case in which $A < 0$ and $C > 0$ is similar). Completing the square, we have

$$A\left(x^2 + \frac{D}{A}x + \frac{D^2}{4A^2}\right) + C\left(y^2 + \frac{E}{C}y + \frac{E^2}{4C^2}\right) \\ = -F + \frac{D^2}{4A} + \frac{E^2}{4C}$$

or

$$A\left(x + \frac{D}{2A}\right)^2 + C\left(y + \frac{E}{2C}\right)^2 = \frac{-4ACF + CD^2 + AE^2}{4AC}.$$

Let

$$G = \frac{-4ACF + CD^2 + AE^2}{4AC}.$$

If $G = 0$, the equation becomes

$$A\left(x + \frac{D}{2A}\right)^2 = -C\left(y + \frac{E}{2C}\right)^2$$

or

$$y + \frac{E}{2C} = \pm\sqrt{\frac{A}{-C}}\left(x + \frac{D}{2A}\right),$$

which is a pair of intersecting lines. (Since $A \neq 0$, the two lines are distinct and have unequal slopes.) If $G > 0$, the equation can be written

$$\frac{(x + D/2A)^2}{G/A} - \frac{(y + E/2C)^2}{-G/C} = 1,$$

where the denominators are positive as required (since they must be squares). If $G < 0$, we write the equation as

$$\frac{(y + E/2C)^2}{G/C} - \frac{(x + D/2A)^2}{-G/A} = 1.$$

51. $(4 + \sqrt{3}, 1 - 4\sqrt{3})$ **53.** $(-4, 0)$
55. $(2\sqrt{2} + \sqrt{3} - 3\sqrt{2} - \sqrt{3}, 2\sqrt{2} - \sqrt{3} + 3\sqrt{2} + \sqrt{3})$
$\approx (2.31, 6.83)$

57. $x'^2 + 4y'^2 = 4$ **59.** $5x'^2 + y'^2 = 4$
61. $y'^2 - 9x'^2 = 36$ **63.** $\theta = \arctan\frac{1}{3}$, $13x'^2 + 3y'^2 = 936$
65. $\theta = 45°$, $y'^2 + 4x' = 0$
67. $\theta = 45°$, $\dfrac{x'^2}{8/3} + \dfrac{y'^2}{8} = 1$ **69.** $\theta = 45°$, $y'^2 = 4\sqrt{2}x'$

CHAPTER 7
REVIEW EXERCISE (pages 340–342)

1. -12, 16, $\frac{4}{3}$ **3.** Vertical line   **5.** True   **7.** False
9. False   **11.** $x - 3y + 11 = 0$ **13.** $x + 3y - 7 = 0$
15. Slope: 2; x-intercept: $\frac{1}{2}$; y-intercept: -1

EXERCISE 8.3 (pages 360–361)

1. $e = 1$, parabola

$r = \dfrac{2}{1 - \sin \theta}$

3. $e = \frac{1}{4}$, ellipse

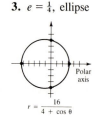

$r = \dfrac{16}{4 + \cos \theta}$

5. $e = 2$, hyperbola

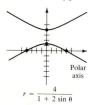

$r = \dfrac{4}{1 + 2 \sin \theta}$

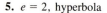

7. $e = 2$, hyperbola

$r = \dfrac{18}{3 - 6 \cos \theta}$

9. $e = 1$, parabola

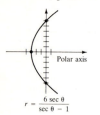

$r = \dfrac{6 \sec \theta}{\sec \theta - 1}$

11. $r = 3/(1 + \cos \theta)$ **13.** $r = 4/(3 - 2 \sin \theta)$
15. $r = 12/(1 + 2 \cos \theta)$
17. $\dfrac{(y - 4)^2}{4} - \dfrac{x^2}{12} = 1$, $e = 2$, $c = 4$, $a = 2$
19. $\dfrac{(x - \frac{24}{5})^2}{1296/25} + \dfrac{y^2}{144/5} = 1$, $e = \frac{2}{3}$, $c = \frac{24}{5}$, $a = \frac{36}{5}$
21. $r = 3/(1 - \sin \theta)$ **23.** $r = 1/(1 - \cos \theta)$
25. $r = 1/(2 - 2 \sin \theta)$
29. The eccentricity $e = c/a$. From Figure 28 and Section 7.4, we have that

$$a = \tfrac{1}{2}(r_a + r_p)$$

and

$$c = a - r_p = \tfrac{1}{2}(r_a + r_p) - r_p$$
$$= \tfrac{1}{2}(r_a - r_p).$$

Therefore,

$$e = c/a = \tfrac{1}{2}(r_a - r_p)/\tfrac{1}{2}(r_a + r_p)$$
$$= (r_a - r_p)/(r_a + r_p).$$

31. $r = 75,000/(4 - \cos \theta)$
33.

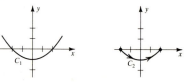

$r = \dfrac{0.7}{1 - 0.7 \cos \theta}$

EXERCISE 8.4 (pages 367–369)

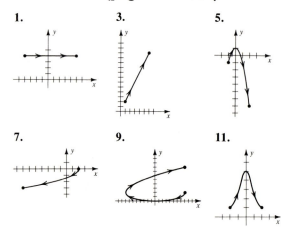

13. $y = 3x + 2$ **15.** $y = x^2 - 3x + 1$, $x \geq 0$
17. $x^2 + y^2 = 36$ **19.** $y = x - 3$, $0 \leq x \leq 2$
21. $x = 1 - 2y^2$, $-1 \leq y \leq 1$
23.

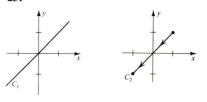

25.

27.

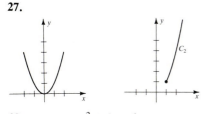

29. $x = t$, $y = t^2 + 4t - 6$, $-\infty < t < \infty$

35. **(a)** Use the trigonometric identities

$$\cos 3t = 4 \cos^3 t - 3 \cos t,$$

$$\sin 3t = -4 \sin^3 t + 3 \sin t.$$

(b) $x^{2/3} + y^{2/3} = b^{2/3}$ **37.** Square both equations and then add.

CHAPTER 8
REVIEW EXERCISE (pages 369–371)

1. True   **3.** False   **5.** $(2, -2)$ **7.** At the rectangular coordinate point $(0, 5)$ **9.** Five   **11.** **(a)** $(2\sqrt{3}, -\pi/6)$; **(b)** $(-2\sqrt{3}, 5\pi/6)$ **13.** $(x^2 + y^2)^2 = 25(x^2 - y^2)$

15. $r^2 \cos 2\theta = 9$

17.

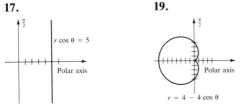

$r \cos \theta = 5$

19.

$r = 4 - 4 \cos \theta$

21.

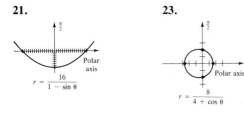

$r = \dfrac{16}{1 - \sin \theta}$

23.

$r = \dfrac{8}{4 + \cos \theta}$

25. $\dfrac{(y - 8/15)^2}{4/225} - \dfrac{x^2}{4/15} = 1$; $a = 2/15$, $c = 8/15$,
$e = c/a = 4$ **27.** $r = a$; $r = 2a \sin \theta$

29. $y^2 = 4x^2(1 - x^2)$ **31.** $\dfrac{(x - 3)^2}{16} - \dfrac{(y + 5)^2}{4} = 1$, $x \geq 7$

33. $y = \ln x$ **35.** Eliminating the parameter gives an equation of a line: $y - y_1 = \left(\dfrac{y_2 - y_1}{x_2 - x_1}\right)(x - x_1)$. For $0 \leq t \leq 1$, x varies from x_1 to x_2.

37.

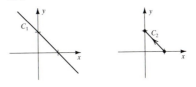

EXERCISE A (pages A-3–A-4)

1. 2.5 cm/sec   **3.** 12 ft/sec   **5.** $0.6/\pi$ m   **7.** 2 radians/sec   **9.** **(a)** $\pi/5$ ft/sec **(b)** 0.4π ft **(c)** $\pi/10$ radian   **11.** **(a)** $2\pi/3$ ft/sec **(b)** $10\pi/3$ ft **(c)** $150°$ or $5\pi/6$ radians   **13.** $5/\pi$ sec   **15.** $25/\pi$ cm   **17.** **(a)** $\pi/10,800$ in./sec **(b)** $\pi/600$ in./sec **(c)** $7\pi/75$ in./sec   **19.** **(a)** $\pi/720$ radian/min **(b)** approximately 1500 ft/sec   **21.** **(a)** $5000\pi/7$ mi/hr $\approx 2.2 \times 10^3$ mi/hr **(b)** $\pi/336$ radian/hr

EXERCISE B (pages A-5–A-7)

3. 19.5 ft   **5.** $3, \dfrac{\pi}{6} + \dfrac{2n\pi}{3} - 0.1133$,
$n = 0, \pm 1, \pm 2, \ldots$ **7.** $500\sqrt{2}\pi/3$ mi/hr ≈ 740 mi/hr   **13.** $\sqrt{58}$ **17.** $62°$ **19.** 69.2 ft

EXERCISE C (page A-16)

1. $m = 0.2430$, $c = 2$ **3.** $m = 0.4786$, $c = -4$ **5.** $m = 0.8344$, $c = 5$ **7.** 3.6749   **9.** -4.2464 **11.** 3.4304   **13.** 1.0285   **15.** 5.8833   **17.** 0.9824   **19.** 2.6742   **21.** -2.9194 **23.** 6.8740   **25.** -1.3509 **27.** $m = 0.3678$, $c = 2$ **29.** $m = 0.4234$, $c = -5$ **31.** $m = 0.7320$, $c = -2$ **33.** $m = 0.6521$, $c = 0$ **35.** $m = 0.7111$, $c = 6$ **37.** $m = 0.2965$, $c = -5$ **39.** 494   **41.** 0.0287   **43.** 0.00000998   **45.** 258.4   **47.** 0.00004084   **49.** 2.542   **51.** 1.7579   **53.** 5.1930   **55.** -0.5655 **57.** 0.9171   **59.** 0.3574   **61.** 0.2022   **63.** 0.6730   **65.** 0.8550   **67.** 0.4122   **69.** $31°23'$ **71.** $68°24'$ **73.** $18°44'$ **75.** $207°$; $333°$

Index

Credits

Chapter 1: page 1, North Wind Picture Archives; page 12, British Museum, Art Resource; page 19, Tim Carlson, Stock, Boston.

Chapter 2: page 59, David Eugene Smith Collection, Rare Book and Manuscript Library, Columbia University; page 61, Dr. Harold E. Edgerton, MIT, Cambridge, Mass.; page 105, Peter Southwick, Stock, Boston.

Chapter 3: page 129, Historical Pictures Service, Inc.; page 144, Ken Karp; page 152, David R. Frazier, Photo Researchers, Inc.; page 163, Smithsonian Institution.

Chapter 4: page 165, Historical Pictures Service, Inc.

Chapter 5: page 207, Smithsonian Institution.

Chapter 6: page 243, Smithsonian Institution; page 249, St. Louis Regional Commerce and Growth Association; page 275, AP/Wide World Photos; page 277, Robert V. Eckart, Jr., The Picture Cube.

Chapter 7: page 283, Columbia University; page 292, Peter Menzel, Stock, Boston.

Chapter 8: page 343, North Wind Picture Archives.

Absolute Value [Section 1.1]

$$|a| = \begin{cases} a & \text{if } a \geqslant 0 \\ -a. & \text{if } a < 0 \end{cases}$$

Distance on the Real Number Line [Section 1.1]

$$d(a, b) = |b - a|$$

Quadratic Formula [Section 1.1]

If $a \neq 0$, then the roots of $ax^2 + bx + c = 0$ are

$$x = \frac{-b \pm \sqrt{b^2 - 4ac}}{2a}$$

Distance Formula [Section 1.2]

For points $P_1(x_1, y_1)$ and $P_2(x_2, y_2)$ in the Cartesian Plane,

$$d(P_1, P_2) = \sqrt{(x_2 - x_1)^2 + (y_2 - y_1)^2}$$

Equation of a Circle [Section 1.3]

$$(x - h)^2 + (y - k)^2 = r^2$$

Slope of a Line [Section 7.1]

Line through $P_1(x_1, y_1)$ and $P_2(x_2, y_2)$, $x_1 \neq x_2$:

$$m = \frac{y_2 - y_1}{x_2 - x_1}$$

Point-Slope Equation of a Line [Section 7.1]

Line through (x_1, y_1) with slope m:

$$y - y_1 = m(x - x_1)$$

Slope-Intercept Equation of a Line [Section 7.1]

Line with slope m and y-intercept b:

$$y = mx + b$$

Function [Section 1.4]

A function from a set X to a set Y is a rule of correspondence that assigns to each element x in X one and only one y in Y.

Even and Odd Functions [Section 1.5]

Even: $f(-x) = f(x)$; Symmetry: y-axis
Odd: $f(-x) = -f(x)$; Symmetry: origin

Shifted Graphs of a Function ($k > 0$) [Section 1.6]

Original graph: $y = f(x)$

$y = f(x) + k$, shifted up k units
$y = f(x) - k$, shifted down k units
$y = f(x + k)$, shifted to left k units
$y = f(x - k)$, shifted to right k units

Inverse Function [Section 1.6]

A function g is the inverse of a function f if $f(g(x)) = x$ and $g(f(x)) = x$.

Exponential Functions [Section 6.1]

$$f(x) = b^x, \, b > 0, \, b \neq 1$$

Logarithmic Functions [Section 6.2]

$$f(x) = \log_b x, \, x > 0, \, b > 0, \, b \neq 1$$

Properties of Logarithms [Section 6.2]

$y = \log_b x$ if and only if $b^y = x$
$\log_b b = 1$ $\log_b 1 = 0$
$b^{\log_b x} = x$ $\log_e x = \ln x$

Laws of Logarithms ($M > 0$, $N > 0$) [Section 6.2]

$$\log_b MN = \log_b M + \log_b N$$

$$\log_b \frac{M}{N} = \log_b M - \log_b N$$

$$\log_b N^c = c \log_b N, \, c \text{ a real number}$$